BIOLOGY OF MAMMALS

BIOLOGY OF MAMMALS

By

Dr. D.R. Khanna
Reader in Zoology
Gurukul Kangri University
Haridwar (Uttaranchal)
&
Dr. P.R. Yadav
Lecturer
Department of Zoology
D.A.V. College
Muzaffarnagar (U.P.)

D P H

DISCOVERY PUBLISHING HOUSE
NEW DELHI-110002

First Published-2005

ISBN 81-7141-934-8

Published by

DISCOVERY PUBLISHING HOUSE

4831/24, Ansari Road, Prahlad Street,
Darya Ganj, New Delhi-110002 (India)
Phone: 23279245 • Fax: 91-11-23253475
E-mail:dphtemp@indiatimes.com

Printed at:

Arora Offset Press
Laxmi Nagar, Delhi 110 092.

Preface

Mammals, our closest relatives in the animal kingdom, are a source of fascination and wonder. Some are the epitome of beauty and grace, others display extraordinary intelligence and have highly developed social systems, and yet more seem to us bizarre in appearance and behaviour.

This thoroughly illustrated title is designed to provide encyclopedic information in easy, simple, clear and lucid language for undergraduate and post-graduate students. Effects have also been made to give an upto date over view of our knowledge of the once-scent lives of mammals.

The present title is designed primarily for undergraduate students though it may also serve an introductory text to those preparing for their master's degree. The title is not intended to be comprehensive, nor could it be at this length, but it concentrates on putting across the basic principles of the subject as briefly and lucidly as possible. It does this with the aid of carefully selected examples, some recent and other classic of the field, and with numerous illustrations.

In the preparation of this book large number of books and research papers have been consulted. So no authenticity is claimed.

Text book can not be written without the support and professional contributions of many people. This book is no exception. The authors are grateful to those teachers and colleagues whose stimulating discussions have clarified certain vague points for him, but all errors, omissions, and corrections needed are solely their responsibility.

The authors tried hard to be accurate and upto date in statement and realises the impossibility of completely avoiding errors therefore, the authors will greatly appreciate having his attention called to any questionable statement.

The authors expresses their gratitute to Mr. Wasan and staff of M/s Discovery Publishing House for their whole hearted co-operation in the publication of this book.

Authors

CONTENTS

1

Introducing Mammals

Scientists divide the animal kingdom into several major groups for classification purposes. By far the largest group is the invertebrates: it contains about 95 percent of the millions of known species of animals, including sponges, mollusks, arthropods, and insects. Groups of vertebrates, or animals with backbones, contain the other 5 percent of known species. They can be divided roughly into fishes, amphibians, reptiles, birds, and mammals. The class Mammalia consists of fewer than 5,000 species, but the sheer diversity of mammals is astonishing. From tiny field-mice to the mighty blue whale, from a hippopotamus to a bat, from armadillos to gorillas, the class Mammalia encompasses some of the best-known and most-studied, as well as some of the least-known, members of the animal kingdom. It also includes human beings. We are mammals, classed along with monkeys, lemurs, and apes in the order Primates.

Features that Link Mammals

Living mammals are warm-blooded animals that suckle their young on milk and have a body covering of hair or fur, prominent external ears, and a mouth armed with teeth. These features—which, incidentally, are not even shared by all mammals—serve to differentiate mammals from all other living vertebrates. For scientists, however, the single feature that defines mammals is the method by which the dentary bone of the lower jaw—the one that houses the teeth—articulates directly with the skull.

Mammals evolved from a group of carnivorous reptiles, and the most primitive mammals were, like their ancestors, small flesh-eaters. As they evolved, mammals spread and adapted to many different habitats and now occupy a wide variety of niches, from small flying insect-eaters to large terrestrial grazers, and from tiny burrowing flesh-eaters to the largest of living animals, the plankton-feeding whales.

Skulls, Jawbones, and Teeth

The mammal-like reptiles from which mammals evolved are known as the therapsids. Many changes in the skull occurred during the evolution of the therapsids, but the mot important concerned jaws. Early during therapsid evolution the upper jaw became firmly attached to the rest of the skull and lost the independent mobility that can still be seen in such reptiles as snakes. Later, there were significant changes to the composition of the lower jaw, the bones housing the teeth became greatly enlarged and the other bones became smaller, until the dentary became the only in the lower jaw. In the earliest mammals, the dentary bones achieved direct contact with the squamosal bone of the skull.

One of the important developments in the mammal-like reptiles was the reduction in the number of bones in the skull and the strengthening of those that remained. The bones around the eye became less important while those surrounding the brain enlarged to accommodate an increasingly large nervous system.

As teeth are the major food-gathering structures of most mammals, they have been subject to a great deal of evolutionary change and have developed a number of distinctive features. They have, for example, different shapes in different parts of the jaw; they are replaced only once, if at all, and not many times as in fish and reptiles; and they are anchored in sockets in the bone.

Skin and Fur

The skin of mammals shows a number of characteristics not found in other backboned animals. These include the continuous growth and replacement of the epidermis, the outermost of the three layers of skin; the presence of hairs; and a number of new types of skin glands. The waterproof epidermis houses the hairs and skin glands, but contains no other structures. The second layer, or dermis, contains supporting collagen fibres, blood vessels, a variety of sensory nerve endings, and muscles that can cause the hairs to become erect. Attached to the inner boundary of the dermis is the sub-dermal fatty layer, which provides insulation.

Hairs, which vary greatly in shape and size, grow from the base of follicles and have a complex structure. Many mammals have at least two main hair types in their body covering; long guard hairs,

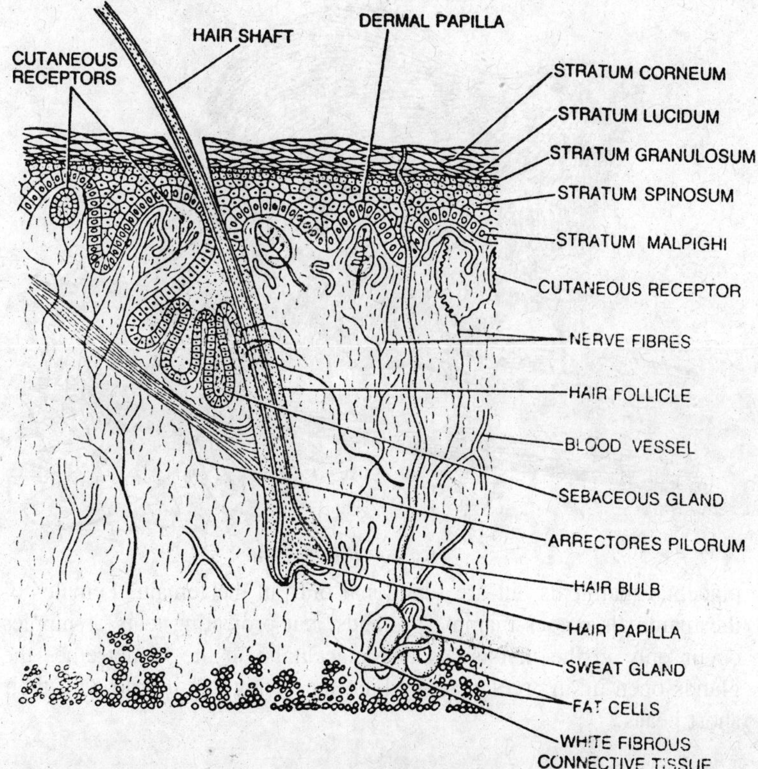

Fig. 1.1. V.S. of mammalian skin.

and shorter hairs or underfur. The combination of these two hair types provides efficient thermal insulation by trapping a layer of still air close to the skin. In some large mammals, hair is sparse or absent, and these species rely on the insulating ability of the skin. Each individual hair is subject to wear and has a limited usable life. Replacement of hairs, or molting, may occur continuously or, in colder climates, at specific times of the year. Some species, such as the ermine, have separate summer and winter coats.

Specialized hairs, or "vibrissae", occur at various points on the body, usually on the head but also on the forearms and feet. Vibrissae have specific sensory cells associated with them and provide important tactile information.

Of the different types of glands occurring in the skin, the most important are the mammary glands. These secrete the milk that nourishes the young. Mammary glands may occur in both sexes in

Fig. 1.2. Beaver.

placental mammals but are functional only in the female. Females of the mammals posses mammary glands, but projecting teats or nipples occur only in the marsupials and placentals; in the monotremes the glands open in an areolar area which the young can nuzzle with their short beaks.

Fig. 1.3. Pika.

Sebaceous glands are present in most mammals, in close association with the hair follicles. They produce an oily secretion that serves to lubricate and protect the hairs. Many mammals have sweat glands, which provide evaporative cooling and elimination of metabolic wastes. Sweat glands may be widely distributed over the body, as in humans, or restricted to particular areas. They many even be completely lacking. Scent glands are highly specialized structures that produce volatile odorous secretions. These are important as a means of communicating information; in some species, they are important as a means of self-defense.

Backbones and Limbs

Living mammals have a double contact between the skull and the atlas—the first cervical (neck) vertebra. Most living mammals have seven cervical vertebrae, although this number does vary among sloths and is reduced to six in manatees. In most mammals, ribs are associated only with the chest vertebrate; the exceptions are the monotremes, which have cervical ribs.

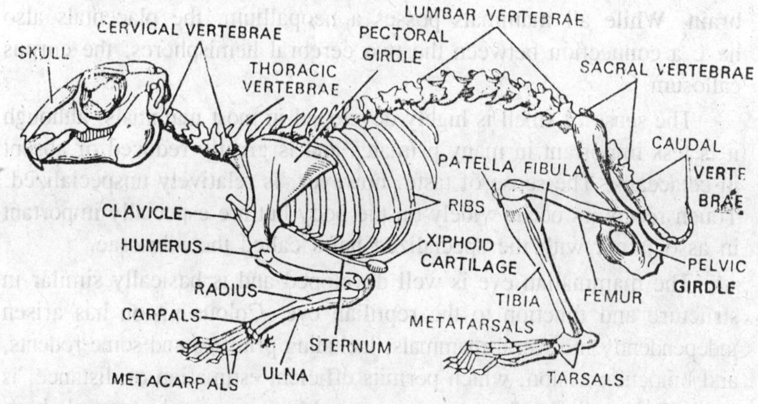

Fig. 1.4. Skeletal framework of rabbit.

The limbs of mammals have retained the basic vertebrate five-fingered plan with few modifications, even in such diverse groups as bats, whose hands have become wings, and whales, whose forelimbs are oar-like. In the hoofed mammals, there has been a progressive reduction in the number of toes from five to two, or even one. The growth of limb bones in mammals differs from that of reptiles in that growth is not indefinite and is confined to two growing points near either end of each long bone.

Monotremes, like a number of early mammal groups, have retained a primitive shoulder girdle with a number of separate bones, but the other living mammals have a shoulder girdle which consists of a scapula (shoulder-blade) and a clavicle (collar-bone). The clavicle is frequently lost in such groups as hoofed mammals, which have evolved extreme flexibility of the limbs to increase their running speed, but is retained and indeed strengthened in those primates that rely on their strong arms for climbing.

The pelvic girdles of mammals consist basically of three fused bones, as in most reptiles, although those of the monotremes and marsupials have an additional bone which provides for the insertion of muscles supporting the ventral body wall.

Brain and Senses

Mammals have much larger brains, relative to body size, than other vertebrates. The increased brain size is due to an expansion of the cerebral hemispheres. A new component of the cerebrum, the neopallium, which occurs as a very small region in some reptiles, has expanded to provide a covering of gray cells over the ancestral reptilian brain. While all mammals posses a neopallium, the placentals also have a connection between the two cerebral hemispheres; the corpus callosum.

The sense of smell is highly developed in most mammals, although it is less important in many primates and is greatly reduced or absent in cetaceans. The sense of taste, however, is relatively unspecialized. Touch receptors occur widely on the body but are especially important in association with the specialized hairs called the vibrissae.

The mammalian eye is well developed and is basically similar in structure and function to the reptilian eye. Colour vision has arisen independently in several mammals, including primates and some rodents, and binocular vision, which permits efficient estimation of distance, is particularly well developed in primates. Many nocturnal mammals have a reflective layer at the back of the eye which increases visual acuity by reflecting weak incoming light back onto the retina.

Hearing is important to most mammals and a number of specializations have evolved, particularly the amplifying function of the middle ear. Most mammals have an external ear that collects sound and concentrates it on the opening to the middle ear. Many mammals can hear sounds of very high pitch, and this has been exploited—by bats, for instance—for echolocation, by which the animal detects obstacles by listening to the reflections of its own sound pulses.

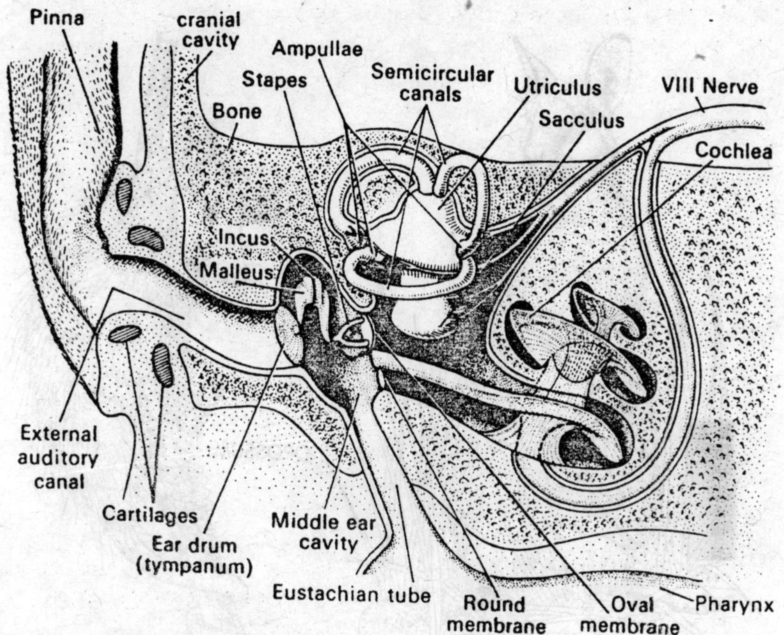

Fig. 1.5. Diagram of the ear region of mammalian head in section.

How Mammals Reproduce

Mammalian reproduction is characterized by internal fertilization which results in an amniote egg. The female reproductive system consists of a pair of ovaries and their associated ducts. Under the influence of hormones produced by the pituitary gland, mature ova are released into the fallopian tubes. The control of ovulation may be environmental, as in many highly seasonal breeders, or it may be induced by copulation, as in cats. Males produce sperm in the testes, which are located within the body cavity or outside in a scrotum, where the sperm can develop at a temperature lower than normal body temperature. After copulation and ejaculation, the sperm migrate rapidly to the upper portion of the female fallopian tubes, where they fertilize the ova.

In monotremes the fertilized egg is surrounded in the uterus by a shell membrane and a leathery shell. It is then incubated in the uterus for about two weeks before being laid and further incubated externally. In marsupials a shell membrane, but no shell, is deposited. In placental mammals shell membranes are absent.

The young of marsupials are born in a very undeveloped state after a very short gestation. The attach firmly to the teat of a mammary

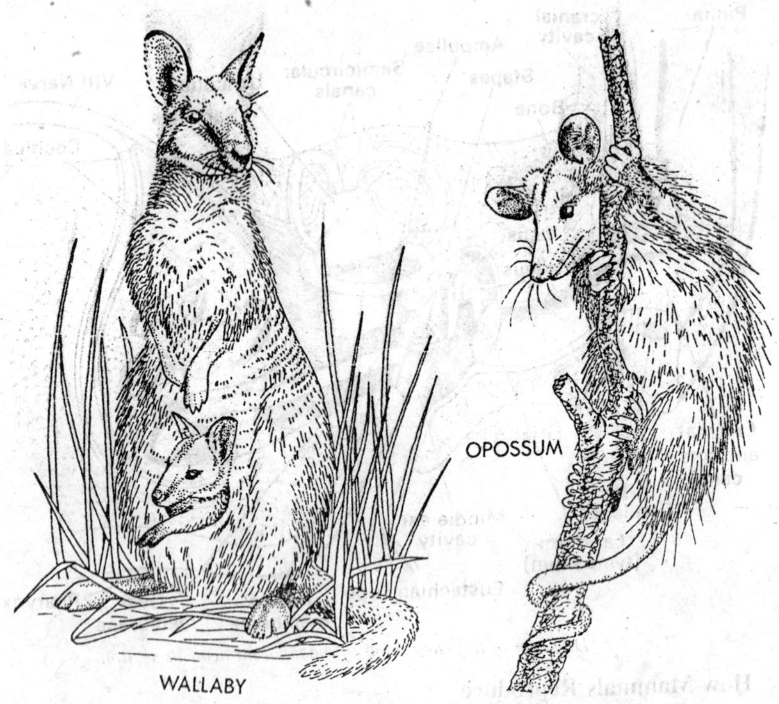

Fig. 1.6. Marsupials: Wallaby and Opossum.

gland which may or may not be located in a pouch on the mother's bell. In placental mammals gestation is relatively longer and the newborn young are in a more advanced state of development, although in some species they are still naked and helpless. A major trend in the evolution of placental mammals has been the lengthening of the gestation period and the birth of more highly developed young. This is usually associated with the production of fewer young in each litter.

Other Organ Systems

In mammals, a muscular diaphragm separates the chest and abdominal cavities. Contraction of the diaphragm, together with the action of the muscles connecting the ribs, draws air into the lungs. The upper respiratory system consists of a trachea which branches out to form the bronchi, leading to the lungs. At the upper end of the trachea is the larynx, where sound is produced.

The circulatory system is dominated by a four-chambered heart, which has a different arrangement from the four-chambered hearts of crocodiles and birds. Separation of pulmonary and systemic circulations is complete in the adult. The red blood cells bind and transport oxygen and carbon dioxide from and to the lungs. A variety of white blood cells perform the housekeeping functions of the body's immune system. All mammals can maintain their body temperature above that of their surroundings, and many are also capable of lowering their body temperature-either on a daily basis or for a prolonged period of torpor or hibernation.

Mammalian kidneys are "kidney-shaped" rather than elongated structures as in other vertebrates. Like many fish, amphibians, and turtles, mammals excrete nitrogenous wastes as urea rather than as uric acid, as most reptiles and all birds do.

The digestive system shows a number of specializations in mammals, particularly among those groups that ferment plant material in a highly modified stomach or in a cecum, which is a sac occurring at the junction of the small intestine and the proximal colon.

2

Classifying Mammals

There are several million different kinds of plants and animals. The purpose of classification is to provide each of these with a unique name by which it may be known; to describe it so that it may be recognized; and to place it in a formal arrangement or hierarchy that will express its relationship to others. The science of arranging plants and animals in this way is called taxonomy or sometimes systematics

Vernacular Names

Many mammals have vernacular or common names, although many do not. These are not used in classification because they many differ from country to country, or from one language to another, and can be ambiguous.

Scientific Names

Scientific names are always in Latin, partly because this was the common language of the early naturalists, but chiefly because Latin is universally accepted by scientists as a means of avoiding the problems of translation and ambiguity.

The scientific name of any kind or species of plant or animal always consists of two words; the first or generic name indicates the relationship of the species to others, the second or specific name signifies the particular species concerned.

Scientific names are often descriptive, drawing attention to some significant feature of their subject, or they may refer to the country or place whence it originated. Sometimes names commemorate a particular person, and some are classical in origin, but they can also be quite arbitrary.

Species

The species is the basic category in classification. About 4,600 species of modern mammals are recognized and although some, such as the giant panda, the lion, or the tiger, are easily seen to the distinct, the majority are much less obviously so and are define by differences in external form and structure, in the structure of the skull and teeth, or even in such attributes as chromosome pattern and structure. The interpretation of similarities and differences is often subjective, and for this reason the total number of mammalian species varies slightly from one authority to another. Also, from time to time new species are discovered and described, even today.

Genera

Species with a number of features in common are grouped together in genera, although occasionally a very strongly characterized species may be the sole member of its own genus. Thus the genus *Vulpes* includes a number of species of fox besides the red or common fox *Vulpes vulpes*, but the Arctic fox is considered sufficiently distinct from these other foxes to be placed by itself in a separate genus as *Alopex lagopus*.

Genera are based on more far-reaching characters than species and are sometimes divided into subgenera to emphasize the features of a particular species or group of species. Again, the interpretation of their characters is subjective and the number of modern genera of mammals is not irrevocably fixed: currently about 1,130 are listed.

Families

In the same way that genera signify groups of species, or sometimes one strongly characteristic species, so families comprise groups of genera that share common characteristics or similarities: a family may include only one very characteristic genus, on occasion with only a single species. There are 136 families of modern mammals.

Orders

Families that share major characteristics are grouped into orders. As a rule these are quite distinctive and separated from others by obvious features, but sometimes the distinctive may be less immediately apparent. For example, elephants occupy an order of their own, the Proboscidea, which can be easily recognized, but the distinction between tree shrews (Scandentia) and elephant shrews (Macroscelidea) is less obvious.

Most orders include several or many families, but a number besides the Proboscidea contain but a single family. One order, the Tubulidentata, has only a single genus and species, the African aardvark or ant bear *Orycteropus afer*. Orders may be subdivided into suborders to emphasize differences among their families. There are 20 orders of modern mammals, the Rodentia or rodents being the largest, with about 40 percent of known mammal species, followed by the Chiroptera or bats, which have about a further 23 percent.

Higher Classification

Mammals belong to the class Mammalia, in which tow major division are recognized. One, the Prototheria, includes only the spiny anteaters of Australia and New Guinea and the Australian duck-billed platypus, which are unique among mammals in that they reproduce by laying eggs. The other, the Theria, includes all other modern mammal but itself has two divisions. One, the Metatheria, is reserved for the order Marsupialia, or marsupials, whose young are born in a very undeveloped state and complete their development in a pouch or fold on the mother's abdomen. The remainder of modern mammals belong to the second group, the Eutheria, whose young develop to a relatively advanced stage within the body of the mother.

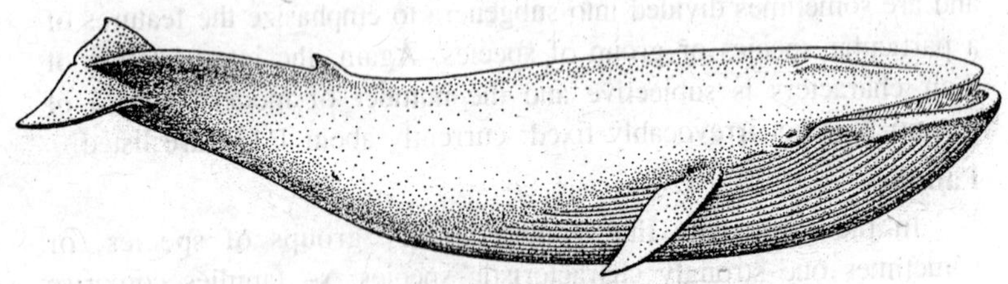

Fig. 2.1. Blue whale.

Fossil Mammals

Modern mammals constitute a very small part of the total kinds that have existed since mammals first evolved 200 million years ago. The most recent fossils may represent genera or even species known today, and many fossil mammals belong to the same orders and families as their modern counterparts. Many fossils, however, represent totally different mammals that have evolved, thrived, and then become extinct during the long period since mammals first appeared.

Changes in Classification

New ways of assessing the similarities and differences between mammals, or of extending the range of characters upon which classification depends beyond those used traditionally, lead from time to time to changes in classification. Also, The interpretation and meaning of their features may differ from one authority to another. There is no immutable classification, and in some instances there is

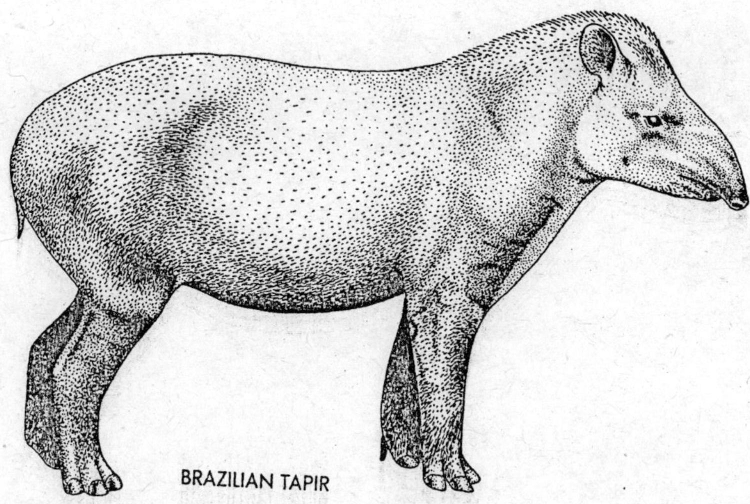

Fig. 2.2. Two perissodactyls.

not complete agreement as to the relationships among and between species, genera, and even families and orders.

Several authors have proposed dividing the marsupials into as many as seven separate orders and these have been accepted by some texts including Wilson & Reeder, 1993. They have not yet been universally accepted, however. As the different lines of evidence point to different interpretations of relationships, yet all marsupials are their own closest relatives, we prefer to follow the conservative course of listing a single order.

Rodents may possibly represent two independent lineages. Even more upsetting would be to place the whales in the order Artiodactyla between pigs and hippos as DNA studies suggest. The unique adaptations of whales to a pelagic life warrant the retention of the order Cetacea. The classifications in the two major references differ in many interpretations. The classification we present here is a compromise erring on the side of caution.

3

Mammals through the Ages

The mammals of the world today are the survivors of a long history that started about 195 million years ago. In rocks of that age, the very first unmistakable mammal fossils occur, as tiny insect-eating animals that look a little like shrews. For almost the first two-thirds of their subsequent history, mammals remained small, inconspicuous animals, probably active only at night. During all this time—the Jurassic and Cretaceous periods of the geologists time chart—they shared their habitat with the dinosaurs. But when the great extinction of the dinosaurs occurred at the end of the Cretaceous, the land appears to have been opened up for exploitation by mammals. From that moment on throughout the following 65 million years to the present, many different kinds of mammals, large and small, carnivorous and herbivorous, terrestrial and aquatic, have evolved, flourished, and disappeared, to be replaced by yet newer kinds.

The Origin of Mammals

Mammals fossilize relatively easily, because their skeletons are generally robust and withstand well the rigors of the fossilization processes. The teeth and jaws in particular are often perfectly preserved, which is doubly fortunate because the exact structure of any particular mammal's teeth tells a great deal about the evolution and biology of that mamma. Consequently the history of the mammals is better known than that of any other comparably sized group of organisms.

Three hundred million years ago the land was populated by primitive amphibians and reptiles, living in and around the extensive wet tropical swamps of the time. Among the reptile fossils, remains of a few forms that were rather larger than the rest, and which had a

Fig. 3.1. Pelicosaurus.

pair of windows, or temporal fenestrae, in the hind part of their bony skulls, have been found. These windows are still found in a modified way in mammals, indicating that these animals were the first of the "mammal-like reptiles", or synapsids as they are correctly called. The importance of this ground is that it includes the animals from which mammals eventually evolved. The early members of the group were not, however, very like mammals, for they still had simple teeth with weak jaws, and clumsy, sprawling limbs; certainly they were not warm-blooded. A well-known example of the early, primitive synapsids is *Dimetrodon*, the finback reptile, which lived over a wide area of what is now North America.

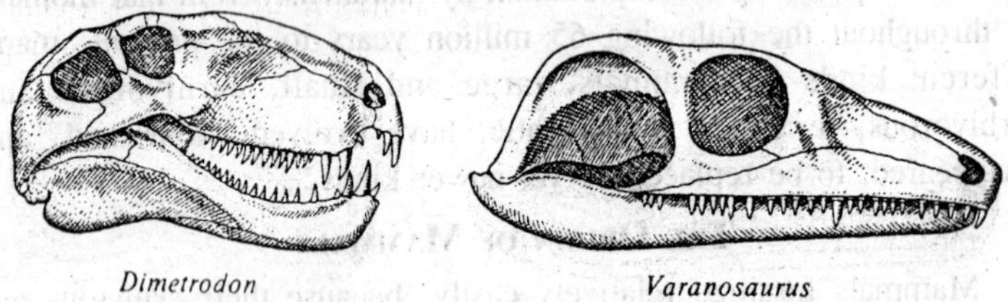

Fig. 3.2. Skull of two early synapsids, that of Dimetrodon and Varanosaurus.

In due course, more advanced kinds of mammal-like reptiles called the Therapsida evolved from *Dimetrodon*-like ancestors. Therapsids had developed more powerful jaw-closing muscles, which they could use with a variety of more elaborate dentitions such as very enlarged canine teeth, or even horny beaks. They also possessed longer, more slender limbs which must have allowed them to run more rapidly and with greater agility in pursuit of their prey, or to escape from predators. Therapsids had probably become more warm-blooded and larger brained as well, although these things are more difficult to tell from fossils.

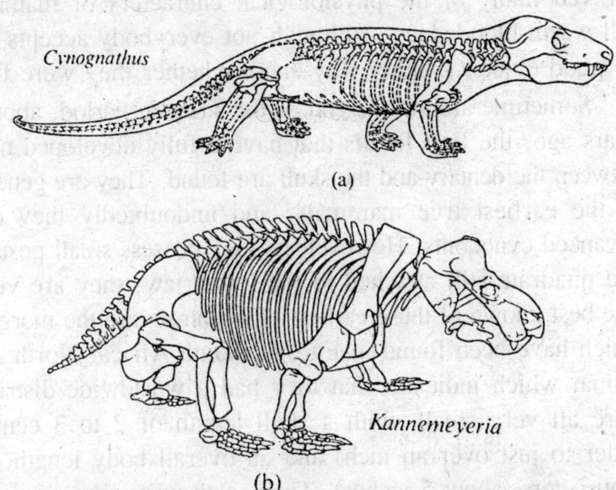

Fig. 3.3. Skeleton of synapsid reptiles: (a) Cynognathus, (b) Kannemeyeria.

One particular group of advanced therapsids, called the Cynodontia, had evolved several other distinctly mammalian characteristics. They had much more mammalian-like teeth, with several cusps, which worked by a shearing and crushing action between the upper and the lower teeth. Along with the modifications to the teeth themselves, very powerful yet also very accurate jaw muscles had to evolve to give these teeth greater biting power. This involved a great enlargement of the bone that carried the lower teeth, the dentary bone, so that it could also carry the attachments of the lower jaw muscles. The other bones of the lower jaw, called the postdentary bones, were correspondingly reduced and weakened. In the cynodonts, these postdentary bones had adopted a new function. Along with the old hinge bone or quadrate of the skull, they transmitted sound waves from a rudimentary ear drum, attached to the jaw, to the auditory region of the animal's braincase. In this way, the cynodonts show an approach towards the fully mammalian arrangement where only the dentary bone is left in the lower jaw, and it has formed a new jaw hinge directly with the squamosal bone of the skull. The mammalian postdentary bones and quadrate have lost their contact with the jaw altogether and have become the sound-conducting ossicles of the middle ear.

Cynodonts also evolved a much more mammalian kind of skeleton, with very slender limbs held much closer to the body, which made for more agile locomotion. It is often thought, too, that the cynodonts had evolved many of the physiological characters of mammals, such as full warm-bloodedness, although not everybody accepts this. There is no good evidence either way about whether they were furry.

Sometime around the end of the Triassic period, about 195 million years ago, the first fossils that have a fully developed new jaw hinge between the dentary and the skull are found. They are generally accepted as the earliest true mammals, and undoubtedly they evolved from advanced cynodonts. However, as they possess small postdentary bones and quadrate still attached to the lower jaw, they are very primitive. The best known of these earliest of mammals are the morganucodontids, which have been found in Europe, South Africa, North America, and China, which indicates that they had a worldwide distribution. They were all very small, with a skull length of 2 to 3 centimeters (just under to just over an inch) and an overall body length of around 12 centimeters (about 5 inches). Their teeth were sharp and multi-cusped, and, from the wear patterns that developed during life, it seems that they were used to capture and chew insects and perhaps other terrestrial invertebrate prey. To judge from this and from their apparently enhanced sense of hearing and smell, morganucodontids are believed to have been adapted to a nocturnal hunting existence.

Fig. 3.4. Morganucodontids.

The Mesozoic Mammals

Morganucodontids and related early mammals were the start of an evolutionary radiation, through the rest of the Mesozoic era, into several different groups. However, without exception they remained very small animals. Even the largest were no larger than a domestic cat, and the great majority were very much smaller than that. Furthermore, most of them remained insectivorous, although one group, the multituberculates, adopted herbivorous habit and looked superficially like the rodents of today.

The most important evolutionary step during the Mesozoic was the development of a more complex kind of tooth, the tribosphenic molar. These teeth have a triangle of three main cusps on both the upper and the lower teeth, while the lower also have an extended basin at the back end. They were more effective in their action, and in the Late Cretaceous, Hound 80 million years ago, two main groups of fossil tribosphenic mammals can already be distinguished. These are the marsupials and the placentals, which were soon to be the dominant mammals of the world. Indeed, even before the end of the Cretaceous several different kinds of placentals had already evolved, and thus the scene was set for the great post-dinosaur radiation of mammals.

Why all the Mesozoic mammals should have remained so small is a mystery. Possibly it was simply because they could not compete with the dinosaurs in the habitats appropriate for large terrestrial animals and therefore had to remain restricted to a nocturnal way of life suitable for small animals. Alternatively, the climate may have been such that the physiology of these early mammals was unsuitable for large animals.

There is also still much argument about what caused the extinction of the dinosaurs and the numerous other groups of animals, and to a lesser extent plants, that suffered at this time. But what is clear is that the mammals not only survived, but were for some reason particularly well placed to take advantage of the relatively empty new world set before them.

The Tertiary Radiations

The evolutionary story of mammals over the last 65 million years to the present day, that is the Tertiary era, began from the few groups of small, insect-eating mammals that survived the extinction of the end of the Cretaceous. The story is very complicated, because as well as the extinction of the dinosaurs, the end of the Cretaceous also saw the beginning of the break-up of the great land mass into separate pieces that proceeded to drift apart, eventually forming the continents familiar today. For a brief time, Australia and South America were connected to one another by Antarctica, but they were shortly to become completely isolated island continents, on which quite different groups of mammals evolved. On the other hand, North America, Europe, and Asia remained effectively connected to one another. A further complicating factor was a series of climatic changes that affected mammal evolution from time to time by causing the extinction of certain groups and allowing other, usually more advanced, groups to survive and flourish. Thus there was an almost continual turnover of species and families as the millions of years passed.

In any event, within no more than 5 million years or so, several important new kinds of mammals had evolved. Looking at the major part of the world, as represented by North America and Eurasia, the early new mammals to evolve from the little insectivorous placental ancestors included the creodonts, which were the first of the large predaceous mammals. These were very like modem carnivores, with specialized shearing molar teeth and powerful sharp-clawed limbs. Several other groups of mammals had evolved into large, browsing herbivores, with various patterns of large, flattened grinding molars, and elongated legs with little hoofs on the ends of the toes to improve their running ability. Some were quite strange looking, such as the uintatheres of North America, which had bony protuberances over the skull, and *Arsinoitherium* from Egypt with its pair of massive horns. Small herbivores, occupying the niches later adopted by the rodents, were represented, surprisingly, by a group of primitive primates called the plesiadapids.

As time passed, new and more modern groups of animals evolved, although often in rather primitive and unfamiliar form. By about 50 million years ago the true Carnivora, and the earliest of the horse family represented by *Hyracotherium*, the eohippus, had appeared, as had the first whales (though these still had serrated teeth) and other groups later to be highly successful, such as the bats, rodents, artiodactyls, elephants, and lemurs.

Around 40 million years ago, the climate deteriorated for a while, causing the extinction of about one-third of the existing mammalian families. The main sufferers were the more archaic, primitive kinds, and consequently the fauna took on an increasingly modem appearance. This phase was followed by a long period of very favourable climate referred to as the Miocene. It was the heyday of mammalian evolution, when more groups of mammals were present than ever before or since.

One of the most significant new developments of the Miocene was the spread of grasses, to form great plains on which herds of grazing animals could thrive. Horses, and even more so artiodactyls— deers, antelopes arid so on—radiated. Although the majority of groups of mammals were those familiar today, they included many bizarre species, particularly among the perissodactyls, such as the chalicotheres, which had longer front than back legs, and *Indricotherium*, the largest land mammal ever to have lived, which stood over 5 meters at its shoulder. Among the carnivores were a number of large sabre-toothed cats.

Fig. 5.5. Indricotherium.

During the Miocene the apes also evolved from more primitive primates, and a little later, about 4 million years ago, the first members of *Australopithecus* had appeared in Africa, as the immediate forerunners of the humans that were to arise 2 million years later.

South American Mammals

During most of the Tertiary, South America was an isolated island continent, and consequently the history of its mammal fauna differed from that of the rest of the world. Several unique orders of herbivorous placental mammals evolved, such as the litopterns which resembled horses but were entirely unrelated to them, and other orders superficially like rhinoceroses, hippopotamuses, and so on. Another important group were the edentates, with such strange animals as the huge armored *Glyptodon*, and the giant ground sloth *Megatherium*. This order still survives today and includes some of the oddest mammals, the armadillos and South American anteaters.

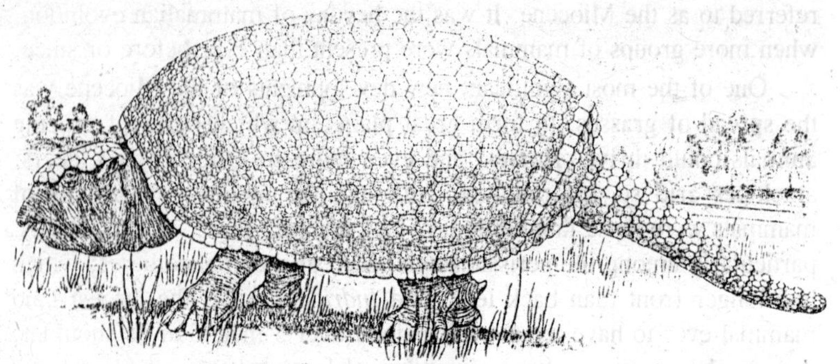

Fig. 3.6. Glyptodon.

The most unexpected feature of South American mammal evolution concerns carnivores, for these consisted exclusively of marsupial mammals, a group absent from the Tertiary of North America, Europe, and Asia. There were smaller forms, such as the didelphids and caenolestids, which have survived to the present. But there were also large predators, the borhyaenids, which played the same role as carnivores in other parts of the world, with forms as large as lions, and even an equivalent to the placental sabre-toothed tiger in the form of *Thylacosmilus*, which had huge upper canine teeth.

The only placental mammals (apart from bats, which easily migrate worldwide) that managed to invade South America during the earlier Tertiary were rodents and primates. They probably came from North America via the Caribbean Islands, and once established they evolved into the characteristic South American groups, the hystricomorph rodents and the platyrrhine or New World monkeys.

Owing to tectonic changes about 3 million years ago, South America eventually drifted northwards to connect with the North American continent. Because of the ensuing climatic changes, there followed what is often called "the Great American Interchange", which resulted in a dramatic alteration to the South American mammalian fauna. The South American herbivore groups disappeared, to be replaced by invaders from the north: representatives of such modern placental groups as horses, tapirs, llamas, and deer. Similarly, the large marsupial carnivores disappeared and in their place came dogs, cats, bears, and so on. On the other hand, many remnants of the old South American fauna do persist, such as edentates, small marsupials, rodents, and monkeys; indeed several of them invaded North America, like the didelphid opossums and the porcupine.

It is particularly clear in South America how the present mammalian fauna is a result of many processes, evolutionary, biogeographic, and climatic, that together caused a long history of change.

Australian Mammals

The early history of Australian mammals is hardly known yet because there are few fossil deposits containing mammals until well into the Tertiary. Apart from a single find of the tooth of a primitive and early herbivore, there is no evidence that placental mammals reached Australia until late in the Tertiary, and then only as a few rodent and bat species from Asia. All other placental species, such as the dingo and rabbit, are human introductions. The egg-laying monotremes diverged quite early from the tribosphenic mammals, the marsupials and placentals. Fossil teeth of a primitive monotreme have been discovered in the Cretaceous of Australia, but the never seem to have been a particularly important group.

With these minor exceptions, all Australian mammals, past as well as present, are marsupials. They radiated into many of the niches associated with placental mammals elsewhere in the world, such as the large carnivorous thylacines, the grazing, herd-living kangaroos, and the insectivore-like opossums.

The Pleistocene Extinctions

The final drama in the history of mammals occurred only 10,000 years ago, in the late Pleistocene epoch. A phase of extinction is shown by the disappearance from the fossil record of many mammal species, but mainly the largest forms. There had been giant members of many groups throughout the world, such as the familiar mammoths, Irish elk, and the sabre-toothed tigers. Less well known but equally remarkable were such animals as giant apes and wart hogs in Africa, a giant lemur in Madagascar, a giant tapir and ground sloth in South America, and giant kangaroo, wombat, and platypus in Australia. All these became extinct at about the same time.

It is much argued whether this was due to some climatic change that made life more difficult for larger species, or whether it was connected with the rapid spread of humans throughout the world at that time. Perhaps this was the first baleful effect of human beings upon the mammals of Earth.

4

HABITATS AND ADAPTATION

Mammals have evolved to successfully occupy nearly every habitat on the face of the Earth. Only the interior of the Antarctic continent remains uninhabited by mammals. To achieve their remarkable success, mammals have evolved various locomotor and feeding adaptations that permit them to survive and prosper in habitats as diverse as tropical rainforests grasslands, tundra, and deserts. Much of the diversity in the external form of mammals reflects their adaptations to specific habitats or environments.

MAMMALS OF THE FOREST

Because trees are a key habitat component of forest-dwelling mammals, particularly as sources of food shelter and routes of travel and escape, many mammals of the forest are excellent climbers and exhibit various degrees of specialization for moving through the trees. In more generalized species, such as many mice and rats, adaptation takes the form of long tails, which act as counterbalances as these mammals move along branches. Other more highly specialized arboreal (tree-dwelling) species have prehensile tails that serve as a "fifth hand". Prehensile tails have evolved independently in forest-dwelling representatives of several groups of mammals, including New World opossums, Australian possums and cuscuses, South American anteaters, pangolins or scaly anteaters in Africa and southern Asia. New World monkeys, South American porcupines and some tropical rats. Opposable big toes, or in some cases thumbs, which aid in grasping branches, arc also specializations for movement through trees and have evolved in several groups of primates, New World opossums, Australian possums and cuscuses, the koala, and the noolbenger or honey possum. Gliding as a means of movement in forests has evolved independently in flying squirrels of North America and Eurasia, African scaly-tailed squirrels, colugos or gliding lemurs in the jungles of Southeast Asia, and in members of the Australian gliding marsupials.

Perhaps the most specialized arboreal mammals are sloths, which spend virtually their entire lives in trees, usually coming to the ground only to defecate, on average about once every seven days. Although almost helpless on the ground, sloths are excellent swimmers, an adaptation to crossing the numerous streams and rivers that characterize their lowland forest habitats in the New World tropics. The minimal level of activity so characteristic of sloths is an adaptation to a diet of leaves containing high concentrations of toxic compounds that are manufactured by plants as a defense against herbivores. The high fiber and cellulose content of these leaves further reduces their nutritional value. By coupling minimal activity with a low metabolic rate and a dense coat for a tropical mammal, sloths substantially reduce their food requirements and consequently their intake of toxic chemicals. In addition, sloths meet some of their energy requirements by basking in the sun in the canopy of trees. The koala, which also subsists on a diet high in secondary compounds, possesses many of the same

Fig. 4.1. Pangolin.

adaptations, including slow deliberate movements and relatively low metabolic rate.

Grassland Mammals

In contrast to forest-dwelling species, mammals living in grasslands often rely on speed, rather than concealment, as a method of escape. The long legs of hoofed mammals of the open plains function to increase their speed as a principal means of predator avoidance, as well as to permit them to efficiently travel long distances in search of food and water. One of the most striking grassland mammals is the maned wolf, a South American carnivore that has evolved extremely long legs to permit it to see over and move with ease through the tall grasses of the pampas.

The open nature of grasslands has promoted the evolution of social behaviour in many species of grassland mammals, such as large ungulates (hoofed mammals), kangaroos, wallabies, ground squirrels, prairie dogs, baboons, and banded mongooses. Individuals in social groups benefit from multiple sense organs (ears, eyes, and noses) which aid in the detection of potential predators at greater distances than would be possible alone. Herding behaviour in grassland ungulates is strongly reinforced by predators who selectively prey on individuals that do not remain part of the herd. Social predators such as lions and cheetahs are also characteristic of grassland habitats. Many smaller grassland mammals are adapted for burrowing and utilize their burrow

Fig. 4.2. Macropus (Kangaroo) female with young.

Fig. 4.3. Didelphys virginianus (Opossum).

systems to escape predators and to provide relief from high temperatures during mid-summer days. Numerous members of the squirrel family, such as North American ground squirrels and prairie dogs, are semi-fossorial, meaning that they utilize extensive burrow systems for shelter and nest sites, but forage on the surface for herbaceous plants and seeds. Several groups of rodents, such as New World pocket gophers, South American tuco-tucos, and African mole-rats, have abandoned life on the surface to exploit rich subterranean plant resources in the form of roots, bulbs, and tubers. Pocket gophers and tucu-tucos utilize powerful forelimbs in burrowing, whereas African mole-rats use their large incisors to excavate burrows. These burrowing or fossorial rodents are characterized by short coats, small cars, small eyes, and short tails.

Mammals of the Tundra

Mammals of the Arctic tundra possess numerous adaptations to permit them to survive the physical rigors of this region. The immense shaggy coats of musk oxen mark this species as a true Ice Age survivor and one of the best adapted of Arctic mammals. One of the most familiar adaptations of Arctic mammals is the ability of some species to turn white in winter. The white winter coat of Arctic hares, Arctic foxes, and collared lemmings provides both camouflage in a white world and excellent insulation.

Locomotion during Arctic winters is often difficult, and tundra mammals possess several distinctive adaptations to increase efficiency. The surface area of the feet of Arctic hares is greatly increased by growth of dense brush-like fur on the pads, The unusually broad hooves of caribou aid locomotion in both winter snow and the marshy ground of summer. As an aid to digging through snow, ice, and frozen ground, the claws of the third and fourth toes on the forefeet of collared lemmings become greatly enlarged during winter.

Fig. 4.4. A—Phoca (Seal); B—Odobenus (Walrus).

Caribou or reindeer are unique among members of the deer family in that females typically possess antlers, albeit smaller than those of males. During winter both males and females use their antlers to scrape away snow and ice to expose vegetation on which they feed Like many large mammals inhabiting grasslands, musk oxen and caribou have responded to living in an open environment by evolving herding behaviour as a defense against predation. This is epitomized by the classic "circle the wagons" defense or musk oxen.

Another anti-predator adaptation in caribou is synchronized calving: 80 to 90 percent or calves in a population are born within a 10-day period in late Mayor early June each year. This brief calving season

Fig. 4.5. Moschus moschiferus (musk deer).

apparently represents a strategy designed to overwhelm and satiate predators with large numbers of young, and thereby allow some of the young to escape predation, The highly precocial newborn calves arc soon able to join the herd and gain its protection. Although synchronized calving in caribou might represent a response to constraints imposed by the very short summer season of the high Arctic, synchronized calving also occurs in some grassland ungulates, such as the wildebeest in Africa, which are not under constraints of an extremely short growing season.

Mammals of the Desert

Desert mammals must be adapted to succeed in harsh physical environments characterized by temperature extremes and shortages of moisture, Most desert mammals minimize the physical stress of these two factors by being active at night (nocturnal) and spending their days in burrows where temperatures are cooler and relative humidities are higher than on the surface. Insects and seeds are extensively utilized by desert mammals both for their nutritional value and as sources of water; the bodies of insects have a high moisture content. The moisture requirements of seed-eating rodents are met largely or entirely from water produced as a byproduct of their metabolism of carbohydrates in seeds, termed metabolic water.

Fig. 4.6. Llama, Lama.

Antelope ground squirrels in the deserts of North America are active during the day and thus must directly confront the challenge of high environmental temperatures. Because lethal body temperatures are quickly reached while foraging, these squirrels periodically return

Fig. 4.7. Antilope cervicarpra (Black buck).

to their burrows to "dump" heat in the cooler environment of the burrow. They resume their foraging after the body temperature returns to normal.

Large desert mammals, such as camels and various species of oryx, cannot escape to burrows and are exposed to daily temperature extremes. One way they survive without expending excessive amounts of water on evaporative cooling is to let their body temperature fluctuate. At night they take advantage of radiational cooling to allow the body temperature to fall below what is normal for mammals. During the day, they permit the body temperature to slowly rise as they are warmed by the sun, and only late in the day as the body temperature approaches lethal limits do they utilize evaporative cooling, thus conserving considerable water. Their pale coat helps both to reflect sunlight and to provide insulation against high environmental temperatures. Oryx also feed at night when the moisture content of plants is higher than during the day.

Convergent Evolution

In examining the adaptations of mammals to various habitats, it is obvious that certain adaptive types appear best suited to particular habitats. As a consequence, we find many examples of unrelated mammals that have evolved remarkably similar adaptations because they live in similar habitats and have similar ecological roles in different parts of the world. This phenomenon is termed convergent evolution.

Among desert mammals a notable example of convergent evolution involves seed-eating rodents, including kangaroo rats in North America, gerbils and jirds in Asia and Africa, jerboas in Asia and Africa, and Australian hopping mice. These rodents exhibit remarkable degrees of convergence in morphology and ecology. All have pale, sand-coloured coats, enlarged hind legs and long tails, and employ bipedal or hopping locomotion as it principal means of escape from predators.

The convergence of seed-eating desert rodents a well as other examples noted in this chapter suggests that there are optimal evolutionary answers to the challenges of adapting to exploit specific types of habitats. Because of their widespread distribution and their adaptability, mammals are excellent examples of convergent evolution.

5

Mammal Behaviour

All animals behave—that is, they react to specific stimuli in specific ways—but mammalian behaviour is characterized by the fact that mammals, with only two exceptions, give birth to living young. In addition, the young of all species are dependent upon their mothers for food, in the form of milk secreted by specialized and modified sweat glands. This necessitates a special type of social behaviour that bonds the mother to its young.

Dependent Mammals

Some types of mammals remain dependent upon their mothers for many months—even years; such is the case with primates (including humans), and with whales and elephants. Others, such as rodents, are weaned in less than three weeks. Whatever the length of this dependence, it is an example of social behaviour that can be defined as an interaction between two individuals of the same species. Even the so-called "solitary" species of mammal, such as the Australian bandicoot or the orang utan, which as adults interact socially only to mate, have a close behavioural interaction with their mothers when they are young. All other behaviours shown by mammals are classified as something other than social. These include behaviours necessary for bodily maintenance, such as feeding, grooming, urinating, defecating, and huddling.

Maintenance Behaviour

It is difficult to dissociate maintenance behaviour completely from social behaviours, since frequently the two are interrelated. For example, passing urine is a biological necessity associated with the physiological processes of nitrogen excretion. Cows, seals, whales, and female dogs and mice simply void urine whenever the distension of the bladder reaches the point where the urination center in the rear of the brain is stimulated. After urination, the stretch receptors in the bladder wall relax, the flow of messages to the brain stops, and the animal no longer feels the urge to urinate.

Adult male dogs, mice, and many other mammals, however, do not urinate in this way. In them the presence of the sex hormone testosterone—produced in the testes—influences behaviour, so that urine is voided at specific points around the animal's environment where other males will encounter it. This is an example of territorial demarcation. So maintenance and social behaviours are sometimes intertwined.

The same is true for feeding behaviour—many carnivorous animals hunt cooperatively in order to fill their bellies—and grooming behaviours which, in primates, serve an important social function. Nevertheless, in all these behaviours the underlying drive is the maintenance of the individual; the social aspect is superimposed upon it.

SOCIAL BEHAVIOUR

Apart from the relatively short Interactions between mothers and their young, social behaviour may he divided into four main categories: play, the acquisition or social dominance, the acquisition of territory, and sexual behaviour. The eventual goal of all of these behaviours is to differentiate the fitter individuals from the less fit, allowing only the fittest to mate and to pass on their genes. The notion of equality has no place in nature.

Play

Play is a special form of social behaviour indulged in by the young of primates and of the carnivorous mammals, especially dogs, cats, weasels, and mongooses. It is noticeably absent in the young of the large herbivores, although occasional glimpses may be seen in lambs, foals, and calves. Play serves to provide the young animal with all opportunity to develop the skills it will later need to support itself in the wild. Young carnivores invest much time in mock fights, during which they practise attacking. Watch two puppies tumble about: they make frequent lunges to the throat and neck area, where a killing bite will be most effective. Primates are the most highly social of all mammals, and among them play serves to teach the young how to form social relationships and how to relate to other individuals. Play also teaches them how to react to signals sent out by another individual which indicate social status and mood.

Acquiring Social Dominance

Many mammals do not live in one place all the time, but migrate from one place to another. Frequently they manifest a social hierarchy, in which one—or a few—individuals are dominant over all the others. In a troop of baboons, for example, the dominant males travel in the center, while younger aspiring males make up the vanguard and rearguard. The dominant males are easily distinguishable by their size and by the thickness of their manes. Some sort of adornment makes dominant members of any group distinctive: it may be visual, as in the baboon or the blackbuck, in which only the dominant buck is black (all other males are light brown in colour); acoustic, as in the South American howler monkey, in which the dominant males have the loudest voices; or olfactory, as in the Australian sugar glider, in which the scent gland on the heads of the most dominant males secrete more copiously than those of the subordinate males. In all cases, the dominants are responsible for most of the matings.

Acquiring Territory

A large number of mammals are territorial, which means that the males compete for, and repel all other males from, a piece of land. The territory may be very large, as is the case with the large cats, containing enough food and other resources for rearing a family, or it may be a token patch barely big enough for the male and his consorts. The latter type of territory is seen in elephant seals and some ungulates, such as the Uganda kob antelope. Territorial lights may be aggressive, but as a rule the loser is not killed. Horns and antlers have evolved to minimize the risk of accidental death through goring. In most cases the territorial sex is the male, who may be substantially larger than the females, in order to give him the best chance of winning his encounter. Almost invariably the males show the visual, acoustic, or olfactory adornment necessary to demarcate and defend the territory against infiltration. Among small and medium-sized mammals, scent marking of obvious points on the boundary and within the hinterland of territory is commonplace. Such scenting places, redolent of the scent signatures of their visitors, act as information exchanges for all who pass by.

Sexual Behaviour

Mammalian sexual behaviour varies from the short and rough, as seen in bandicoots, to the highly stylized and ritualized performances of creatures such as the Uganda kob antelope. Whether mating is promiscuous, as in these two examples, or whether male and female mate for life, as in many of the small forest-dwelling antelopes, or whether it is something in between, as in most mammals, depends on many complex and intertwined environmental factors. Non-promiscuous species show a degree of courtship, during which a male and female annexe themselves from the group. In a number of species, the close proximity of the male acts to bring the female on heat, and only at this time will he attempt to mate. It is thought that pheromones, or signalling odors, in the male's urine act to stimulate the female's reproductive system. In some species—the domestic dog and the hamster are familiar examples—the female may emit a special odor which lures males from afar. She may then choose with whom she will mate. In those species in which both sexes need to help with rearing the young, for instance most primates and many carnivores, courtship will be of sufficient duration for the offspring of any previous alliance to become apparent before mating occurs. If the female is pregnant, a pair-bond may not form. In this way the male will be assured that the subsequent offspring, which he is helping to rear, is truly his own. Courtship in these species is characterized by subtle and covert cues, not intended for a general audience.

In stark contrast, those species that mate promiscuously advertise their sexual readiness in flamboyant behaviours. Female chimpanzees, who are strongly promiscuous, are adorned with a large patch of white sexual skin in the perivaginal area which inflates massively during the heat. Males are strongly attracted to this and line up patiently waiting their turn in the mating queue. Interestingly there is little interference and aggression seen in this whole bizarre process.

The Survival of the Species

Mammalian behaviour is not only complex, but also highly adaptive—that is, it contributes to the survival of the species just as much as any other biological attribute. Research into mammalian groups that occupy a wide range of habitats, such as primates or antelopes, is revealing that the ultimate factor that governs which type or behaviour evolves is the environment. Thus the strange behaviour of marmosets, in which the female always has twins but is large enough to carry only one youngster, may be explained by considering their territories. These are so large in relation to the tiny size of the animal that they can only be adequately defended by both the male and the female—a female on her own is unable to rear a family and protect the territory. Natural selection has evolved a reproductive strategy in which, if the male does not remain with the female to help carry and protect the young—and incidentally help defend the territory—his genetic investment in the next generation will be impaired. This apparent act of genetic blackmail on the part of the female is no more than the influence of the environment on aspects of the species biology that lead to increased chances of survival.

Attack, Defense and Survival

The battle for life makes it necessary for all mammals to have offensive and defensive adaptations. In predators the power to attack and kill is essential to survival; in their prey there is an equally urgent need to avoid or repel the hunter. The same capacities are needed in the often fierce competition for mates. Efficiency in attack and defense is, in fact, one of the most fundamental requirements of any animal.

In mammals, attack and defense are exemplified in many different forms. First there are the actual physical structures—teeth and claws, for example, or armor and quills—which have obvious so offensive or defensive functions. Less obvious physical specializations are such devices as the foul-smelling glandular secretions of skunks, the capacity of certain mammals to shed part of their tails, as lizards do, if these are grasped by a predator, and various camouflaging or warning colours. Yet a third expression of a mammals's offensive or defensive abilities is not connected with physical structure at all, but with behaviour. Here we can cite such reactions as flight, "freezing" into a rigid position, "playing possum" and on Many mammals are equipped with specializations in all three of these categories.

The horns of antelopes and the antlers of deer are particularly obvious examples of weapons possessed by mammals. Although comparable in function, they differ considerably in structure. Horns, usually possessed by both sexes, are permanent features which continue to grow throughout the animal's life. They consist of bony processes jutting out from the skull, covered with a hard substance known as keratin, which is actually tougher than bone. Keratin is also found in other mammalian structures such as the skin, claws, hoofs, hair and fingernails.

Antlers, by contrast, are composed of pure bone and are shed and regenerated seasonally. With the exception of reindeer and caribou, in which they occur in both sexes, they occur in both sexes. They are normally grown only my male deer. The main function of antlers seems to be connected with sexual selection, for they are at the peak of their efficiency during the mating season, when rival males are fighting for the possession of females. Antlers vary greatly in size and design, ranging from the elaborate branching structures grown by moose and wapiti to the much simpler antlers characteristic of many smaller species. In fact, antler size seems to be very largely correlated with the body size of the animal. Such small deer as the pudu of the Chilean Andes and the tufted deer of China have antlers which are only half-inch spikes, while Chinese water deer have no antlers at all. They are equipped instead with greatly elongated upper canine teeth, or tusks.

Fig. 5.1. The growth of horns.

There is considerable variation in the horns of wild cattle, sheep and goats. Cape buffalo, for example, have massive side-sweeping horns that curve outward and then upward at the ends. The musk ox has hooked horns that sweep downward and then up, while the little Celebes anoa has almost straight horns, which are directed backward. The wild sheep usually have horns that curve in a tight spiral, those of the Rocky Mountain bighorn often making a complete circle or more. The horns of wild goats are generally simple curves, but they often feature twists or spirals as well.

Even greater variations of design are found in the horns of antelopes. There are the rapierlike horns of the oryx, the sweeping, crescent-shaped horns of the giant sable, the double-curved horns of the impala, the little spiky horns of the duiker and the exceptionally large corkscrew horns of the kudu. Many other antelopes possess horns quite different from any of those mentioned.

Quite apart from their use to their owners, horns and antlers are of great interest in themselves. The yearly growth of antlers is a particularly remarkable phenomenon which imposes a great strain on the animals producing them. The structures develop from two projections of the frontal bone, known as pedicels, which are normally concealed by the skin of the head, but in such deer as the muntjac they protrude in prominent fashion. The growing process takes up to five months and resembles that of typical bone formation, but because it has to take place so much more quickly, the animal utilizes far greater quantities of lime and phosphoric acid than would normally be the case.

Starting as simple swellings that bud and push out from the bony pedicels of the skull, the antlers rapidly increase ins size and in many species they branch repeatedly as they grow. At this time the antlers may be bent or deformed very easily, for they are composed of flexible bony material enclosed in a covering of soft skin covered with fine hair, called "velvet."

Fig. 5.2. The growth of antlers.

A network of blood vessels in the velvet nourishes the antlers until they reach full growth and then ossification proceeds from the base until the complete structures become rigid and formidable symbols of their owner's virility. Finally, the velvet peels off, a process which the stage hastens by rubbing his antlers against trees and rocks or even against the ground. A little blood may appear at this time, derived from the blood vessels lying beneath the velvet. It is the drying up of these blood vessels which causes the velvet to peel, and their former channels can be seen as well-defined grooves in the completed antler.

After the breeding season, thè antlers are shed. A little blood usually appears again when each antler drops from its pedicel, but this quickly dries and the surface heals over. Within several months, new buds appear and the next set of antlers begins its grown.

With each successive growth, the size and elaboration of the antlers increases for a number of years. Thus the first pair of antlers grown by a yearling white-tailed deer are small and simple spikes. The following year they are bigger and more complex. In the European red deer, nine successive growths and sheddings are sometimes necessary before the antlers appear in their full glory; even the smaller roe and fallow deer need from five to six years for their antlers to become fully mature. For several years thereafter, provided the animals have an equivalent food supply each year, the antlers remain fairly constant in size and shape. But when the animal passes his full prime of sexual vigor, the antlers begin to degenerate in size and quality.

Besides such weapons as horns and antlers, the teeth of mammals are widely used for attack and defense. Mammalian teeth differ greatly from those of the reptiles from which they evolved. Reptiles characteristically possess a battery of peglike teeth, each of which—except in such cases as the fangs of poisonous snakes—is generally little differentiated from the next. Mammals, by contrast, have evolved much more specialized teeth, which as stated earlier, can be broadly classified into three groups: incisors, canines and cheek teeth. The use of these teeth in eating has already been discussed, but in many cases they are of at least equal value as offensive and defensive weapons. The general value of teeth for almost any animal in attack and defense is too obvious to emphasize, but some of the adaptations they have acquired for these activities are of particular interest.

The most typical offensive specialization in mammalian teeth is represented by the canines. There are usually two of these sharp, elongated teeth, both in the upper and the lower jaw, and they are excellent natural daggers. Canines are essential to such typical predators as lions and tigers in dealing with their prey and are also well developed as a deterrent to aggressors in non predators such as the gorilla. One of their most extreme specializations occurred in the extinct saber-toothed cats. But various mammals living today have canine teeth which are just as remarkable.

The tusks of the walrus are simple, highly developed canines. Apart from being used for digging up mollusks from the bed of the sea, these teeth are used by both sexes in defending themselves against such predators as polar bears and by the bull walruses in fighting rivals at mating time. The hides of old specimens are often formidably scarred by the tusk marks inflicted during these ferocious combats.

Another instance of teeth which have become specialized into tusks occurs in the elephant. But unlike the walrus' tusks, the tusks of elephants have not been developed from the canines, but from the incisors. Elephants when fighting are capable of inflicting dreadful wounds with their tusks, even to the extent of disemboweling their opponents. The tusks are also of value in protecting young members of ahead from the large cats, which are always ready to attack an elephant call if opportunity offers. The only creature against whom the tusks can offer no defense is man. With his powerful rifles he can kill an elephant from a distance of several hundred yards, and between 2,000 and 5,000 elephants have been destroyed annually in recent times in Kenya alone by sportsmen and ivory poachers.

As a final example of mammals possessing specialized weapon teeth, we may cite two members of the pig family: the wart hog and the babirussa. In the male, or board, wart hog the canines of the upper jaw turn upward and may attain a length of 8 to 10 inches. With this equipment it can defend itself and its family from the attack of all but the largest and most formidable predators. The babirussa, from the Celebes and parts of the Moluccas, is even better equipped, for the canines of the lower jaw are also well developed and grow in the same direction the upturned canines of the upper jaw. These upper canines actually grow through the roof of the snout. Besides the aggressive possibilities of this battery of specialized teeth, they also provide a protective cage which effectively screens the upper part of the face.

As a purely defensive structure, armor is particularly effective, and the bodies of various mammals have a protective covering which makes it difficult for predators to get at the vital organs. The thick skins of elephants and rhinoceroses are armor of a king, but the very bulk of the creatures also plays a major part in discouraging attack. More highly specialized armor is found in several smaller mammals, especially the armadillos and pangolins. Covered with thick plates, some armadillos can roll themselves into a ball so that their soft part are virtually inaccessible even to the most persistent predators. Others ward off attacks by pressing themselves flat on the ground beneath the arch of rigid plates which covers their backs, One of the most remarkable armadillos is scientifically known as *Chlamyphorus truncatus*. In addition to the armored "cloak" that extends along the upper part of its body, this creatures has its tail and hindquarters covered with a thick bony shield. If threatened, it bolts into its burrow and stops the entrance with its armored rear. The pangolins likewise have a protective covering of tough overlapping scales, or plates and can roll themselves up at will. The female Indian pangolin has even been seen to envelop her young in her own protective covering when danger threatens.

Fig. 5.3. An armor-plated rump.

Spines are another valuable means of defense, and the prickly surface of a rolled-up hedgehog is enough to deter most predators. Porcupines, however, possess the most formidable defensive structures of this kind. These animals, which may attain a length of two or more feet, live in both the Old and New Worlds. All are characterized by an array of quills on their backs, which can be raised by powerful muscles lying directly beneath the surface of the skin. This arrangement is most highly developed in the Old World forms. The common African porcupine, for instance, sometimes has quills as long as 15 inches.

When disturbed by a possible enemy, the African porcupine raises its quills and rubs them together to make a threatening sound. At the same time it makes grunting and snorting noises to show that it is not to be trifled with. If the annoyance persists, the porcupine charges swiftly backwards at its foe, with its quills held out like a battery of lances. If they make contact, many of the quills may become detached from the porcupine's back and remain embedded in the wounds they have made. This can have disastrous consequences to even the most powerful predator. There is one case on record of a tiger, almost fully adult, that was killed by a porcupine whose quills perforated its liver and lungs during one of these aggressive-defensive maneuvers.

The quills of the Canadian porcupine are not so long, but they are just as effective, for they have barbed tips, which enable them—like fishhooks—to work their way deep into the victim's flesh. The average Canadian porcupine has some 30,000 of these barbed spikes, and any that become lost or broken are replaced. The mountain lion and fisher successfully prey on the porcupine by bowling the animal over and attacking its soft underbelly. Few other experienced predators venture to attack it. Dogs are frequent victims and so are the inexperienced young of other meat eaters.

Fig. 5.4. A battery of quills.

An animal unfortunate enough to get a number of porcupine quills lodged in its flesh will be greatly distressed and may eventually die, for the quills continually work themselves deeper and deeper into the tissues. This occurs by muscular action on the barbs of the quills. The rate of penetration can be quite astonishing. A case is cited in the American *Journal of Mammalogy* of a quill fragment three quarters of an inch long which advanced almost two inches in just over a day.

Reinforcing the effect of offensive weapons and defensive armament is the concealing or warning coloration adopted by mammals. Rudyard Kipling in his story "How the Leopard Got His Spots" gives a delightful account of the principles of protective coloration in nature. In this story the giraffe, the zebra and the leopard, among other animals, were all reputed to live on the High Veldt and to be "sclusively sandy-yellow-brownish all over." But then, to avoid the leopard, the giraffe and the zebra retreated to the forests and acquired blotches and stripes from the broken shadows caused by the sun shining through the leaves. When the leopard followed them they were so perfectly camouflaged that they just melted into the forest with he words: "One-two-three! And where's your breakfast?" The leopard himself was then forced to acquire spots before he had a chance of hunting them on equal terms.

It would be wise not to trust Kipling too deeply about the way in which these protective colours were acquired, but his story does show how important camouflage can be to both predator and prey. The spotted pattern of the leopard does indeed conceal it most effectively in the wooded regions where it chiefly hunts. And many writers have commented on the way in which the zebra's stripes and the giraffe's blotches make these animals almost invisible in the thin cover where they habitually live, especially at dusk when an attack is most likely to occur. Thus one traveller, E. White Stewart, tells how on numerous occasions both he and his native tracker failed to observe zebra in scrub at distances of no more than 40 or 50 yards, and only detected them when they actually moved or switched their tails. Another traveler, Vaughn Kirby writes of the giraffe: "The ghostly manner of their disappearance is most remarkable. I have often been riding up to them, when some other object has drawn my attention off for a few moments; glance again, and thy are gone! Not merely gone to a distance though still in sight, but gone utterly, vanished like a mist-wreath at sunrise."

A striking example of the protective value of colour is found in the variations which occur between local races of the same species of mammals. This subject has been studied in detail by Hugh Cott of Cambridge University, England, one of the world's leading authorities an animal camouflage. Cott has brought together a large body of evidence to show that the colour variation in mammalian races is almost certainly due to the elimination by predators of individuals that do not match their backgrounds as effectively as others. Thus in desert regions, many small mammals are a much lighter and sandier colour than their close cousins, which may live in a darker environment. One race of pocket mice, which lives on the gypsum sand dunes of New Mexico's Tularosa Basin, is almost white. Another race of a closely related species inhabits dark lava flows in the same area and is almost totally black.

It seems certain that natural selection has been at work in such cases as these, eliminating the light forms from the dark-colored areas, and vice versa. Scientists have proved the point in controlled

Fig. 5.5. Seasonal colour changes of a weasel's coat.

experiments, using different varieties of deer mice on different backgrounds. The barn owls and long-eared owls that were released in the rooms with them did indeed take more of the mice which did not blend with their backgrounds.

Certain arctic animals have year-round white coats which match the perpetual snow and ice of their environment. These include the polar bear, the northern races of the arctic hare and arctic world, and the Peary caribou of Ellesmere and other Canadian arctic islands. With various other northern mammals, such as the arctic fox, the varying hare and the short-tailed weasel, colour variations are seasonal with the same individuals. These mammals often have a white coat in winter, but in summer assume a darker one.

The period during which the two coats are grown seems to be very closely correlated with the duration of total snow cover in the region. Thus the typical arctic fox of the far north alternates light and dark coats with the seasons, while members of the same species living in the less severe climatic environment of Iceland normally retain a dark coat the year round. Conversely, the local variant of the arctic hare which occurs on Ellesmere Island has a light coat throughout the year, while its cousin from the Hudson Bay region assumes a dark coat for they year, while its cousin from the Hudson Bay region assumes at a dark coat for some nine weeks in late summer when the snow is in retreat.

Although heredity plays a part in determining such changes of coat, experiments made with the snowshoe rabbit suggest that the length of daily illumination also has an important effect. The amount of light received through the eyes probably influences the pituitary gland, which in turn affects the hormones controlling the molting cycle.

But it is not only in the Arctic that mammals show a seasonal change in protective coloration. Another region where conditions vary greatly from one part of the year to the other is in the deciduous forests. Here is summer there is a thick canopy of leaves and a dappled pattern of sunshine and shadow is cast on the ground below. In winter this canopy disappears and a much more uniform light streams through the bare branches overhead. As we might expect, several mammals show protective adaptations to these changing conditions. To take only one example, the common fallow deer of Europe has a white-spotted coat in summer, which provides it with the same kind of concealment in this habitat as we have already described for the leopard. In winter, however, when spots would only make it more conspicuous, this coat is lost and the fur becomes a uniform grayish brown. The correlation of this species' seasonal coat changes with changes in the environment is supported by the fact that spotted deer living in tropical and subtropical evergreen forests do not go through such transformations. Thus the axis deer of Asia keeps its spots all the year round as a protective adaptation to the permanent leaf cover in its forest home.

In some mammals, bold warning colours rather than camouflage may serve as an aid to survival and this is often achieved by the presence of conspicuous colour patterns which contrast with the green and brown background of nature. Such patterns are generally associated with some dangerous or repellent characteristic and therefore act as a warning that aggression will bring unpleasant consequences.

The skunks are striking example of this particular defensive device. These familiar American mammals are represented by three distinct genera, all of which possess coats of striking black-and-white patterns. All are also well known for their ability to emit a powerful odor when angry or frightened. This originates in their anal glands, which secrete a foul-smelling substance that can be projected with remarkable accuracy to a distance of 10 to 12 feet.

It seems quite clear that the skunk's specialized and particularly unpleasant defensive weapon is closely correlated with its conspicuous markings. These are simply a warning advertisement which says, in effect, to a possible predator, "attack me at your peril." Moreover, when first disturbed a skunk will go through an elaborate aggressive display, designed to bring these warning markings in the most effective way to the attention of the animal threatening it. These displays vary with different species.

The common striped skunk, *Mephitis mephitis*, raises its plumed tail over its back and stamps its feet threateningly on the ground. The little spotted skunk, *Spilogale putorius*, is even more emphatic. Doing a handstand, this species supports itself on its forelimbs, with its hind limbs raised vertically in the air so that the full extent of its spotted back is shown to the aggressor. It can maintain this attitude for five or six seconds at a time and it is a very rash or inexperienced predator that will not in such circumstances hastily decamp. If it fails to do so, the skunk will next discharge its anal glans with great precision in the direction of its foe. Although not fatal, the results are so demoralizing that the attacker, if intelligent, will tend in the future to give the conspicuous black-and-white markings a wide berth.

As a final example of a physical defensive mechanism in mammals, a word must be said about self-amputation of the tail. This reaction to attack is, of course, very well known in many lizards which, if grasped by the tail or even it excited to extreme fear without being touched, will abandon part of it before dashing away into cover. The fact that certain mammals practice a similar defensive technique is not nearly so well known. Yet several small rodents, particularly dormice and field mice, are quite as capable of shedding part of their tails in emergency as lizards are, although the means by which this is achieved is somewhat different. Whereas the lizard tail drops away as a result of a clean break occurring in the middle of one of the vertebrae, a mouse only sheds part of the tissue in which its tail is encased. The tail is not thereby shortened, but a predator who grasps it is left only with its outer cover, while the mouse, with the inner part of the tail, escapes. Again in contrast to lizards, which can regenerate their tails at least partially, the denuded portion of the mouse's tail withers away in a few days and drops off. The value of this adaptation is suggested by the observations of field workers who have reported seeing more mice with mutilated than with intact tails.

Apart from the physical aspects of attack and defense, what are the patterns of behaviour adopted by mammals to achieve their aggressive or defensive ends ? The majority of these, as might be expected, are closely related to movement although some species also find safety in cooperation and numbers. Musk oxen, for example, if attacked by wolves, form a defensive circle, with the calves and young animals inside. Prairie dogs exhibit another type of defensive cooperation. Individuals sighting danger utter shrill cries which warn their neighbours to take cover. In similar circumstances a beaver will slap the water with its tail, signaling to the others in the colony that an enemy is approaching.

The most obvious method by which an individual animal can escape from a threatening predator is by running way. Flight is, indeed, the most common expedient adopted by those mammals which are the hunted rather than the hunter. Flight, however, is obviously useless if the pursuer can run faster than the pursued—unless by chance there is nearby cover which can be reached in time. Because of this, several mammals have adopted other techniques which either are used from the beginning or are substituted for flight when this proves to be useless.

What often happens is as follows. First, the predator begins to stalk its victim, which at this time may not be unduly disturbed. But when the distance between the two animals is narrowed to a certain point, the quarry is stimulated to flight. The predator must then decide either to dash directly after the quarry or allow time for its intended victim to settle down again so that it may make a new, and perhaps closer, approach. Even in the second case the moment comes when the predator must decide that it is the time to make a now-or-never attempt to reach its prey. Abandoning stealth, it rushes forward and the quarry dashes off at full speed.

But if the quarry realizes (although whether consciously or instinctively it would be difficult to say) that flight is useless, if often adopts a quite different behaviour pattern. In a last desperate effort to save itself, it may turn to face its persecutor and place itself in an aggressive attitude. Sometimes this sudden change of technique so surprises the predator that it will be shocked into temporary confusion, allowing the quarry a new chance to escape. A rabbit will sometimes do this, at the last gasp turning on the pursuing hound. In a somewhat different but quite comparable context, the weak boy at school may, in desperation, suddenly face the bully who is tormenting him and by a lucky blow put the larger boy to flight.

Another rather special defensive technique utilized by mammals is sheer immobility. The rabbit or deer "freezes" in its tracks at the sight of an enemy and perhaps escapes detection. The habit of freezing is also used by predators, especially by members of the cat family stalking their prey. Anyone who has watched a domestic cat stalking a bird feeding on the ground will have observed how the cat advances in a series of short quick run, pausing in between in a completely rigid attitude.

The classic example of the use of immobility as a survival technique is the death feigning of the American opossum, and the term "playing possum" has passed into general usage to describe such behaviour. This animal, if attacked or frightened, sometimes falls down on its side in a position that most realistically resembles death. The attacking animal, after a few sniffs at the prostrate body, will generally move away. A few minutes later, however, when the coast is clear, the "corpse" will come to life and assume its normal activities. This behaviour, incidentally, is probably physiological, for the sham is almost certainly a shock reaction to a crisis situation.

The Quick and the Dead

All of the survival devices developed by other animals have been adopted by mammals: various tactics of pursuit, group cooperation, and such weaponry as teeth and claws, spines, armor, scent and poison glands. Even the habit of feigning death is not unique. But mammals are more intelligent: many can quickly change their usual modes of attack and defense to cope with new circumstances.

Cooperative Hunting of the Pack

Hunting in groups is a particularly mammalian trait, depending on a high order of cooperation and intelligence. The wolves and their allies regularly run in packs and are able to bring down swift and resourceful, as well as larger and stronger, prey.

The relationship between wolves and moose has been situated in detail on Isle Royale in Lake. Superior. In summer the wolves' favorite fare of moose is supplemented by beaver and small game, but with the coming of freeze-up these disappear, and the long big moose is the prime target. The wolves range single file for miles until they sight one, then spread out downwind and stand still as pointers. Then they make a sudden rush, one large wolf heading for the quarry while the rest string out behind. If the moose runs, they close in, snapping at its rump and flanks. It may escape, but once a vital spot is hit, it dies in minutes. A moose that stands its ground, however, can always hold off the wolves until they slink away.

Learning to Cooperate with Other Mammals

The first lessons in self-sufficiency begin when the parent of a carnivore brings disabled but living prey—a small ground squirrel, mouse or rat—for the young to practice hunting and killing. The capture of larger food animals requires more patience and skill. Young wolves on the Alaskan tundra learn how to do this by accompanying and watching older members of the pack. When half-grown, they help in running down a caribou, but do not touch it until a more experienced wolf has made the kill. Later they will cooperate fully in stalking, attacking, killing and eating the prey. The learning process never stops, for every hunt poses some new problems. Occasionally a member of the dog of weasel family is so adaptable that it learns to travel and hunt with an individual of another species. This kind of association, based on self-interest, lasts as long as it is of some benefit to one or perhaps both of the partners.

The Strategy of Stalking

Unlike the members of the dog family, cats depend on stealth and the quick dash to secure their prey, and usually hunt alone or in pairs. Some exceptions are the Canada lynx, whose young accompany the mother through the first winter, and the lions on the African plains that travel in family groups, or "pride." Although they do not run as a pack, three or four members of the pride occasionally cooperate in hunting, and when they do, their strategy is deliberate and organized. An experienced lion generally moves upwind of a herd and, by growling and giving the uneasy animals its scent, causes them to stampede toward the hunting party. A lioness usually makes the kill, but the old males eat first, followed by the females and young. The remains of the carcass ordinarily fall to hyenas, jackals and vultures.

When close enough, a lion simply stalks and then rushes at its quarry. If it can get within 30 yards of an antelope, it can usually pull it down in a few bounds. The maximum speed of the charge is about 35 miles an hour. However, a lion is not so well adapted for the chase as the lithe cheetah, which is capable of running down its victim at 65 to 70 miles an hour. It can maintain that pace for a quarter mile.

The Tactics of Bluff

When an animal cannot get away, it may try to appear larger and fiercer than it really is, it may fight back in desperation or even feign death. The smaller and weaker mammals display a variety of these defensive reactions. Certain shrews hump their backs, grate their teeth and utter high-pitched squeaks; then, as an ultimate gesture, they collapse on their backs with spread and waving feet, repeating staccato notes. Other rodents, such as the house mouse and the European field mouse, rear up, bit and spar silently with their front feet, and the ordinarily inoffensive cottontail rabbit has been known to kick a bobcat in the face with its strong hind legs. Although opossums may try to bluff larger animals by hissing and baring their teeth, they are the best of the pretenders among the higher animals. A pursuing dog can easily catch a running opossum and kill it, but if it "plays dead," it usually not be molested. When it does this, the opossum may be pretending, or actually scared into a state of catalepsy, in which its heart rate slows down and it can stay motionless for a span of as little as eight seconds to as long as six hours. The jackal, honey badger and stiped hyena may also faint or feign death.

Sharp Quills and Hard-to-Crack Defenses

Some mammals combine special external protections with the ability to roll into a defensive ball, which makes their capture difficult or unrewarding. The pangolin and various armadillos, which are armored, all roll up. Porcupines, however, are often overtly aggressive.

They advertise their presence loudly and, when irritated, erect their quills. A swat from a porcupine tail will embed hundreds of quills in a predator. The European hedgehog is also noisy in its movements, but it usually curls up when threatened. An active climber, it occasionally falls from heights, and its quills act as a cushion on landing.

"This shall End When One is Dead"

Encounters between two animals that are both well equipped to fight also occur in nature, and when they do there may be a prolonged struggle to the death. The Indian mongoose has, through many centuries, earned the reputation of a snake killer. Its prowess is celebrated in a jungle story by Rudyard Kipling in which a pet mongoose, Rikki-Tikki-Tavi, battles and kills a pair of cobras that attack him in defense of their nest. It is not true, however, that "all a grown mongoose's business in life is to fight and kill snakes." It preys on nearly anything it can get, including grasshoppers, frogs, lizards, nesting birds, rats and some of the smaller snakes, rarely attacking large poisonous snakes. But when mongoose and cobra, for whatever reason, become engaged at close quarters, there is no escape for either. The fatal dance must go on, snake's blow against mongoose's jump for an hour or more. The cobra usually loses because it cannot strike or retract fast enough and is not adapted to stab, but must bite to inject venom. In a fight with a constrictor or a pit viper, the mammals would be at a disadvantage.

6

Appearance of Mammals

Mammals are warmblooded animals whose body is more or less clothed with hair. Whose young are produced alive, except the egg-laying monotermes, and are nourished for a while after birth by the secretions of milk (mammary) glands. The skeleton shows several import and distinctions from most reptiles and birds, in having double occipital condyle-the articular facets which unite the skull with the first cervical vertebra-and having a simple lower jaw. In reptiles and birds, on the other hand, the condole generally single and the jaw is a bony complex. There is also an intervening bone, the quadrate, between the jaw and the skull, which is lacking in the mammals. A further mammalian distinction lies in the fact that the vertebrae and limb-bone ossify from three separate bone-forming centers: the body or shaft, as the case may be band the articular ends or epiphyses.

The mammalian dentition is peculiar in the local differentiation of the teeth (heterodonty) into incisors, canines, premolars, and molars and also in that there are but two sets in series, the milks or lacteal teeth and the permanent ones; never does one see anything comparable to the amazing successional teeth of the predentate dinosaurs, for instance. Secondarily acquired simplicity of the teeth may occur as in the toothed whales, and the number of teeth may be increased or reduced.

The mammals are certainly the highest class of vertebrates from many standpoints, in some ways, however, this place may be disputed by the birds but he latter represent culmination of the one line of ascent and the mammals another.

ORIGIN OF MAMMALS

Stock

At least two views have been held as to the origin of mammals. The older one, that advocated by Huxely in 1880, would derive, them from the much in Huxlye's theory is undoubtedly correct, except that authorities do not now believe that the Amphibia represent the next lower stage, but that there was an intervening condition, one in which gill-breathing had been lost but truly mammalian characters had not yet appeared, although some of them were already foreshadowed.

There are found in Triassic rocks in South Africa a group of reptiles known as the Cynodontia, in allusion to their dog-like teeth, and there seems to be a large body evidence in favour of the view that out of this group, although no known member of it the mammals have been derived. The cynodonts resemble the mammals in the possession of a heterodont dentition, in that the teeth are clearly divisible into incisors, canines, and molars, and in the paired occipital condyles in addition to which there are similarities of construction. Structurally the cynodonts bridge the gap between reptiles and mammals because while the dentary, the single bone of the mammation lower jaw is large and important, the jaw is nevertheless complex in that it possesses the several bones typical of the reptile. There are other reptilian characters as well all, of which may reasonably be expected in the remote ancestors of the Mammalia. These cynodont reptiles are low in the reptilian scale, far removed from the dinosaurs and birds with which we have been concerned, although capable of as high a degree of specialization along other lines.

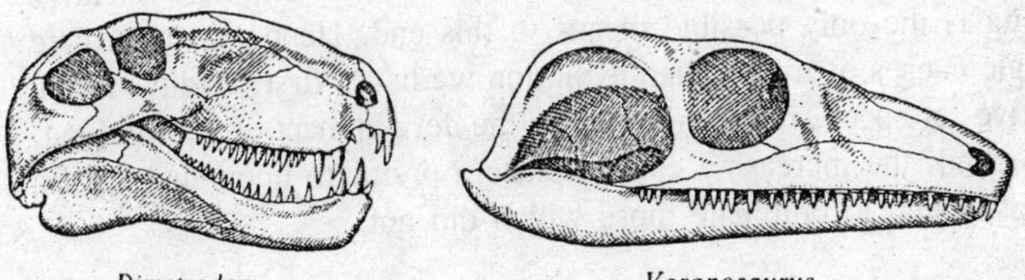

Fig. 6.1. Skulls of two early syanpsids, that of Dimetrodon was about 40 cm long and of Varanosaurus 23 cm.

Place of Origin

While we find members of this order in both North America and Africa, although they must have been much more widely diffused, it was possibly in the latter place that the mammals arose, probably due not so much to the potentiality possessed by the reptiles of one place over the other as to a happy combination in Africa of a potent stock and an impelling cause.

Cause

Therein is emphasized the statement made by Broom that all of the characters wherein a mammal differ from a reptile are the result of increased activity, for the says that when the cynodont took to walking with feet underneath and body off the ground, it first became possible for it to become warm-bloded animals. But back of this lay impelling geologic causes which Brom does not even hint at. These were, first increasing aridity of climate which from Permian time well into the Jurassic seems to have been characteristic of all lands, as the extensive series of red sediments may imply. And aridity has been found to be a great stimulative to speed. Add to this the evidence, especially in the southern hemisphere between latitudes 20° and 35°, of extensive glaciations-greater even than in the more familiar Plesitoncene Glacial period and we have a high incentive to the retention of bodily heat. For a while, perhaps, there were recurring warm seasons, sufficiently long and frequent so that a normal reptilian life could still be led, the creature hibernating when the weather reptilian life could still be led, the creature hibernating when the weather became to severe. But as in the case of the aestivation lung-breathing fishes, sufficient time must still be had for the active portion of the creature's career, so that, although in the origin of terrestrial forms premium would be placed upon the capability of atmospheric respiration, here it would be upon ability to withstand the cold and yet remain active and the acquisition of warm blood and heat retentive clothing is the only possible means to this end. Hence as immediate geologic causes of mammalian evolution we have, first, aridity, in the incentive for speed, rendering *possible* the development of warm blood, and second, the increasing cold to place a premium upon such as did develop it and to eliminate those which did not.

Time

The time of mammalian evolution can only be fixed within certain limits. Geological evidence, if we have read the cause aright, points to its inception in Permian time for "there is now undeniable evidence that in middle Permian times most of the lands in the southern hemisphere were subjected to as severe a glacial climate as that of the present polar areas, and that this Permian glaciation was likewise interrupted by times of warmth" (Schuchert).

On the other hand, geological record seems to point of the Triassic at any rate of the time of *culmination* of the evolutionary movement, for here are found for the first time by cynodont reptiles and the actual relics of the mammals themselves. But the cause must always precede the effect and it may be that the known cynodonts were persistent reptilian survivors of the group out of which the mammals actually sprang, and that the earliest known mammals themselves had already had a long transitional period. It seems reasonable to suppose, therefore that the time of mammalian origin was not later then Middle Triassic nor earlier than Lower Permian.

Mammals of the Mesozoic

It spite of all the uncertainties of the fossil record it is now possible to follow the history of the Mammalia back to their origin from cotylosaurian reptiles of the Permian, more than 225 Ma ago. Sufficient information is available for us to be quite sure that some population of early anapsid reptiles such as *solenodonsaurus* of the Carboniferous and Permian times, besides giving rise to all the modern reptiles, to the dinosaurs, and to the birds, also produced the mammals. The evidence for this connection rests on a most interesting series of fossils, together with some 'living fossils' the monotremes of Austrialia. The fossil history is not at all times equally clear. The mammalian stock first became distinct in late Carboniferous times, as a special type of cotylosaurian reptile, with a tympanum placed behind the jaw and later a lateral temporal aperture. This stock quickly became very abundant and successful as the pelycosaurs and therapsides of Permian and Triassic times, but then nearly died out in the Jurassic, during which period we know of the mammals only from fragmentary remains of a few small animals. Then, in the Cretaceous, some of these small forms became more numerous from them arose, before the Eocene, a variety of different types of mammal, from, which the histories of the modern orders can be followed in some detail.

We have, therefore, abundant evidence of the earliest stages of mammalian evolution, say 200 Ma ago from fossils in the Permian Red Beds of Texas. For the next following 40 Ma or so we have also rich material from the Upper Permian and Triassic Karroo beds of South Africa. In these early times the mammal line was a flourishing one, more so indeed than the diapsids or any other of the descendants of the cotylosaurs. The animals were mostly carnivorous, though there were also herbivorous types. Many of these early mammal-like forms became quite large and numerous. In fact this stock dominated the land scene in the Permian. In the later Triassic, however, if our fossil evidence is a safe guide, their numbers became reduced, perhaps the carnivorous dinosaurs took their place. Already at this time many of the essential features of mammalian organization had been developed, so far as these can be judged from the bones. Possibly the soft part of these Permian forms were also mammal-like, the animals may have been active and 'intelligent', have possessed hair and warm blood. But their brains were small and reptilian in structure.

Throughout the succeeding 100 Ma of Jurassic and Cretaceous times our knowledge of the mammalian stock is dim. Mammals of a some what rodent-like type, the multituberculates, became quite numerous and produced some larger forms, but then became extinct in the Eocene. Unless we have been singularly unlucky in the preservation of mammalian remains, no other mammal-like animals large than a polecat existed throughout this long period. Since we possess detailed information based on numerous fossible, about scores of large and small diapsid reptilian types, it can hardly be only an accident that mammal-like fossils are so rare.

We must conclude that the mammalians organization, after an initial success in the Permian and Triassic, was almost supplanted in the Jurassic and Cretaceous by the various diapsid creatures. Only a few are fossils, mostly of teeth and lower jaw bones, to give us some idea of the nature of the animals that carried out type of organization through the Jurassic period. Then from the top of the Cretaceous, about 70 Ma ago come the first fossil remains of mammals similar to those alive today; rare, insectivorous creatures, not widely different from modern hedgehogs. This was the time of the beginning of an astonishing revolution, which completely altered the life on the face of the earth. The descendants of these shrew-like animals multiplied considerably in the new conditions, and by the earliest Palaeocene times, 65 million years ago, they had produced recognizable ancestors of most of the modern orders of mammals.

Our knowledge of the origins of mammals derived from fossils is supplemented by certain surviving mammals, whose structure shows them to have diverged rather early from the main stock, especially the monotremes (duck-billed platypus and spiny anteater) and the marsupials (kangaroos, opossums, etc). Unfortunately it is still not clear exactly how these surviviours are related to the main stocks as Pleistocence fossils; so, tantalizibgly, we know only little about the affinities of this ancient group. The characteristics of their bony skeleton show that they must have diverged from other mammals well back in Mesozoic times. For instance, the pelvic and pectoral gridles are very 'reptilian' in structure. It is therefore not wild speculation to use the character of the soft parts of monotrems to deduce those of the mammalian stock in late Triassic or earky Jurassic times, say, 195 Ma ago. The marsupials appear as fossils in the late Createaceous. Some of the earliest forms are very like modern opossums, and we may conclude that the condition of modern marsupials throws some light on the probable condition of the soft part of mammals 70 or 80 Ma ago. However, it has been realized in recent years that many features in marsupial organization are special developments, not 'primitive' characteristics carried over from the ancestral condition.

The piecing together of all this evidence to give a reliable picture of the history of mammals is a valuable exercise in zoological and geological method, as well as a means of becoming famillar with a fascinating story. We may now returns to the Carboniferous times to examine the evidence more in detail.

Mammal-like Reptiles, Synapsida (joined arches)

All our evidence about the origin of mammals must of course be based upon the study of hard parts, which can become fossilized, and in particular on the skull. The characteristic feature of the skull in the populations that led to the mammals is the development of a hole in the roof, low down on the side of the temporal region, the lower temporal fossa. This hole was at first bounded above by the post-

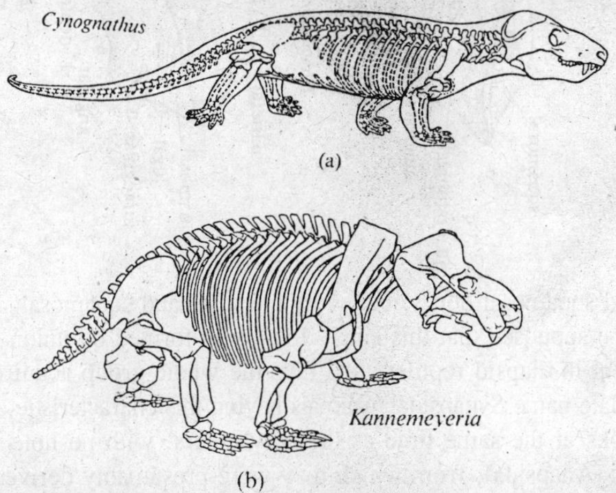

Fig. 6.2. Skeleton of synapsid reptiles. (a) Cynognathus, (b) Kannemeyeria.

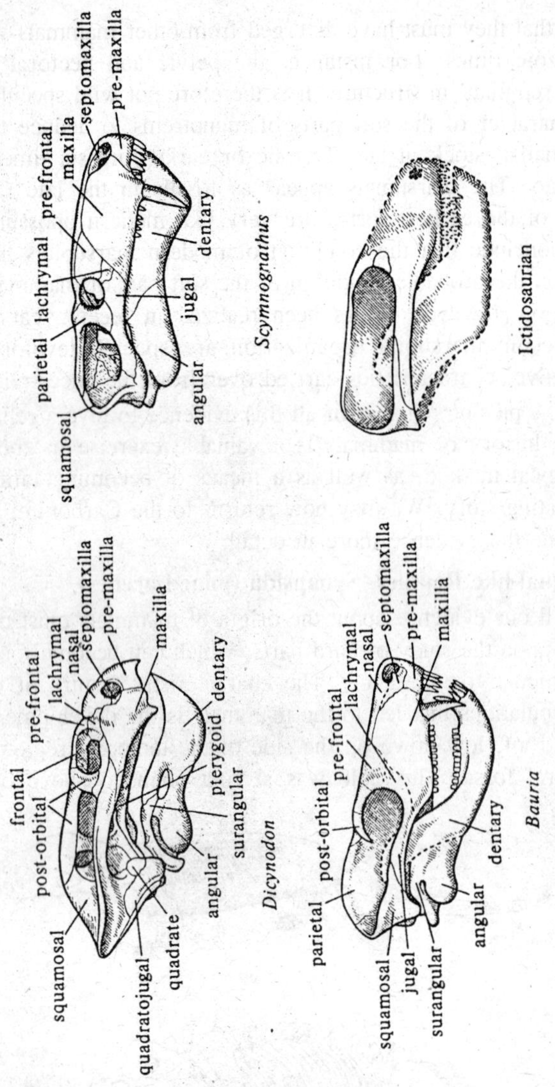

Fig. 6.3. Skull of various synapsid reptiles.

orbital and squamosal bones, below by the jugal and squamosal. It was at one time supposed that this single fossa was formed by union of the two present in diapsid reptiles, and thus the whole group acquired the inappropriate name Synapsida. Animals having this characteristic appear in the rocks at the same time as the cotylosaurs, with no hole in the skull roof (Anapsida), from which they were presumably derived. We have only fragmentary knowledge of anapsids from the carboniferous;

most of our information comes from lower Permian forms, such as *Solenodonsaurus*, yet there are quite well preserved synapsids of Upper Carboniferous date. Therefore we have no complete series of successive types to show how the earliest mammal-like types arose; nevertheless we can see something of the probable stages of this progress within the Permian Anapsida. One of the characteristics of the skull of amphibians was the presence of an oticnotch, in which lay the tympanic membrane, at the back of the skull. In *Captorhinus* and similar anapsid fossils from, the Lower Permian this notch had disappeared. The tympanum apparently lay behind the quadrate, leaving the whole side of the skull as a rigid support for the jaw. A long series of subsequent evolutionary changes let to one of the most characteristic features of the skeleton of mammals, namely, the conversion of the hinder jaw bones (quadrate and articular) into ossicles for the transmission of vibrations to the ear.

Captorhinus and its allies were 30cm or more in length and showed some development of the limbs towards the mammalian condition, in that the bones were moderately elongated and slender and the limbs perhaps tended to be held vertically under the body. However, there was no great development of a back-wardly directed elbow and forward knee. The teeth were considerably specialized, possibly for eating molluscs, and there were several rows of crushing teeth on the edge of the jaw, and long overhanging ones in the premaxillae.

Perlycosaurs

Anatomy of Early Mammal-like Reptiles

Pelycosaurs are recognized by fragmentary remains as early as the Lower Pennsylvanian. By the Upper Pennsylvanian, they constituted approximately 50 percent of the described reptilian genera. Archaeothyris, from the middle Pennsylvanian provides the earliest evidence of the configuration of the pelycosaur skull and gives some information regarding body proportions and details of vertebral and limb structure.

In their general appearance, early pelycosaurs would have resembled large living lizards such as iguanids and varanids although the pelycosaur's limbs were relatively shorter. Middle Pennsylvanian pelycoasaurs were at least twice the length of contemporary protorothyrids, and they slow progressive size increase into the early Permian.

The original divergence of the pelycoaurs primitive amniotes may have resulted form an adapoptation toward feeding on large prey. Not only is the entire body much larger, but the size of the skull increases disproportionately. The canine teeth, which are present in the most primitive amniotes, are accentuated, and the remaining teeth are also enlarged relative to hose of the insectivorous protorothyrids. In relation to feeding on larger prey, the postorbital region of the skull is shorter but higher than in other primitive amniotes. There is a shorter area for the adductor muscles to attach to the lower jaw, but the individual muscle fibers were longer. Both features permit a wider gape and a rapid, but less powerful, jaw-closing action. Pelycosaurs were the first carnivorous (as opposed to insectivorous) amniotes.

Pelycosaurs also appear to be the first major group to diverge from the primitive amniote stock, as can be seen by the retention of several primitive features that are lost in all other groups of primitive reptiles. These features include the presence of two, rather than a single, coronoid bone in the lower jaw and the large size of the medial central of the tarsus.

On the other hand, even the earliest pelycosaurs are advanced in a number of cranial features that characterize later synapsids and provide the basis for the origin of mammalian anatomy. The most important of these features is a lateral temporal opening where the postorbital, squamosal, and juggal met in early anapsids. In contrast with early diapsids, the quadratojugal does not form the ventral margin of this opening in primitive pelycoasurs. In contrast with protorothyrids and early diapsids, the supratemporal and postorbital meet above the temporal opening and separate the squamosal and parietal superficially. The arrangement of bones is somewhat similar to that seen in anthracosaurian labyrinthodonts, which may be close to the ancestry of reptiles. The pattern in pelycosaurs might represent the primitive condition for amniotes, but the great posterior extension of the postorbital can also be interpreted as a specialized feature that was developed to strengthen the skull as the temporal opening developed. All the bones of the skull seen in other early amniotes are retained, but the postparietals fuse at the midling to form a single median element.

The skull of early pelycosaurs is also distinguished by the configuration of occuiput. It is essentially vertical in other groups of early amniotes. In pelycosurs, it slopes posteriorly from the skull table to the area of the jaw articulation. As in protorothyrids, the braincase in primitive pelycosaurs is not solidly integrated with the dermal bones that surround the occiput. The middle Pennsylvanian genus *Archaeothyris*

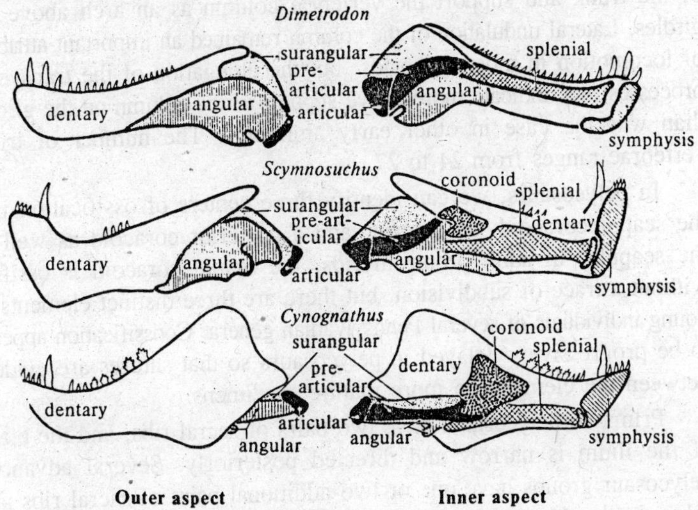

Fig. 6.4. Reduction of the articular and other bones and increase in the dentary in therapsid reptiles.

retains a discrete supraoccipitall that was not suturally attached to the skull table or firmly connected to either the tabular or the otic capsule. The otic capsule is small and does not contact the cheek. This pattern is retained in the primitive Lower Permian genus *Ophiacodon*. Other pelycosaurs integrate the bones of the occiput into a platelike structure that incorporates the supraoccipital, broad paroccipital processes, and the dermal bones of the margin of the skull. The stapes, which may have assisted in the support of the baincase in early pelycosaurs, remains a large element throughout the group, although it no longer served a structural role after the solidification of the occiput.

Postcranially, primitive pelycosaurs differ from the early protorothyrids primarily in features that may be associated with their greater body weight. The limb bones are more massive, with large and mode complex areas for muscle attachment, and the metapodials and phalanges are relatively shorter.

Like the protorothyrids, the pelycosaurs retion narrow and arches, with the zygapophyses close to the midline. In primitive species, the neural spines are low and triangular, but they become greatly elongated in several families and the transverse processes also become longer. These changes may be related to an early stage in reorganization of the trunk musculature. In mammals, the elaboration of axial musculature above and below the transverse processes helps limit lateral undulation of the trunk and support the vertebral column as an arch above the girdles. Lateral undulation of the column remained an important attribute of locomotion in most synapsids, but the elongation of the transverse processes may indicate more active support of the column on the girdles than was the case in other early ammiotes. The number of trunk vertebrae ranges from 24 to 27.

In pelycosaurs, we can identify three centers of ossification into the scapulocoracoid: an anterior and a posterior coracoid as well as the scapula. In adult protorothyrids, the scapulocoracoid is ossified without a trace of subdivision, but there are three distinct elements in young individuals of several Pennsylvanian genera. Coossification appears to be progressively delayed in pelycosaurs so that sutures are evident between the elements in more mature specimens.

Primitive pelycosaurs have two pairs of sacral ribs, and the blade of the ilium is narrow and directed posteriorly. Several advanced pelycosaur groups have one or two additional pairs of sacral ribs and a broad iliac blade to support their greater body size.

Diversity of Pelycosaurs

By the early Permian, pelycosaurs made up 70 percent of the known ammniote genera and had diversified into a number of distinct families. Unfortunately, remains from the Carboniferous, when these groups underwent initial differentiation, are still poorly known, and specific interrelationships of these families continue to be in dispute.

Romer and Price (1940) provided the most comprehensive anatomical and taxonomic review of the pelycosaurs. They recognized three suborders: the Ophiacodontia, which include a primitive ancestral assemblage and large piscivorous genera from the Lower Permian; the Sphenacodontia, which include large carnivorous genera; and the Edaphosauria, which consist of two groups of herbivores. In their recent taxonomic revisions, Reisz (1980, 1986) and Brinkman and Eberth (1983) questioned the validity of these suborders but recognize most of the family units that Romer and Price had described.

Ophiacodontiade

The oldest-known and most primitive pelycoasaurs are included in the family Ophiacodontidae. A side from primitive features, this family is characterized by a long narrow snout and relatively low skull table. It is represented by several genera in the Pennsylvanian and remains common in the Lower Permian. The largest and best-known genus is *Ophiacodon*, which was more than 4 meters long. The carpals and tarsals were poorly ossified, which led Romer and Price to suggest that this group was semi-aquatic and that the long narrow snout was an adaptation to feeding on fish.

Sphenacodontidae

Phylogenetically, the most important of the pelycosaur families are the sphenacodonts. They appear in the middle or late Pennsylvanian and are the dominant terrestrial carnivores in the Lower Permian. At some stage in their history they gave rise to the therapsids. The best-known sphenacodonts are the Lower Permian genera *Dimetrodo* and *Sphenacodon*, which exceed 3 meters in length. The skull is typically much deeper than in the ophiacodonts and the canine teeth are greatly enlarged. To support the dentition, the maxilla extends far up on the snout and separates the lachrymal bone from the margin of the external nares.

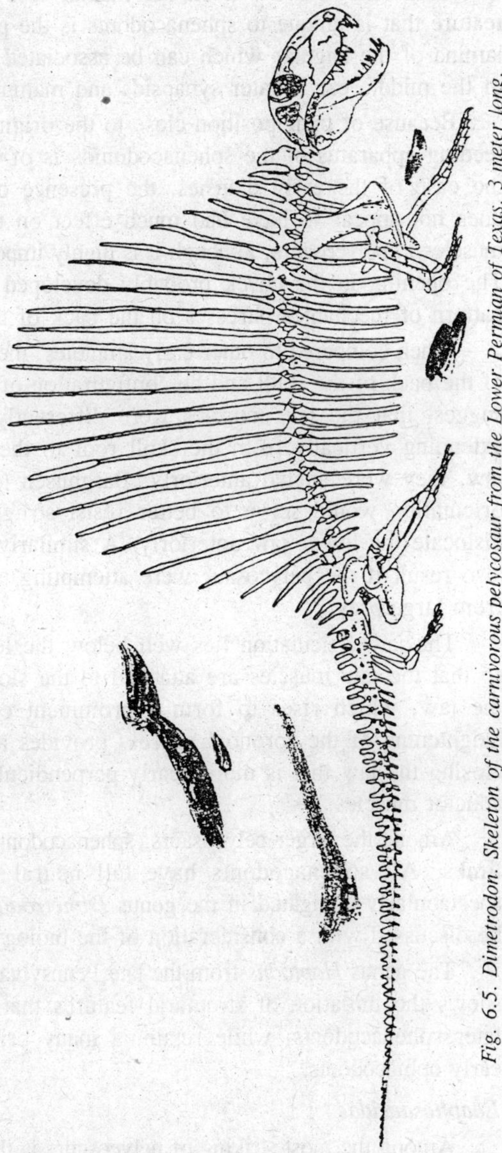

Fig. 6.5. Dimetrodon. Skeleton of the carnivorous pelycosaur from the Lower Permian of Texas, 3 meter long.

The jaw articulation is well below the level of the tooth row in sphenacodonts, the articulato is greatly enlarged, and the articulating surface extends far down on the medial face of the lower jaw. A feature that is unique to sphenacodonts is the presence of a reflected lamina of the angular which can be associated with the development of the middle ear in later synapsids and mammals.

Because of their position close to the origin of the therapists, the feeding apparatus of the sphenacodonids is of special interest. As in the case of the early diapsies, the presence of a temporal opening does not appear to have had much effect on the pattern of the jaw muscles in pelycosaurs, although it is highly important in the therapisds. The opening in the cheek probably developed in relationship to the pattern of mechanical stresses on the back of the skull.

When compared to other early amniotes, the changes in the shoape of the back of the skull and he configuration of the adductor chamber suggest that the jaw muscles were differently oriented. Instead of extending vertically from the skull roof to the dorsal margin of the jaw, they were angled anteriorly. Barghusen (1973) argued that this orientation would serve to better resist struggling prey that might dislocate the lower jaw anteriorly. A similarly directed force would also result if the pelycosaur were attempting to tear chunks of meat from large prey.

The jaw articulation lies well below the level of the tooth row, so that the jaw muscles are attached to the sloping dorsal margin of the jaw, which rises to form a prominent coronoid process. The heightening of the coronoid process provides a longer lever arm for closing the jaw that is more nearly perpendicular to the fibers of the aductor muscles.

Among the larger pelycosaurs, sphenacodonts have the most gracile limbs. All sphenacodonts have tall neural spines, but they are spectacularly elongated in the genus *Dimetrodon*. Their function will be discussed with a consideration of the biology of the pelycosaurs.

The genus *Haptodus* from the late Pennsylvanian and early Permian shows the initiation of structural features that are elaborated in the later sphenacodonts, while retaining many primitive features of the early ophiacodonts.

Edaphosauridae

Among the most striking of pelycosaurs is the genus *Edaphosauris*, which is known from the late Pennsylvanian and early Permian. Like Dimetrodon, it has greatly elongated neural spines, but they are further characterized by the presence of short, transversely directed cross bars. The skull is markedly different from that of sphenacodonts in being very short and broad. The teeth are not long piercing and slashing structures but short pegs of nearly uniform length. The palate and inside surface of the lower jaw are covered with additional teeth of similar size that form a large biting surface. The individual teeth are not expanded or otherwise specialized for crushing hard food and Romer and Price suggest that they May have fed on relatively soft plant material. The limbs are shorter than those of sphenacodonts, and the rib cage is long and laterally expanded as an one might expect in a herbivore.

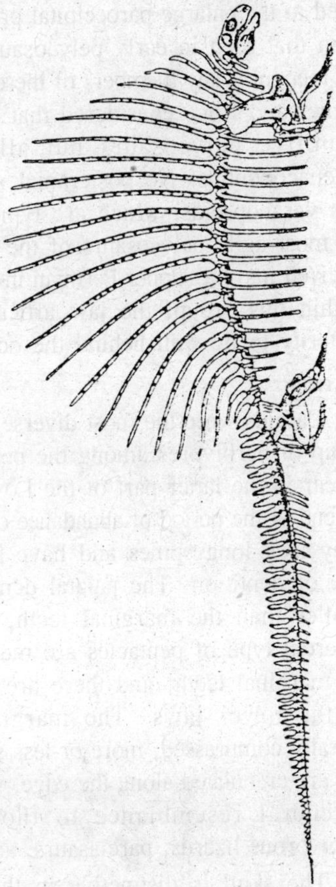

Fig. 6.6. Skeleton of Edaphosaurus.

A newly described edaphosaur from the Upper Pennsylvanian, Ianthosaurus, lacks tooth plates on the palate and lower jaw, and the marginal dentition resembles that of primitive sphenacodonts. Reisz and Breman (1986) argue that Ianthosaurus may demonstrate a close link between these families. The separation of these two groups must have occurred before the evolution of the distinctive sphenacodont reflected lamina of the angular that is not observed in the edaphosaurs.

Eothyrids and Varanopseids

The phylogenetic position of two families of smaller pelycosaurs, the eothyrids and varanopseids, remains in doubt. Their skulls appear primitive relative to those of other pelycosaurs in having a short antorbital region and a lacnmal bone that reaches the narial opening. The tooth row is horizontal and at the same level as the jaw articulation. They seem advanced in the large size of the temporal opening, which is bordered ventral by a narrow bar made up of a long quadratojugal that meets the maxilla and separates the jugal from the margin of the skull. The ocfciput is formed by alarge platelike supraoccipital that is fused to the enlarge paroccipital processes. This pattern is quite distinct from that seen in early pelycosaurs and genus *Ophiacodon*. We don not recognize any members of these families before the Lower Permian. Romer and Price considered that the eothyrids were among the most primitive pelycosaurs but allied the varanopseids with the sphenacodontids. No postcranial material is known of the eothyrids. The varanopseids, which are typically about meter long, are among the most agile pelycosaurs of the late Lower Permian and may have survived into the Upper Permian in southern Africa. Within this group, the jaw articulation extended posterity to lie well behind the occipital condyle.

Caseidae

Caseids were the most diverse and widespread group of herbivores among the pelycosaurs. They appear in the latter part of the Lower Permian, at the end of the period of abundance of *Edaphosaurus*. They lack long spines and have a very different type of dentition. The palatal denticles are much smaller than the marginal teeth, and have very different type of pentacles are much smaller than the marginal teeth, and there are no tooth plates on the lower jaws. The marginal teeth have laterally compressed, more-or-less spatulate crowns that are crenulated along the edge, which give them a general resemblance to those of modern herbivorous lizards, pareiasaurs, and plateosaurs.

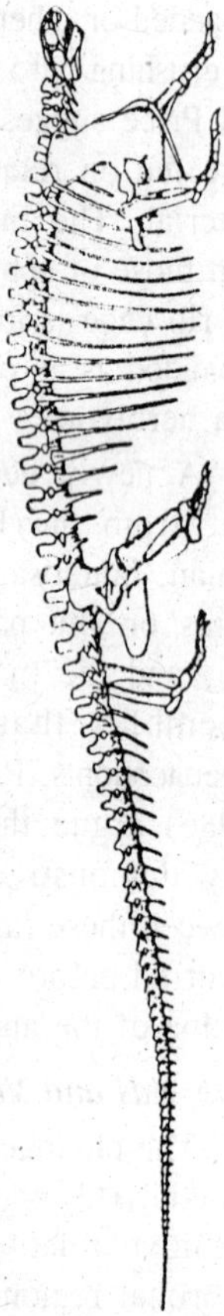

Fig. 6.7. Skeleton of Cotylorhynchus.

The skull is distinctive in the large size of the temporal opening and the relatively enormous external narial opening. The surface of the skull is sculptured with a scattering of rounded pits. The oldest-know genus, Casea, is little more than 1 meter long, but *Cotylorhynchus* reached 3 meters and weighed at least 600 kilograms. In most genera, then skull is extremely small relative to the trunk length, and the rib cage is enormously expanded to surround a great mass of food undergoing digestion. The number of sacral ribs increases from three to four pairs within the group. As in edaphosaurs, the supinator process has expanded distally to surround completely the ectepicondylar foramen. The hands and feet are short and wide (a feature elaborated progressively within the group). The phalangeal count is always less than in other pelycosaurs and may be as little as 2, 2, 3, 3, 2.

Romer and Price placed edaphosaurs and caseids in the same suborder. Both share a number of derived features, but most can be associated with their large size. The dental specialization toward herbivory is entirely different in the two groups. If they did share a common ancestry, it must have been at the level of primitive, camivorous pelycosaurs. Reisz pointed our that eothyrids and caseids a are unique among pelycosaurs in the possession of an anteroposteriorly elongated external narial opening, a pointed rostrum, and the entrance of the maxilla into the ventral margin of the orbit. Both also possess a short face and a low platelike occiput, although these features are not unique to these groups.

As Olson pointed out, the eothrids do not closely approach the caseids in the nature of their dentition. When they first appear in the fossil record, caseids are very distinct from all other groups of pelycosaurs. Fossils of caseids do not occur with the normal Permian fauna in Texas and Oklahoma and they may have lived in a distinct environment. Caseids are represented in the middle Permian of France and Russia but are otherwise limited to North America.

Biology of Pelycosaurs

There is only one skeletal feature of pelycosaurs that shows unequivocally that they are related to the ancestry of mammals – the presence of a lateral temporal opening of a pattern known in no other early tetrapods. There is no feature that suggests any tendency in the direction of a high metabolic rate or other physiological characteristic that typifies mammals. Pelycosaurs probably had a physiology that was very similar to modern reptiles such as lizards, turtles, and crocodiles.

One feature that indicates temperature control is the tall neural spines of the sphenacodonts. Their high degree of articulation in the fossils, even when individual spines have been broken shows that they were embedded in a single sheet of tissue or "sail" in the living animal. Romer (1948) calculated that the length of the spines in *Dimetrodon* increased disproportionately relative to other linear dimensions, so that the area of the sail varied in proportion to the volume of the body. Grooves at the base of the spines probably accommodated blood vessels. A vascular supply to the sail would permit it to function as a rapid means of absorbing or radiating heat from the body.

According to Bramwell and Fellgett (1973), a specimen of *Dimetrodon* weighing about 200 kilograms could warm up from 26°C to 32°C in 205 minutes without the sail or 80 minutes with the sail. The sail would have enabled them to be active much earlier in the day than were other predators or prey of comparable size. While the presence of a said argues strongly for a selection regime favoring temperature control, it also demonstrates that these animals functioned primiarlu as ectotherms rather than endotherms.

THERAPSIDS

Origin of Therapsids

Pelycosaurs are known primarily from North America and Europe, with a few late-surviving genera in sourthern Africa and Russia. The most complete fossil record is from the Lower Permian Redbeds sequence of the south western United States. These deposits continue into the Upper Permian, but the fauna becomes progressively depauperate as a result of increasing aridity. Olson described fossils from the early Upper Permian that have some features in common with therapsids, but their remains are to fragmentary to provide information regarding the transition between the two groups. The upper beds in the southwestern United States are sometimes referred to informally as middle Permian in age, as are the Russian beds in which we find the earliest therapsids. However, only Lower and Upper Permian are recognized as formal geological units.

The oldest unquestioned therapisids come from deposits in European Russia at the base of the Upper Permian. Several distinct groups can be recognized, which suggests a significant period of prior evolution. The early therapsids overlap in time with the youngest caseid pelycosaur from Russia, *Ennatosaurus*, but their remains come from different localities, which leads us to believe that they lived in different environmental separation, the early therapsids probably had adapted to a way of life that was distinct from that of the pelyconsaurs, perhaps because of differences in their physiology. These differences may account for the rapid radiation of therapisids that quickly reached a level of diversity far greater than that of their ancestors.

The closest affinities of the thereapisids lie among the sphenacodont pelycosaurs, based specifically on the common presence of a reflected lamina of the angular. Dimetrodon and other members of the subfamily Sphenacodontinae appear too specialized in the great length of the neural spines and the reduced number of premaxillary teeth to the direct ancestors of therapsids. However, the more primitive genus *Haptodus* could have filled this role. This lineage leading to therapsids may have diverged from animals that were similar to *Haptodus* at any time between the late Pennsylvanian and the middle Permian, a period of at least 25 million years.

The therapsids are clearly advanced over the pelycosaurs when they appear in the Upper Permian, particularly in the specializations of the postcranial skeleton.

Early Therapsids

Eotitanosuchia (Phthinosuchia)

Therapsids are typically grouped in two major assemblages, the carnivorous theriodonts and the herbivorous anomodonts. They share a common ancestry among primitive therapsids, which also include both carnivorous and herbivorous genera.

Among the earliest and most primitive therapsids are the biarmosuchids from the Ocher (Ezhovo) locality at the base of the Russian sequence. The skull of *Biarmosuchus* resembles that of the sphenacodonts in most features. The temporal musculature is largely confined to the inner surface of the skull. The occiput is inclined slightly anteroventrally, the reverse of the orientation seen in pelycosaurs, but has a similar platelike configuration. The supratemporal bone is lost. The canine tooth is especially prominent, but there is not an anterior "step" in the maxilla, as in *Dimetrodon*. Like the advanced sphenacodontids, the maxilla extends dorsally to the nasal and separates the lacrimal from the narial opening. Unlike any pelycosaur, the septomaxilla has a long exposure on the surface of the skull.

The palate retains most of the primitive features of sphenacodontids, including the presence of teeth on the transverse flange of the pterygoid, but the vomers are recessed above the level of the internalnares and are partially fused. The palatines and anterior portion of the pterygoids are arched above the midline, which suggests the presence of a narrow air passage above the remainder of the denticulate palate. The stapes is a large bone that is oriented transversely and has a very large stapedial foramen. The reflected lamina of the angular is clearly separated from the remainder of the bone and is characterized by radiating ridges.

Much of the postcranial skeleton is known, although no restoration has been published. The vertebrae resemble of primitive sphenacodontids, without the exaggerated neural spines of the Sphenacodontinae. The structure of the girdles and limb indicates a posture much advanced above the level of the pelycosaurs. The glenoid and acetablulum both open more ventrally, and the femur may have assumed a semierect posture that we comparable to moderately advanced the above the level of the pelycosaurs. The glenoid and acetabulum both open more ventral, and the femur may have assumed a semierect posture that was comparable to moderately advanced thecodonts. The blade of the scapula is much narrower than that of pelycosaurs, but the clavicle and interclavicle remain large elements. The outline of the pelvis remains primitive in the presence of platelike pubis and ischium.

The humerus retains a primitive configuration with both ends widely expanded, but the femur has a sigmoid curvature and inturned head that is comparable to that of a crocodile. The hands and feet are more symmetrical than those of pelycosaurs, which indicates that they faced more directly forward throughout the stride, and the length of some phalanges is greatly reduced, which presages the reduction in phalangeal number that is typical of later therapsid and mammals.

Dinocephalians

The dinocephalians were contemporary with the primitive carnivorous biarmosuchids and include some of the earliest herbivorous

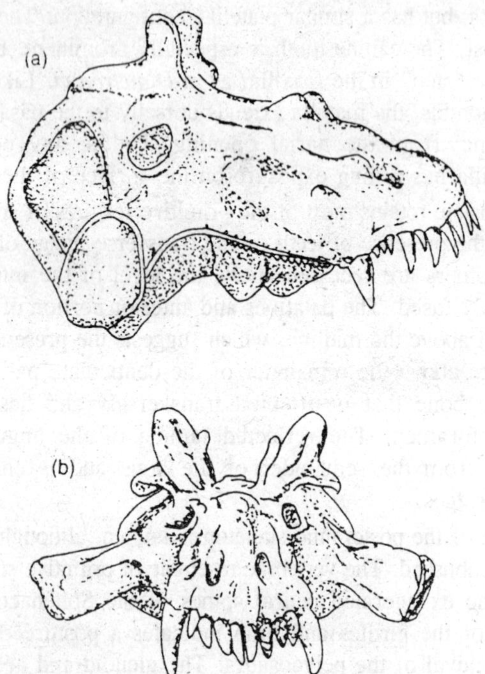

Fig. 6.8. The skull of Estemmenosuchus. (a) Lateral and (b) anterior views.

therapsids. They are confined to the Upper Permian but are known in sourthern Africa as well as Russia. They left no descendants.

Among the most specialized of the dinocephalians is the genus *Estemmenosuchus* from the lowest Upper Permian zone in Russia. The skull is large and massive, with hornlike protuberances on the maxillae, frontals, parietals, and jugals. The estemmnosuchids had large canine teeth but tiny cheek teeth. Like other dinocephalians, the incisors are large and have a specific interdigitating arrangement.

Better known are the brithopodids from the Rusian Zone II, including Titanophoneus. The limbs show a more upright posture, but the postcranial skeleton retains the general pattern of sphenacodont pelycosuars with a very long tail. The digital formula is reduced to 2, 3, 3, 4, 5 in the manus and 2,3, 3, 3,3 in the pes. The temporal opening is greatly enlarged dorsoventrally compared with the pattern of biarmosuchids. Dorsally, there are sharp crests near the midline that indicate the great dorsal extent of the jaw musculature. The canines are long, and the piercing cheek teeth suggest that the brithopodids had a primarily carnivorous diet. The incisors are large and interdigitating; they bear a narrow shelf at their base, which ensures a close occlusion that would have enabled them to cut pieces of flesh neatly from their prey.

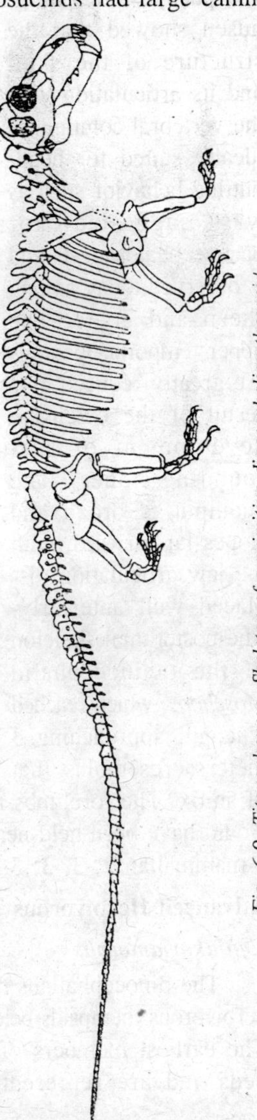

Fig. 6.9. Titanophoneus. Skeleton of a primitive dinocephalian theapsid.

We find more advanced dinocephalians in southern Africa. They include the titanosuchids and tapinocephalids, in which the cheek teeth are reduced in size but increased in number and evolved a chisel shape like that of the incisors. The canine teeth are indistinguishable from the remainder of the dentition. These animals were surely herbivores. In aolvanced genera, the snout region is broad and low.

Among the titanosuchids and especially the tapinocephalids, the bones at the back of the skull are extremely thick (up to 10 centimeters) and pachyostotic. Barghusen showed that the structure of the skull and its articulation with the vertebral column are ideally suited for head-butting behavior such as was suggested for pachycephalosaurids and is observed today among sheep and goats. The upper temporal openings are greatly reduced as a result of the extensive development of surrounding bones. The occuiput is broad and slopes far forward, with the jaw articulation displaced well anteriorly. The postcranial skeleton of the tapinocephalid *Moschops*, which reached a length approaching 3 meters, resembles that of an ox. The forelimbs retain a sprawling posture, but the hind limbs would have been held nearly erect. The phalangeal count is reduced to a mammalian 2, 3, 3, 3,3.

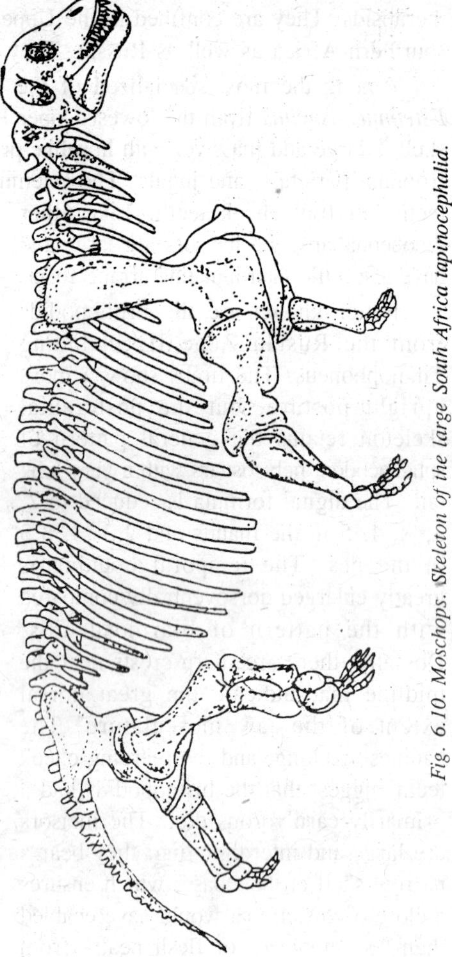

Fig. 6.10. Moschops. Skeleton of the large South Africa tapinocephalid.

Advanced Herbivorous Therapsids : The Anomodonts

Venjukoviamorphs

The dinocephalians did not survive the end of the Permian. Most herbivorous therapsids belong to a distinct assemblage, the Anomodontia. The earliest members of this group are known in the early Russian beds and are represented by the genus Otsheria in Zone I and

Venjukovia in Zone II. The skulls of these genera are small and short, superficially resembling the herbivorous pelycosaurs. The zygomatic arch is located high on the cheek and exposes the adductor musculature widely between the cheek and the lower jaw. The temporal openings have expanded posteriorly beyond the level of the occipital condyle, but they do not extend as far dorsally as those of early dinocephalians. The canine teeth are reduced and the cheek teeth are short and blunt.

The premaillae form a broad shelf of bone anterior to the internal nares. The palate is primitive in retaining a movable basicranial articulation and a distinct transverse flanges of the pterygoid, but it has lost the denticles that occupied the margin of the flange in the more primitive therapsids.

The premaxillae form a broad shelf of bone anterior to the internal nares. The palate is primitive in retaining a movable basicranial articulation and a distinct transverse flanges of the pterygoid, but it has lost the denticles that occupied the margin of the flange in the more primitive therapsids.

Like the latter anomodonts, the coronoid bone is lost from the lower jaw and a large mandibular foramen is evident. The dentary bones in *Venjukovia* are fused at the symphysis.

As herbivorous forms, it was long thought that the venjukoviamorphs might form a link between the primitive dinocephalians and later anomodonts. Barghusen (1976) pointed out that *Venjukovia* is more primitive than the earliest known dinocephalians in not having the adductor musculature expanded over the dorsal skull table behind the orbits. In contrast, *Venjukovia* initiated the spread of a lateral sheet of the adductor mandbulae externus, which orginates on the lateral surface of the squamosal and inserts on the lateral surface of the lower jaw. Many of the specialization's of a herbivorous diet are clearly different in these two groups and evolved separately. There is no evidence to ally the anomodonts with any other particular group of primitive therapsids.

Dromasaurs

Three genera that are represented by only four specimens are the sole members of a second group of anomodonts, the dromasaurs, which are known only from the Upper Permian of southern Africa. The skull of dromasaurs appears superficially like that of the venjukoviamorphs, with a high and narrow zygomatic arch, narrow postorbital bar, and narrow squamosal. It is relatively short and there is a partial secondary palate that is formed by the premaxillae. Canine teeth are not prominent and premaxillary teeth are progressively lost within the group. In contrast with venjukoviamorphs and dicynodints, the postoribital does not extend back to the squamosal. The adductor musculature expands out of the termporal opening dorsally, and the lateral surface of the lower jaw is grooved to suggest an area for insertion of a superficial layer of the external adductor.

The limbs are long and slender, in contrast with the dicynodonts, and dromasaurs lack the key changes in the jaw apparatus that characterize that group.

Dicynodonts

The most numerous, diverse, and long-lived anomodonts were the dicynodonts, which appeared at the very be base of the Upper Permian sequence in southern Africa, became worldwide in distribution in the Lower Triassic, and continued to the end of the Triassic. They were by far the most abundant of terrestrial vertebrates in the late Permian and early Triassic but lost ground by the mid-Triassic, perhaps in competition with rhynchosaurs and herbivorous cynodonts.

The early dicynodonts were small in comparison with the contemporary dinocephalians ans pareisaurs of the Upper Permian, but later genera ranged up to the size of an ox. The body of dicynodonts is short and stout. Inlarge genera, the rear limbs were held erect, but the fore limbs were sharply bent at the elbow. In all members, the phalangeal count was reduced to 2, 3, 3, 3, 3. The postcranial skeleton varied in relationship to body weight and to different ways of life that ranged from semiaquatic to subterranean, but the skull shows a remarkable consistency in its proportions that is related to a unique specialization of the feeding apparatus.

The genus *Dicynodon* shows the cranial pattern that is typical of the group. Dicynodonts differe from venjukoviamorphs in lengthening the temporal region and elaboration of the squamosal, which results in a broad plate of bone on the lateral surface. The teeth are greatly reduced, typically leaving only a pair of canines, each fitting into a massive canine boss at the front of the maxilla. Even these teeth are lost in some genera. Their presence or absence in some species may reflect sexual dimorphism.

The premaxillae as well as the dentaries are fused at the midline in all but the most primitive genus, Eodicynodon. The bone at the front of the upper and lower jaws is marked by tiny nutritive foramina that resemble the surfaces that underlie the horny beak of birds and turtles. It is logical to assume that dicynodonnts had a similar covering that formed a beaklike structure. The palate is advanced in having the braincase sutured to the pterygoids, rather than retaining a movable basicranial joint. The premaxillae, maxillae, and palatine form a long secondary palate, with the internal nares opening behind the midpoint of the skull. The temporal fossae are enormous, which indicates the great mass of the adductor musculature. The area of the pterygoid that served for origin of the pterygoideus musculature extends anteriorly rather than laterall, so that the conspicuous transverse flangs of the pterygold, which is present in *Otsheria*, is no longer evident.

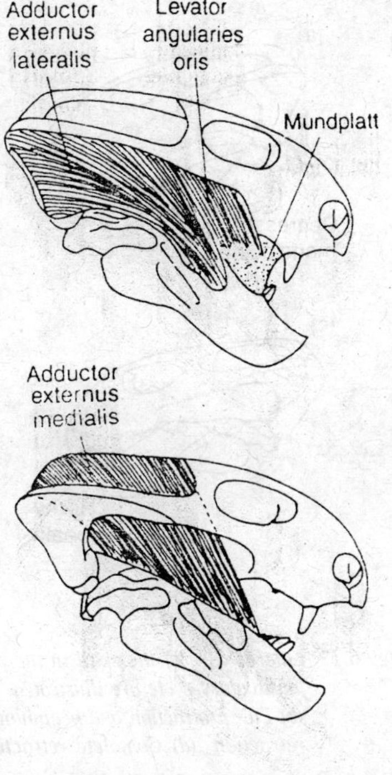

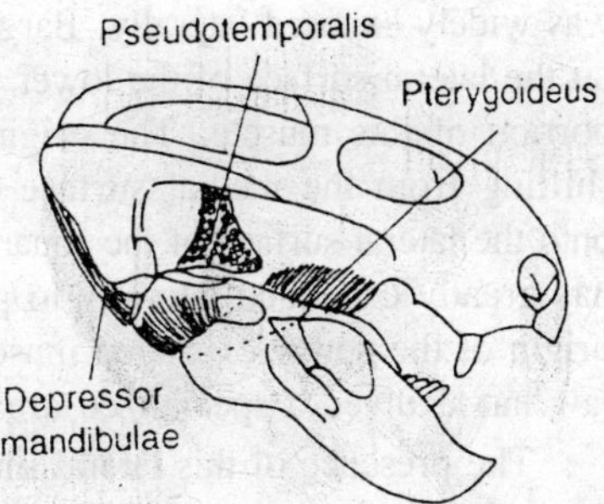

Fig. 6.11. Emydops. Reconstruction of the jaw-closing muscles in the primitive dicynodont at progressively deeper levels.

The configuration of the jaw articulation differs significantly from that of venjukoviamorphs. The articularting surface of the articular is approximately twice as long as that of the quadrate. The surface of both bones is convex is lateral view but grooved longitudinally, which indicates that the articulating surcace of the lowe jaw was translated across that of the quadrate, in contrast with the simple hings joint of most tetrapods.

The shape of the jaws and temporal region shows that the major muscles that opened and closed the mouth were oriented nearly horizontally, which suggests that anterior and posterior movements of the jaw were of great importance.

Within the anomodonts, a major new muscle mas was evolved. Already in the venjuikoviamorphs, the adductor mandibulae externus

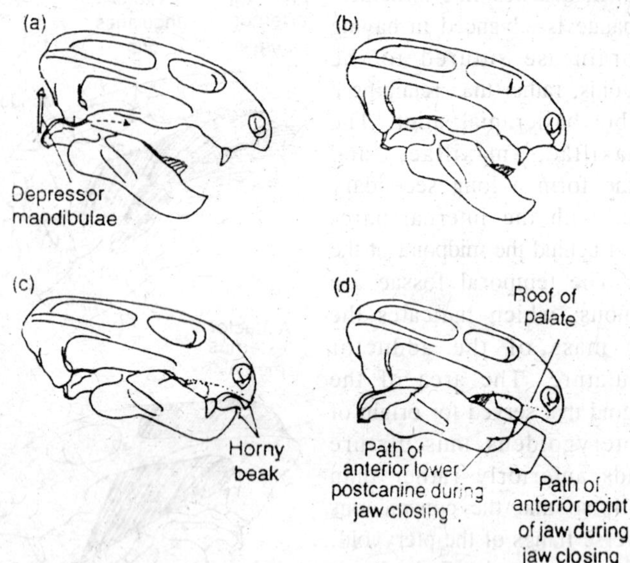

Fig. 6.12. Lateral view of the skull in the primitive Dicynodont Emylops. Stages in the masticatory cycle are illustrated. (a) Depression and beginning of protraction. (b) Full protraction and beginning of elevation. (c) Beak bite and beginning of retraction. (d) Complete retraction.

was widely exposed laterally. Barghusen (1976) identified a depression on the lateral surface of the lower jaw into which inserted a superficial portion of this muscle. The origin may have been in the process of shifting from the medial surface of the back of the zygomatic arch onto the lateral surface of the squamosal. In dicynodonts, the squamosal has greatly expanded laterally to provide a very large surface for the origin of the newly expanded muscle. The lateral surface of the lower jaw has evolved a specialized area for its insertion, the lateral shelf.

The presence of this lateral muscle would have balanced the more medially directed force of the remaining elements of the adductor mass. This balancing force may have made it possible for the reduction of the transverse flange of the pterygoid, which in more primitive synapsids prevented the lower jaw from moving medially. The change in orientation of the transverse falnge of the pterygoid increased the area of the subtemporal fenestra and permitted the anterior expansion of the adductor muscles.

Cromption and Hotton demonstrated that the major force for breaking up food resulted from retraction of the lower jaw when it was nearly closed. The jaw opened from its maximally retracted

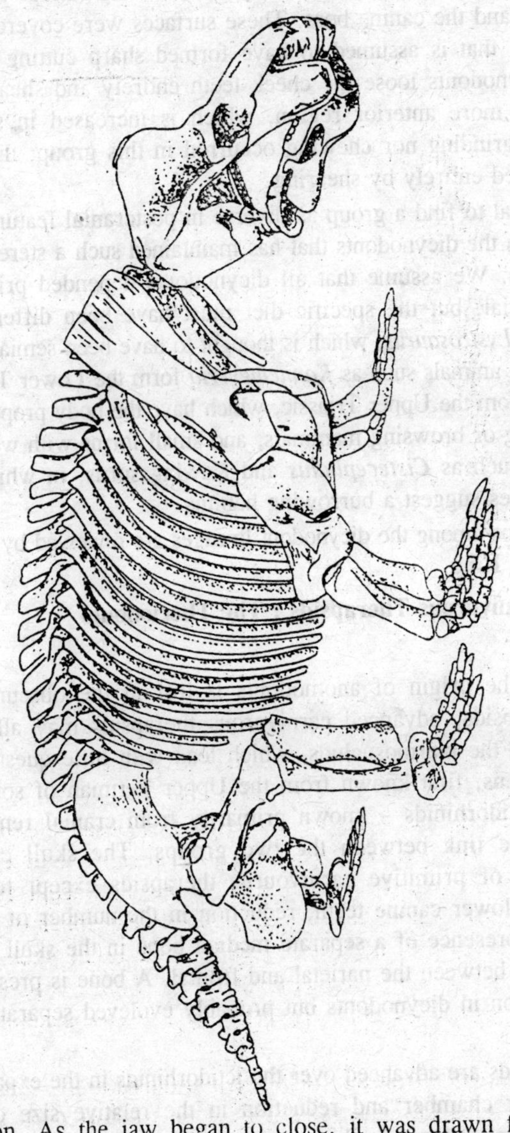

Fig. 6.13. Skeleton of the large Dicynodont Kannemayeria.

position. As the jaw began to close, it was drawn forward by the pterygoideus muscle. It then closed and retracted simultaneously, with the force of the adductors originating from the back of the temporal region.

In primitive dicynodonts, this jaw action produced shearing at two points, posteriorly between the remaining cheek teeth of the lower jaw, maxilla, and palate and anteriorly between the lateral edge of the lower jaw and the canine boss. These surfaces were covered with horny material that is assumed to have formed sharp cutting edges. Advanced dicynodonts loose the cheek teeth entirely and shearing is limited to the more anterior region, which is increased in surface area. Neither grinding nor chewing occurred in this group; the food was comminuted entirely by shearing.

It is unusual to find a group as diverse in postcranial features and as long lived as the dicynodonts that has maintained such a stereotyped feeding pattern. We assume that all dicynodonts depended primarily on plant material, but the specific diet must have been different in genera such as *Lystrosaurus*, which is thought to have been semiaquatic, huge terrestirial animals such as *Kannemeyeria* form the Lower Triassic and Placerias from the Upper Triassic, which have the body proportions of large grazing or browsing mammals, and small forms with wedged-shaped skulls such as *Cistecephalus* and Kawingasurus, in which the limbs and girdles suggest a burrowing habitus.

Relationships among the dicynodont lineages are reviewed by Kemp and Cluver and King.

Advanced Carnivorous Therapsids : The Theriodonts

Gorgonopsinans

Although the origin of anomodonts is lost in the obscurity of ancestral therapsids, advanced carnivorous therapsids may all their origin to forms the biarmosuchids, which lead with little question to the gorgonopsians, first known from the Upper Permian of southern Africa. The ictidorhinids – known primarily from cranial remain – provide a close link between the two groups. The skull closely resembles that of primitive carnivourus therapsids except for the prominence of lower canine teeth, reduction in the number of cheek teeth, and the presence of a separate median bone in the skull table, the preparietal, between the parietal and frontal. A bone is present in a similar position in dicynodonts but probably evoleved separately in the two groups.

Gorgonopsids are advanced over the ictidorhinids in the expansion of the adductor chamber and reduction in the relative size of the orbit. The canine teeth are further emphasized and the cheek dentition reduced. The skull is massively constructed and reaches a length of 45 centimeters in one geenus. We find gorgonopsids primarily southern Africa, but they also appear in Russia toward the end of the Permian. Twenty-two genera have been described; they were the dominant carnivores in the later Permian and probably preyed on the large pareiasaurs and dinocephalians.

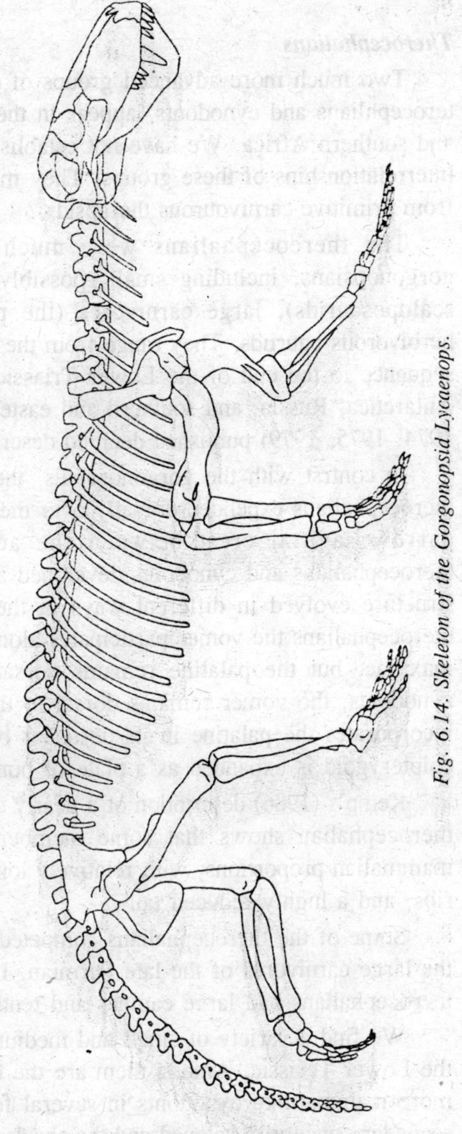

Fig. 6.14. Skeleton of the Gorgonopsid Lycaenops.

The postcranial skeleton, which is well known in the genus *Lycaenops* superficially resembles that of cursorial mammals. The posture of the forelimb remains primitive, with the humerus held essentially horizontal. According to Kemp (1982), the structure of the femur is comparable with that of crocodiles and indicated a similar capacity for two styoles of locomotion—one in which the limb retains a sprawling posture like that of primitive amniotes and the other in which the femur was held at about a 45-degree angel, which enabled the lower limb to move in a parasagittal plane. Changes in the structure of the ankle and foot that distinguish advanced theriodonts are little evident in gorgonopsids, which retain the primitive phalangeal count 2, 3, 4, 5, 3, with only a slight diminution in the lengths of some phalanges.

Gorgonopsians did not survive past the end of the Permian and are not closely related to the origin of either of the more advanced groups of camivorous therapsids. Sigogneau (1970) published the most recent review of this group.

Therocephalians

Two much more advanced groups of carnivorous therapsids, the terocephlians and cynodonts, appear in the Upper Permian of Russia and southern Africa. We have not established the specific origin and interrelationships of these groups. They may have evolved separately from primitive carnivourous therapsids.

The thereocephalians were much more diverse than the gorgonopsians, including small, possibly insectivorous forms (the scaloposaurids), large carnivores (the pristerognathids), and the herbivorous bauriids. They range from the base of the Upper Permian sequence to the end of the Lower Triassic and are known in China, Antarctica, Russia, and southern and eastern Africa. Mendrez (1972, 1974, 1975, 1979) published detailed descriptions of several genera.

In contrst with the gorgonopsians, the jaw musculature in early therocephalians expended dorsally over the branincase, leaving only a narrow sagittal crest between the adductor chambers. Both therocephalians and cynodonts developed a secondary palate, but this structure evolved in different ways in the two groups. In primitive therocephalians the vomer participates along with the premaxillac and maxillae, but the palatine remains dorsal in position. In primitive cynodonts, the vomer remains dorsal to the secondary palate, which incorporates the palatine in its posterior border. In both groups, the epipterygoid is expanded as a plate of bone lateral to the braincase.

Kemp's (1986) description of a nearly complete skeleton of a small therocephalian shows that some members of this group had very mammalian proportions, with relatively long limbs, attenuated lumbar ribs, and a highly reduced tail.

Some of the therocephalians competed with the gorgonopsians as the large carnivores of the late Permian. Like that group, these early therocephalians had large canines and tended to lose the cheek teeth.

We find a variety of small and medium-sized thereocephalians in the Lower Teiassic. Among them are the bauriids, which parallel the more mammal-like cynodonts in several features. A nearly complete secondary plate is developed and the cheek teeth have expanded crowns to crush and grind food. The postorbital bar is no longer complete. The dentary bone is enlarged, but the other bones of the lower jaw are nor significantly reduced. Paleontologists once thought that thereocephalians related to *bauria* might be close to the ancestry of at least some mammals, but increased knowledge of both thereocephalians an cynodonts indicates that only the latter groups shows the specialize features of the dentition, braincase, and lower jaw that are expected in mammalian ancestors.

Cynodonts and the Origin of Mammals

Therocephalians and gorgonopsians were the dominant reptilian carnivores in the late Permian. Within the early Triassic they were supplanted by the most advanced thereodonts, the cynodonts. Although some features that we associate with mammals were achieved in other therapsid groups, only the cynodonts show a significant approach to the mammalian conolition their general morphology. Cynodonts are represented in the latest permain of both Russia and southern Africa by the family Procynosuchidae. The skull of Procynosuchus is advanced over other early therapsids in a complex of features that presages the mammalian condition. To appreciate the importance of these features it is necessary to consider some of the basic differences between primitive reptiles and mammals.

We tend to think of large brain size and live birth as being particularly important attributes of modern mammals, but both were achieved relatively late in mammalian evolution. The placental pattern of live birth was certainly not achieved until late in the Cretaceous and large brain size evolved only within the Tertiary.

A much more fundamental mammalian feature is their high metabolic rate. Modern mammals require approximately ten times more food and oxygen than do reptiles of comparable size. The higher metabolic rate allows mammals to be active more continuously and to maintain a high, constant body temperature that is independent of the environment. Such a radical difference in metabolic rate affects nearly all the systems of the body and is responsible, directly or indirectly, for nearly all the differences we observe between reptiles (be they modern or Paleozoic) and mammals.

The initial change may have been a shift in the quantitative importance of fermentative importance of fermentative and oxidative metabolism of the voluntary muscles. Reptiles rely primarily on fermentative, anaerobic metabolism of glucose to power muscle contractions. Glucose is present within the muscles and the response can be immediate and is little affected by temperature. By relying on anaerobic metabolism, reptiles can run short distances as rapidly as mammals. A critical drawback of anaerobic metabolism is the accumulation of lactic acid, which results in muscle fatigue.

CYNODONTS

Cynodonts were the last of the major groups of therapsids to evolve. By the end of the Traiassic they included some very mammal-like groups and had given rise to the mammals themselves. A marked reduction in body size characterized the cynodont lineages. Some early cynodonts were as big as large dogs, but by the middle Triassic, cynodonts had skulls only 70 millimeters long, and in the total body length of the earliest mammals was less than 100 millimeters.

Both carnivorous and herbivorous forms are represented among the cynodonts. The carnivorous cynodonts include the early Triassic cynognathids, the middle Triassic chiniquodontids, and the late Triassic trithelodontids. These animals had well-developed canine teeth and incisors, like the earlier groups of carnivorous therapsids, but in advanced cynodonts the post-canine teeth of *Cynognathus* indicate that the upper and lower teeth had a tearing action aided by a cutting function of serrations on the cusps of the teeth. This arrangement would have allowed cynognathids to use the cheek teeth to cut off bite-zie pieces of meat. In contrast, the carnivorous dinocephalians and gorgonopsids had poorly developed cheek teeth and apparently used the incisors to nip off bits of flesh.

Surprisingly, these changes in dentition may have reflected selective forces that acted initially on the ability of cynodonts to hear. They jaw muscles of cynodonts showed progressive differentiation of the superficial masseter muscle in association with the development of a pronounced coronoid process of the dentary bone. These chariges probably improved the bite force of the jaws, and also allowed a reduction in the size of the articular and quadrate of cynodonts are homologous with the malleus and incus of the middle ear of modern mammals, and it is likely that at the cynodont stage they already played a role in sound transmission. If this was the case, lighter bones would be more effective than heavy bones at transmitting vibrations picked up by the tympanic membrane, but a reduction in the size of the bones would have reduced the strength and load-bearing capacity of the jaw joint. Thus, the reorganization of the jaw muscles and the development of the coronoid process of the dentary bone seen in cynodonts could have been advantageous because they articular and quadrate bones to be small. In this analysis, selective pressures to improve hearing are seen as beng the important factors in the reorganization of the mammalian jaw.

However, the improvement in hearing would have some cost associated with it. A weak jaw would have limited the ability of cynodonts to chew their food: The canines and incisors could puncture and tear prey, but the cheek teeth could not process food effectively because those chewing motions (especially unilateral chewing) would exert too much stress on the jaw joint. Ultimately the cynodont lineage evolved a new load-bearing jaw joint between the dentary and squamosal bones that was strong biting. This is the mammalian type of jaw suspension. A classic example of an evolutionary intermediate is provided by the trithelodontid *Diarthrognathus*. This animal gets its name from its double jaw joint (*di* = two, *arthro* = joint, *gnath* = jaw). It has both the ancestral articular-quadrate joint and, next to it, the mammalian dentary-squamosal articulation.

The herbivorous cynodonts includes the early Triassic diademodontids, the middle Triassic traversodontids, and the late Triassic tritylodontids. As the names suggest, features of their teeth are particularly significant in these groups. The surfaces of the cheek teeth developed patterns of cusps associated with crushing, cutting, and grinding plant material. The arrangement of teeth in the jaws of herbivorous cynodonts was also quite mammalian. For example, the tritylodont *Oligokyphus* had a large diastema separating the anterior teeth from the cheek teeth. The postcanine teeth were multirooted like the cheek teeth of mammals, and had crescentic cusps on their surfaces forming opposing pairs of cutting edges suitable for shredding plant material. *Oligokyphus* was about 50 centimeters in total length, and nearly half of that was tail.

The postcranial skeleton of the cynodonts was also verging on the mammalian condition. The structure of the pelvis and femur indicates that the hindlimb was held most directly beneath the body and moved in a parasagittal plane (parallel to the long axis of the body). The ability to shift between a sprawling and an upright posture that was present in gorgonopsids had been lost in cynodonts. The forelimbs of cynognathids and diademodontids appear to have retained a widespread position, but changes in the shape of the scapula suggest that the speed of the stride was increased. In the tritylodont *Oligokyphus* the shoulder girdle was very like that of mammals, and the forelimb probably moved in a parasagittal plane. Taken together, these changes indicate increased power and agility in the cynodonts.

Some features of the soft anatomy of therapsids can be inferred from traces left on the bones. The smooth surfaces of the premaxilla,

maxilla, and dentary adjacent to the teeth suggest that cynodonts had muscular lips like those of primitive mammals. The lips of therapsids could have functioned in conjuction with the cusped cheek teeth and, probably, a muscular tongue to manipulate food as it was chewed.

An abrupt change in the size and shape of ribs in the lumbar region suggests that cynodonts had a muscular diaphragm like that of mammals. Furthermore, ridges in the nasal passages suggest that turbinal bones may have been present. These features are likely to have been associated with ventilation of the lungs by large volumes of air, and that inference in turn suggests that cynodonts were capable of high rates of metabolism. In mammals the turbinal bones support tissues that, in addition to having olfactory cells, warm and humidify inhaled air before it reaches the lungs and then recover some of that heat and water when the air is exhaled. This countercurrent recycling system can conserve enough heat to be an important part of the energy budget of a mammal. The presence of turbinal bones in cynodont therapsids would suggest that their thermoregulation was basically like that of mammals.

First Mammals

The origin of mammals is often treated as if it were a discrete event that happened suddenly sometimes in the late Triassic or early Jurassic, but this is an oversimplified view. The animals we know as mammals are the products of an evolutionary lineage that stretches back to the division between diapsids and synapsids in the Carboniferous. The anatomical structures that characterize mammals evolved in mosaic fashion through the Permian and Trassic, and their original functions were not necessarily the same as those they have today. The biology of animals in the synapsid lineage was also changing during the late Paleozoic and early Mesozoic. We have traced the evidence for increasing locomotor agility and increased specialization of the jaws and teeth in cynodonts, and the most advanced cynodonts are little different from the earliest mammals. A list of the derived characters of mammals is useful as a summary, but it emphasizes the differences between mammalian and nonmammalian synapsids, whereas a review of their evolution stresses the continuous process of transition.

Trithelodontids : The Sister Group of Mammals?

The Tritheledontidae (also known as the Diarthrognathiae) is a poorly known group of small cynodonts from the middle and late Triassic of South Africa. Tritheledontids may be the cynodonts most closely related to mammals. *Diarthrognathus*, the best known of the tritheledontids, possessed the mammalian dentary-squamosal joint and also the nonmammalian articular-quadrate joint. The postcranial skeleton of tritheledontids is not known, but the skull and teeth showed derived mammalian characters: The postorbital and prefrontal bones were absent and the zygomatic arch was slender. Furthermore, tritheledontids are the only cynodonts known in which the teeth were covered with prismatic enamel like that of mammals. Thus, the tritheledontids may be the sister group of mammals, but that relationship is by no means certain and two other groups might be more closely related to mammals than are the tritheledontids. The tritylodontids have derived mammalian characters in the skull and postcranial skeleton, and *probianognathus*, a cynodont from the middle Triassic of South America, has cheek teeth very like those of the earliest mammals (the Triconodonta, represented by the family Morganucodontidae). We have placed the tritheledontids as the sister group of mammals, largely on the basis of the prismatic enamel on their teeth, but a strong case can be made for giving that position to the tritylodontids.

Nontherian and Therian Mammals

The difficulty in identifying the sister group of mammals is a reflection of the fragility and small size of the skeletons of these small animals. Most fossil material consists only of scraps, and the postcranial skeletons of several groups are unknown. The fossils are often so small that they must be examined under a microscope, and they are found by sifting and washing tons of dirt and rock. In fact, searching for fossils of early mammals is very much like searching for gold.

Teeth are hard structures that are more likely to be fossilized than other bones, and much of the phylogeny of mammals is based on the patterns of tooth cusps. Differentiation of the teeth into incisors, canines, and molariform teeth was evident in sphenacodont pelycosaurs and continued in the cynodonts. The molar teeth of the advanced cynodonts were not simple conical pegs as they were in pelycosaurs and early therapsids. They had various complex patterns of cusps and double or triple roots. The interlocking cusps guided the closing jaws into alignment and sheared food, rather than tearing it as had the teeth of early cynodonts. Tooth morphology is complex and difficult to visualize, but it is basic to understanding the classification and appreciating the life-styles of early mammals.

The study of teeth has played a large role in understanding mammalian evolution, in large part because little material except for teeth has been available. Mammals with teeth that have three cusps aligned anteroposteriorly (the *triconodont* pattern) have been distinguished from forms that have three cusps in a triangular (*tribosphenic*) pattern. The tribosphenic molar differs from the ancestral mammalian molar in having a much-enlarged talanid process on the posterior margin of the lower molars. The talonid process contains a basin that receiver the inner cusps (the protocone) of the corresponding upper molar. Together they mash and grind food caught between them.

The cynodont-mammals transition appeared to incorporate a dichotomous radiation with the morganucodontids (triconodont molars) closely related to nontherian mammals, and the kuehneotheriids (tribosphenic molars) related to therian mammals. New discoveries of fossil mammals are suggesting that this scheme may be oversimplified and a new classification will be needed when more information is available. Material from early Jurassic deposits in Arizona (*Dinnetherium*), Switzeland (*Hallautherium*), and China (*Sinocondon*) has revealed unsuspected diversity. None of these fossils appears to have been closely related to the morganucodontids or kuethneotheriids. Apparently, mammals had undergone an extensive radiation before the earliest mammals we have yet found. For the present, however, a distinction between therian and nontherian mammals persists in the paleontological literature.

Nontherian mammals

Morganucodontid mammals (triconodonts) and the poorly known docodonts had three cusps on the molar teeth arranged in a line from anterior to posterior; the middle cusp was the largest. When the jaws closed the inner faces of the cusps of the upper molars slipped down over the external faces of the cusps of the lower molars and produced a shearing action like that of the blades of a pair of scissors. A piece of food could be processed by shearing it repeatedly, cutting it into smaller and smaller bits. The postcranial skeleton of morganucodontids had many mammalian features. The joint between the atlas and axis was apparently capable of as muchratation as that of modern mammals, and the joint between the atlas and the double occipital condyles allowed flexion and extension. The cervical, thoracic, and lumbar vertebrae was well differentiated, and the cervicothoracic flexure indicates that the head and neck were held in an upright posture. The lumbar vertebrae lacked ribs, as they do in modern mammals. The anterior neural spines were angled posteriorly, and the posterior spines were angled anteriorly. Three transitional vertebrae are evident in the posterior thoracic region. This pattern of vertebrae in modern mammals is associated with flexion and extension of the trunk in a vertical plane in contrast to horizontal flexion. The pelvic girdle of *Morganucodon* was fully mammalian: It had a narrow, rodlike ilium that was directed anterodorsally, a large obturator foramen, and a reduced pubis. The form of the limbs suggests that morganucodontids were active foragers with climbing abilities similar to those of many small mammals today.

The kuehneotheriids are probably included among the symmetrodonta of the middle Jurassic and early Cretaceous. These fossils are known largely from teeth, which have three cusps arranged in the triangular tribosphenic pattern. The symmetrodonts appear to be the primitive sister group of the therian mammals.

Therian mammals

Multituberculates were a diverse and successful group of mammals. They included nearly 50 genera and survival for 100 million years, from the late Jurassic into the Oligocene. Some multituberculates were the size of small mice and others grew as large as a woodrodent-like appearance that is produced by a pair of long incisors separated by a diastema from the cheek teeth. The lower premolars of most multituberculates were specialized as large shearing blades (*plagiaul-acoid* premolars); similar premolars are found among some living marsupials. Multituberculates were probably omnivorous, much like modern rodents that include both plant and animal material in their diet.

The living mammals Monotremata (the duckbilled platypus and echidnas), Marsupialia (the marsupials), and Eutheria (placental mammals) have long been considered to represent surviving lineages of nontherians (the monotremes) and therians (the marsupials and euthorians). However, echidnas have extremely reduced teeth and the duck-billed platypus lacks teeth altogether, so it has not been possible to link the tooth structure of modern monotremes with the triconodont pattern of nontherian fossil mammals. The recent discovery of a fossil monotreme from Cretaceous deposits in Australia indicates that early monotremes had a triangular arrangement of tooth cusps. The teeth are not true tribosphenic molars, but they seem most likely to be derived from the pattern of tooth cusps seen in therian mammals. Furthermore, a second character used to distinguish nontherian and therian mammals, the construction of the sidewall of the braincase, now appears less convincing than it was once considered. In this context, monotremes seem likely to be their mammals.

REGULATION OF BODY TEMPERATURE

The essence of being a mammal or bird is expressed in endothermic regulation of constant and high body temperatures, that is, endothermal homeothermy. This "warm-bloodedness" opened nocturnal niches to the earliest mammals, for in equable climates of the Mesozoic, the ability to be active at night was probably the key factor that allowed them to persist in an ecosystem dominated by reptiles. Preadaptations to homeothermy in fishes, reptiles, and birds have already been discussed. What are the advantages of regulating body temperature (T_b) at a high level? How do mammals regulate body temperature? What might the ecological consequences of endothermal homeothermy be?

Advantages of a High Body temperature (T_b)

Mammals and birds have high metabolic rates, at least six times the SMR of poililotherms. Although it is energetically expensive, there are benefits to regulating T_b at a high rather than a low level. The biochemistry of vertebrates involves thousands of interacting enzyme-catalyzed reactions, most of which are temperature sensitive. A constant internal temperature is required to obtain maximum chemical coordination among these reactions. In addition, the higher the T_b the more rapid the response of cells to organism needs. Certainly complex systems like the CNS, although capable of acting at very cold temperatures (as in arctic fishes), function more effectively at high temperatures. As one example neurotransmitters like acetylcholine and norepinephrine act by diffusing from their site of release across the synaptic junction to the postsynaptic receptor surface. Because diffusion is a physical process its rate increases as temperature increases. A high T_b therefore enhances the rate of information processing, a competitive advantage too often neglected when considering the success of mammals and birds. Very rapid the same neurological benefits when they are warm, but very rapid responses on cool nights can occur only in endothermal homeotherms. In addition, muscle viscosity declines at high temperatures. This reduction in internal friction results in more rapid, forceful contraction and faster response times.

Endothermal homeothermy provides an additional advantage not enjoyed by ectotherms: great independence of environmental temperatures, or ambient temperature (T_b). In particular, endothermy allows animals to be active at night when solar energy is absent, and in cold climates where solar energy is not sufficient to warm an ectothermal vertebrate.

Mechanisms of Temperature Regulation

In general the T_b of mammals and birds is at or above 36°C. Why was T_b set so high in the evolution of homeothermy? Certainly the influences of a high T_b on activity and reaction times was beneficial and these are part of the reason. The mechanisms used to regulate T_b are important too, for they impose physiological and ecological limitations on homeotherms. Because it is easier to balance heat loss against gain in a cool environment ($T_b > T_a$) than in a warm environment ($T_b < T_a$), high internal body temperatures are ecologically beneficial. Heart flows from higher to lower temperatures, thus when T_b is greater than T_b excess body heat can be dissipated freely to the environment. If T_a is greater than T_b, excess body heat can be lost only by costly metabolic processes. To understand this more fully let us consider how a mammal regulates its temperatures as T_b varies below and above T_b.

When the environment is cooler than the body, heat loss may be increased by increasing blood flow to the skin. Areas of skin with reduced fur (feet, ears, abdomen, and son on) are highly vascularized and excess heat is quickly dissipated across these surfaces. When the environmental temperature falls, several physiological responses take place, usually in sequence, all of which lead to heat conservation and maintenance of T_b. Peripheral vasoconstriction reduces skin temperatures and, therefore, the rate of heat loss. Pilerection increases fur thickness and the amount of insulation. Finally, shivering, an asynchronous contraction of skeletal muscles, begins and increases metabolic heat production.

When T_a exceeds T_b, the temperature gradient is reversed and heat flows into the animal; unless checked, T_b will rise. This is much more difficult problem for a vertebrate. Erecting the fur increases insulation and reduces the rate of heat gain across the skin, but metabolic heat production causes the body temperature to rise. Mammals and birds must resort to the evaporation of body water to combat the rise in T_b. Its high heat of vaporization makes water evaporation an effective cooling mechanism. Many mammals cool themselves by sweating: secreting water through pores onto the body surface. Other mammals, devoid of sweat glands, evaporate water across the nasal, lung, and oral surfaces. They pant, ventilating rapidly, when warm. Panting, however, requires muscular activity that increases metabolic heat production. This heat compounds the problem and requires further loss of water.

Most habitats seldom have air temperatures that exceed 35 to 40°C. Even the tropics have average yearly temperatures below 30°C. Only in open situations where solar radiation impinges directly on a mammal or a bird can the environment reverse the normal heat gradient. Thus the high body temperatures maintained by mammals ensure that in most situations the heat gradient is from animal to environment. Still higher body temperatures, around 50°C for example, could ensure that mammals were always warmer than their environment. There are definite upper limits to the body temperatures that are feasible, however. Many proteins denature near 50°C. During heat stress, some birds and mammals may tolerate body temperatures of 45-46°C for a few hours. Only some bacteria, algae, and a few invertebrates exist at higher temperatures. This is another case in which the direction of vertebrate evolution has been established by a balance between biotic needs and physicochemical realities.

Although endotherms have achieved independence from the constraints imposed by environmental temperatures, the mechanism used has, in its turn, imposed a new set of limitations upon them. because of the high metabolic rate needed to maintain a high body temperature, endotherms consume energy much faster than ectotherms. Energy comes from the food animals eat; therefore, endotherms must eat more, and more frequently, than ectotherms. Well-known examples illustrate these differences. Shrews may starve to death if they cannot consume their own weight in prey daily, but snakes may fast for several months after capture and emerge from their fast little the worse for wear.

Ecological Consequences of Endothermy: Cost-Benefit Analysis

A clearer understanding of the costs, the ecological and evolutionary effects, of endothermy can be obtained by constructing energy budgets for a particular species. An energy budget, like a financial budget, shows income and expenditure using units of energy as currency. The energy of different activities and alternative behaviors can be evaluated. The importance of adaptations that allow a species to increase energy consumption or reduce energy expenditure become clear.

Recent studies of vampire bats by Brian McNab have revealed a clear-cut relationship between energy intake and expenditure and the species' geographic range. These bats of the suborder Mirochiroptera inhabit the neotropics and are specialized to feed exclusively on blood. If one is willin to be immundized against rabies before starting to work with the bats, and can stand the smell of their caves, vampires are ideal for calculations of energy budgets. Their daily pattern of activity is simple: they spend about 22 hr in their caves, fly out a night to a feeding site, and return after they have fed. Typically, a vampire flies about 10 km round trip at 20 km/hr to find a meal. Thus, a bat spends half as hour per day in flight. The remaining hour and a half outside the cave may be sent in actual feeding.

The vampire's food is as convenient for energetic calculations as its daily schedule, because of the relatively constant caloric content of blood.

I = ingested energy (blood). The blood a bat drinks must provide the energy needed for all of its life processes: maintenance, activity, growth, and reproduction.

E = excreted energy. As in all animals, not all of the food ingested is digested and taken up by the bat. The energy contained in the feces and urine is lost.

I - E = assimilated energy. This is the energy actually taken into the bat's body.

M = metabolism. This can be subdivided into M_i (Metabolism while the bat is inside the cave) and M_o (nonflight metabolism while the bat is outside the cave).

A = the cost of activity, a half hour of flight per day.

P = the biomass increase. This term is the "profit" a bat shows in its energy budget. It may be stored as fat or used for growth or for reproduction (production of gametes, growth of a fetus, or nursing a baby).

In its simplest form, the energy budgets is

energy in = energy out ± biomass change

The biomass term appears as ± because an animal metabolizes some body tissues when its energy expenditures exceed its energy intake. This is what every dieter hopes to do in order to lose weight.

Translating this general equation into the terms defined gives

$$I - E = M_i + M_o + A \pm B$$

All these terms can be measured and expressed as kilocalories per bat per day (kcal/bat • day). These calculations are based on McNab's studies and apply to a Brazilian vampire bat weighing 42 g.

Ingested energy

In a single feeding a vampire can consume 57 percent of its body weight in blood, which contains 1.1 kcal/g. Thus the ingested energy is

$$42 \times 57\% \times 1.1 \text{ kcal/g blood} = 26.3 \text{ kcal}$$

Excreted energy

A vampire excretes 0.24 g urea in the urine plus 0.95 g of feces daily. Urea contains 2.5 kcal/g and the feces contain 5.7 kcal/g. Thus the excreted energy is

$$0.24 \text{ g urea} \times 2.5 \text{ kcal/g} + 0.95 \text{ g feces} \times 5.7 \text{ kcal/g} = 0.6 \text{ kcal} + 5.4 \text{ kcal} = 6.0 \text{ kcal}$$

Assimilated energy

The energy the bat actually assimilates from blood equals the energy ingested (26.3 kcal) minus that excreted (6.0 kcal).

Thus a vampire's energy "income" is 20.3 kcal/day

$$26.3 \text{ kcal} - 6.0 \text{ kcal} = 20.3 \text{ kcal}$$

Metabolism

In a tropical habitat, 20°C is a reasonable approximation of the temperature bat experiences both inside and outside the cave. While at rest in the laboratory at 20°C a vampire's metabolic rate is 3.8 ml O_2/g • hr. As a result, the term for metabolism can be calculated and converted to calories using the caloric equivalent of oxygen (4.8 cal/ ml O_2).

$$M_i = 42 \text{ g} \times 3.8 \text{ ml } O_2/\text{g} \bullet \text{hr} \times 4.8 \text{ cal/ml } O_2 \times 22 \text{ hr/day} = 16.9 \text{ kcal}$$

$$M_O = 42 \text{ g} \times 3.8 \text{ ml } O_2/\text{g} \bullet \text{hr} \times 4.8 \text{ cal/ml } O_2 \times 1.5 \text{ hr/day} = 1.1 \text{ kcal}$$

Activity

The metabolism of a bat flying at 20 km/hr is three times its resting metabolic rate (3 × 3.8 ml O_2/g • hr × 11.4 ml O_2/g • hr). The cost of the round trip from the cave to the feeding site is

$$A = 42 \text{ g} \times 11.4 \text{ ml } O_2/\text{g} \bullet \text{hr} \times 4.8 \text{ cal/ml } O_2 \times 0.5 \text{ hr} = 1.2 \text{ kcal}$$

Biomass change

The quantities calculated so far are fixed values that the bat cannot avoid. The biomass change is a variable value. If the assimilated energy is greater than the fixed costs, this energy "profit" can go to biomass increase. Fixed costs which exceed the assimilated energy are feflected as a loss of biomass. For the situation described there is an energy "profit":

$$I - E = M_i + M_O + A \pm B$$

$$26.3 \text{ kcal} - 6.0 \text{ kcal} = 16.9 + 1.1 \text{ kcal} + 1.2 \text{ kcal} \pm B$$

$$B = +1.1 \text{ kcal/bat} \bullet \text{day}$$

These calculations show that vampires can live and grow under the conditions assumed. What happens if we change some of the assumptions? Professor McNab points out that the northern and southern geographic limits of vampires conforms closely to the winter isotherms of 10°C. That is, the minimum temperature outside the cave during the coldest month of the year is 10°C; the bats do not occur in regions where the minimum temperature is lower. Is this coincidence, or is 10°C the lowest the bats can withstand? Calculating an energy budget for a vampire under these colder conditions provides an answer.

Inside the bats' caves the temperature remains constant at 20°C, so only the conditions a bat encounters outside the cave are altered. Because of limitations of stomach capacity ingestion cannot increase beyond 57 percent of body weight, the value assumed in the previous calculation. Therefore we need reconsider only M_0, A, and B.

Metabolism outside

At 10°C a bat must increase its metabolic rate to maintain its body temperature, and laboratory measurements indicate the resting metabolic rate increases to 6.3 O2/g • hr.

$$M_0 = 42 \text{ g} \times 6.3 \text{ ml } O_2/\text{g} \bullet \text{hr} \times 4.8 \text{ cal/ml } O_2 \times 1.5 \text{ hr/day} = 1.9 \text{ kcal}$$

Activity

The cost of activity will not change because the metabolic rate of the bat during flight (11.4 ml O_2/g • hr) is higher than the resting metabolic rate needed to keep it warm (6.3 ml O_2/g • hr). Only the term M_0 changes, increasing from 1.1 to 1.9 kcal, and the sum of the energy costs becomes 20.0 kcal/bat • day.

Because the assimilated energy remains at 20.3 kcal/bat • day, only 0.3 kcal is available for biomass increase. The assumptions in these calculations introduce a degree of uncertainty, and probably 0.3 kcal is not different from 0 kcal. In other words, at 10°C a bat uses all its energy staying alive; it does not have an energy profit for growth or reproduction and can tolerate these conditions for only a short period. The 10°C January isotherm runs south from the vicinity of Brownsville, Tesas, along the Gulf Coast of Mexico and cuts across the middle of the Florida penisula. Most of the Gulf Coast of the United States is north of this isotherm and has too harsh a climate for vampires.

Additional calculations reveal more about the selective forces that shape the life of vampire bats. For example, a bat's stomach can hold a volume of blood equal to 57 percent of its body weight, but a bat cannot fly with that load. The maximum load with which a vampire can fly is 43 percent of its body weight. Before it can take off to start the flight back to its cave, therefore, a bat must reduce the weight gained from its meal. For example, if a bat weighting 42 g takes in 23.9 g of blood, it must reduce this load to 18.1 g before it can fly. Vampires do this by rapidly excreting water. Within 2 min after it begins to eat, a vampire starts to emit a stream of dilute urine. Experiments performed at Cornell University has revealed that a vampire produces urine at a maximum rate of 0.24 ml/g body weight • hr. Thus, in the hour and a half the bat may spend in feeding, it could excrete as much as 15 g of water-more than enough to allow it to fly.

Although rapid excretion of water solves the bat's immediate problem, it introduces another. The bat is left with a stomach full of protein-rich food that will yield a large amount of urea. To excrete this urea, the bat needs water to form urine. By the time a vampire gets back to its cave it is facing a water shortage instead of a water excess. Unlike many mammals, vampires seldom if ever drink water to compensate for water loss. Instead, they depend on blood for their water requirements. Like other mammals adapted to conditions of water scarcity, vampire bats have kidneys capable of producing very concentrated urine to conserve water. As a result of its unusual ecology and behaviour, a vampire bat can be considered to live in a "desert" of its own making in the midst of a tropical forest.

As can be seen from studies of the vampire bat, consideration of the energy requirements of alternative behaviours, each of which could satisfy the needs of an organism, help us understand why vertebrates behave as consistently as they do. Application of energy-budget analysis to variety of vertebrates is an exciting approach to behaviour evolution and ecology that has barely begun.

Mammalian Integument

We have established the selective advantage of having a high and constant body temperature despite its energy costs. It should have also become clear that any energy lost can be critical to the endothermic homeotherm. While behavioral microhabitat selection helps to curtail energy loss, most mammalian thermoregulation would be impossible without the characteristics and dynamic reactions of the integument. A great deal of the ability of mammals to individually and evolutionarily invade new habitats is attributable to the properties of their integument.

The integument of mammals is a single orga, the largest of the body; it includes a series of adaptive structures that are outgrowths of the skin. In mammals, as in other vertebrates, the skin functions to protect against onslaughts from, and to receive information from the outside world. The skin, its hair, glands, and sense cells interact with every other aspect of mammalian life in such significant ways that each deserves special consideration.

Skin

The major divisions of the vertebrates skin are the epidermis (the superficial cell layer derived from embryonic ectoderm and the dermis (the deeper cell layer of mesodermal orgain). Both rest on the subcutaneous tissue (hypodermis) that overlies the muscles and bones. The uniqueness of the mammalian integument lies in its derivatives: growing replaceable hair; lubricant - and oil-producing holocrine (sebaceous) glands; aprocrine and exocrine glands that secrete volatiles, water, and ions; the unique mammary glands; scales, nails, claws, and hoofs; and horns. The integument is also a contributing factor in the growth of antlers. Not every species has each class of integumentary appendage, but the various types often occur in unrelated mammals, attesting to their broad importance.

Epidermis

The epidermis is an avascular cell layer with active and proliferating cells in its deepest part but dead cells near the surface that are regularly shed. The deep, active cells (*stratum germinativum*) obtain their metabolic needs from the underlying vascular dermis. Epithelial cell divisions at this interface force daughter cells toward the skin's surface. These daughter cells produce and accumulate proteinaceous keratohylain granules and related substances, which slowly change chemically as the cell dies and is forced nearer to the skin's surface. By the time these cells reach the surface they are adherent platelet like sacs of *soft kerain*, a tough, pliable hydrophobic protein.

Delicate balances between rates of cell proliferation, production of keratohyaline granules, "maturation," and the shedding of surface cells produce regional and species differences in skin thickness and texture. Some small rodents have exceedingly delicate epidermis only a few cells thick. Human epidermis varies from a few dozen cells thick over much of the body to over a hundred cells thick on the plams and soles. Elephants, rhinoceroses, hippotoamuses, tapirs, and pigs were formerly classified as "pachyderms" because their epidermis is several hundreds of cells thick. The texture of the external surface of the epidermis varies from smooth (in fur-covered skins and the hairless skin of cetaceans) to rough, dry, and crinkled (many hairless terrestrial mammals). The tail of many rodents it textured with epidermal scales very like those of reptiles. Without doubt the most adaptive character of the mammalian epidermis is cohesiveness between cells during all stages of development and degeneration. This imparts marvelous characteristics to the tissue: It is living, growing, regenerative – and at the same time dry, abrasive, resistant, and expendable.

Dermis

The dermis, unlike the epidermis, is primarily composed of extracellular products. The dermis is laid down in two layers whose mechanical properties dominate and determine those of the skin as a whole. Over joints and in areas of mobility, the dermis is elastic, loose, and thin. On the back and areas of continual friction with the outside world (the soles, prehensile tails, flippers, and so on), the dermis is thick, firm, and rather immobile. Its deepest layer is rich with interlaced bundles and sheets of collagen fibers produced by sparsely scattered cells (fibroblasts). The interweaving of this dermal fabric and the amount of elastic fibers determine the extraordinary tensile strength of the mammalian skin. It is primarily this layer of mammalian skin that is used to make leather. Occasionally smooth muscle fibers occur in this area. Their contraction produces a wrinkling of the skin such as that of the nipple of the mammary glands or the male scrotum.

Twings of blood vessels and nerves course through the lower dermis to their destination in the superficial dermis. The latter is a thin layer composed of more delicate and loosely arranged collagen and elastic fibers. It is a "reverse impression" of the complex deep surface of the epidermis. Here arterioles break up to form and enormous, complex web of capillaries closely applied to the dermis-epidermis junction. These fuse again as venules, which exit from the superficial dermis often in parallel with an incoming arteriole. In addition there may occur direct connections between afferent and efferent vessels (arteriovenous anastomoses). Two or more flat networks of vessels occur at different depths within the dermis and hypodermis parallel to the body surface. These vascular networks are best developed in mammals lacking thick fur (for example, swine and humans) or in naked regions of hairier mammals (the perineum, that is, around the urogenital/anal openings; the scrotum; the nostrils, and so on). They have a special function in body colling. The amount of blood and rate of flow within these vessels can be varied by central neural and hormonal control or directly by local temperature. Under conditions of excess heat, the vessel plexus dilates and warm blood is cooled by its close contact with the skin surface. When chilled, all but a portion of these capillaries constrict, minimizing blood flow to the skin. The close juxtaposition of afferent arterioles and efferent venules, which continue to carry small amounts of life-supporting blood to the epidermis, acts to conserve heat by countercurrent exchange.

The dermis houses the majority of sensory structures and nerves associated with the sensations of temperature, pressure, and pain. In addition, many motor nerves run to the vascular, muscular, and secretory structure of the skin. Some free neve endings penetrate the epithelium, but the majority terminate in the dermis as specialized end organs. Generally these organs are made up of connective tissue capsules of varying thickness.

Skin colour resides in the dermis, as well as the epidermis. Some neural crest cells, the melanocytes, come to rest in the dermis, in the deepest layers of the epidermis, and in the interface between the dermis and epidermis. They produce granules of melanin, which may be yellow, rusty red, brown, or black in colour. Through long, narrow, cytoplasmic extensions of the melanocytes, melanin is injected into adjacent cells that lack the melanin producing enzymes. Additional hues may be produced by structural colours overlying melanin, and by the vascularization of the skin to produce shades from pink to scarlet. Skin colouration, which incorporates the vascular system, has the advantage of being rapidly variable so that red display areas may become brilliant in excited individuals.

Hypodermis

The subscutaneous tissue or hypodermis is not functionally a part of the skin but interfaces between the skin and the deep-lying muscles and bones. Collagenous and elastic fibers are prevalent, but much of the hypodermis is unstructured. This region often contains a major energy store of the body: subcutaneous fat. Fat storage reaches its maximum development in pinnipeds and cetaceans in the form of insulative blubber. Many mammals have extensive subcutaneous muscles that move the skin relative to underlying tissues. The fly-disturbing jiggle of a horse's skin is an example. In many carnivores, ungulates, and especially in man and the other higher primates, subcutaneous muscle reaches exceptional development in the facial region. Known as the platysma, this muscle can produce facial expressions, movement of the external ears, eyelid opening and closure, and the ability to purse the lips and suck – a vital attribute to early mammalian life.

7

MAMMALIAN EVOLUTION

Little more than a century ago, few people realized that the diverse array of mammals we have just described were not only related to one another but to all other living things. Religion taught that each individual type of animal had been created by God to fulfill a special role in the word. The idea that the mighty elephant, the tiny shrew, the lithe and graceful panther, the monkeys, apes, and even man were in a very real sense cousins would have been regarded at best as a mad delusion and at worst as blasphemy. Yet we now know beyond a doubt that all these creatures can their ancestry back to a common stock.

The revelation of this fact is, of course, due to the discovery of the principles of organic evolution. The idea of the evolution of all life was discussed by the ancient Greeks, but after them it was forgotten for more than 2,000 years and men's minds were preoccupied with more magical interpretations of nature. It was not until the middle of the 18th century that the idea was reborn, this time with new vitality. A century later, in 1859, Charles Darwin's *On the Origin of Species by Means of Natural Selection* was published. This classic book presented the story of evolution so clearly that it could no longer be rationally denied, and later work has served only to confirm its basic truth.

Very briefly, the theory of evolution comprises the idea that all living things belong to one great family and that later and more complex forms have developed from simpler forms that preceded them. The various members of this family became differentiated over many millions of years by the process known as "natural selection." This process can operate because individual organisms from time to time produce definite heritable variations, known as "mutations," in the germ cells that give rise to the next generation. Thus in any given environment, there will be individuals who, through favourable mutation, have become better adapted to their way of life than their fellows. These individuals will be more likely to remain alive and reproduce themselves, and nature can therefore be said to have "selected" that type for survival from among its less well-equipped rivals. This process is repeated in each generation, with different qualities selected in different environments. Thus different "species" are eventually established, each with its own adaptations to its particular circumstances.

The history of the earth is very long, at least 4.5 billion years, according to the findings of the latest research. However, no more than half of this immense span of time has been characterized by the evolution of living organisms. Through chemical reactions on the shores of the very early rivers or seas, the first tiny one-celled organisms known as "protistans" probably developed. As the generations passed, life increased in variety and complexity, and higher and higher kinds of organisms evolved. First there were soft-bodied creatures such as sponges and jellyfish, then creatures with protective shells. After these came a sequence of backboned animals, or "vertebrates," represented in ascending order by fishes, amphibians, reptiles and, finally, mammals and birds. The full story of this wonderful process and how it has been pieced together from the study of fossils is told in another book in this series, *Evolution*, by Ruth Moore.

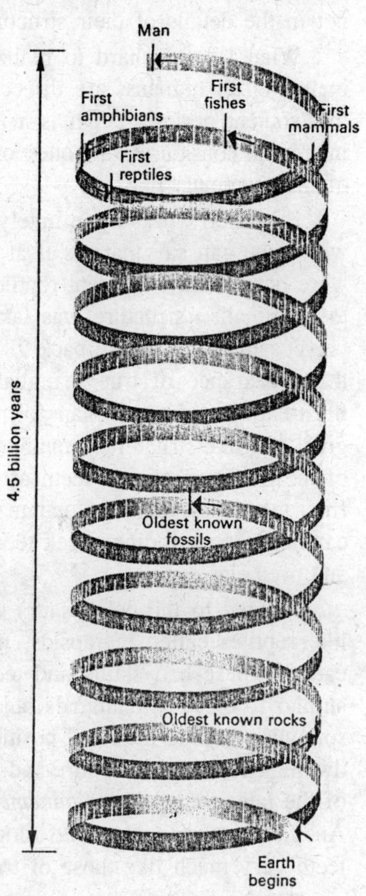

Fig. 7.1. The spiraling tape depicted here represents four and a half billion years of earth history, starting at bottom with the formation of the planet and proceeding upward to the present.

Here, where we are concerned only with the mammals, our tale begins some 180 million years ago. At that remote time, which is yet recent when we consider the total age of the earth, our planet was dominated by reptiles. Giant dinosaurs ranged over the land, huge aquatic reptiles known as ichthyosaurs and plesiosaurs swam in the seas, and weird flying reptiles known as pterosaurs took to the air. These last and membranous "wings" comparable to function, though not in the details of their structure, to those of living bats.

What is often hard to realize is that all mammals living today, including all humans, are direct descendants of reptiles that lived in this ancient period. Yet it is a fact—and this fact gives us an added interest in considering the question : What were these ancient ancestors of the mammals like?

Evolution works by infinitely slow stages, so there is no one point where we can say that, as if at the blowing of a trumpet, mammals were born from the parent reptile stock. An important early approach to mammalian structure was taken by lizardlike creatures known as pelycosaurs, which date back 280 million years or more—long before the appearance of true mammals. These were represented by both meat-eating and vegetarian forms, some of which were typified by great sail-like structures running the length of their backs. But instead of the uniform rows of teeth of other reptiles, many pelycosaurs had front teeth somewhat like mammalian incisors. Behind these were large canines, then grinding cheek teeth comparable to a mammal's molars and premolars.

Related to the pelycosaurs was a group of even more mammal-like reptiles called therapsids, most of which were aggressive meat eaters. These had skulls and teeth which in many ways were quite similar to those of mammals, and limbs which were shifting from the sprawling "out-to-the-side" position of early amphibian and some land-dwelling reptiles to the fore-and-aft position of typical mammals. One of the later therapsids, *Kannemeyeria*, had a strange beak-like snout. Another, called *Cynognathus* (dog jaw), was a powerful carnivore with teeth very much like those of true mammals.

The structure of the teeth, among other features, shows that pelycosaurs and therapsids, although not necessarily the direct ancestors of mammals, may certainly be regarded as their ancient uncles and aunts. The fossil record does not tell us whether therapsids were warm-blooded, whether they had hair instead of scales and whether they nursed their young. But judging from the progressive modifications of the hard parts, they may possibly have been developing these typically mammalian characteristics.

Fossils, unfortunately, reveal very little about the creatures which we consider the first true mammals. From the new remains which have been discovered—mainly teeth and jaws—we know that they were mostly tiny animals no bigger than rats and mice. They exploited different ways of life from those practiced by the reptiles. Some of them hunted the insects that dwelt in the undergrowth of tropical jungles, while others took to the trees to avoid direct competition with their ground-dwelling contemporaries. In the unstable environment of the tree-tops, which makes such rigorous demands on the coordination of the muscles controlling locomotion, these ancient ancestors of monkeys, apes and men began to develop the complex brains that still distinguish the primates today.

Fig. 7.2. The earliest mammal.

Now all these events were occurring at time when reptiles still dominated the earth. It was not until some 60 million years ago that the pattern of life changed and the true mammalian descendants of the mammal-like reptiles finally began to assert their pre-eminence. For reasons which are still not fully understood, almost all the great reptiles which had ruled the earth for some 200 million years became extinct.

One the main causes of the mass extinction was probably a radical change in climate, related to geological and cosmological events. The fulling reptiles, with their limited tolerance of external temperatures and their reliance on specialized types of food which no longer flourished in their former abundance, were ruthlessly eliminated. Unable to adapt, they perished. The disappearance of these formerly dominant creatures created a kind of biological vacuum, which the more adaptable mammals were quick to exploit. And with their internal thermostats, more advanced brains, and an improved method of bearing and raising their young, they were able to exploit it very efficiently indeed.

The next 60 million-odd years of earth history have been aptly termed the Age of Mammals—the great period of mammalian expansion and differentiation when all the creatures mentioned in the previous chapter went through the major phases of their evolution and became established, with suitable physical and mental specializations, in the different regions they occupy today. In a sense we still live in the Age of Mammals, although the past million years in particular have been a period of rapid decline for the group as a whole. This has been attended by the spectacular rise of one particular mammal, man, whose evolving mind has enabled him to establish a position of unchallenged supremacy over his mammalian cousins. Meanwhile, there is an exciting story to tell about the time when the group as a whole was enjoying its heyday.

For convenience, students of the earth's past divide the Age of Mammals into seven epochs. The age as a whole is known as the Cenozoic, from two Greek words, *kainos*, meaning "new," or "recent," and *zoikos*, "pertaining to life." The dating of the different epochs is determined by a variety of scientific tests applied to the rocks in which their respective fossils are embedded. The fossils them-selves are the keys to the whole story. What do they tell us about the early mammals whose descendants for the past 60 million years have been such as important part of life on our planet?

Fig. 7.3. Alticamelus.

To answer this question it will be easiest to consider the procession of ancient mammalian life as it appeared in the successive epochs.

Some of the mammals of which fossil remains have been discovered are probably direct ancestors of mammals living today. But others, often very bizarre and intriguing, represent early mammalian forms that long ago reached the ends of their lines and diet out. These fossils give us evidence of evolutionary enterprises which never came to fruition; of creatures which, through some limitation of brain power. some narrow specialization of physique or because of a sudden environmental change to which they were unable to adapt, were thrown on the scrap heap of nature's discarded experiments. The study of these ancient mammals is, however, essential to any proper understanding of the role of mammals in the world today.

The story begins in the Paleocene, which lasted from 63 million to 58 million years ago, the first epoch in the Age of Mammals. Whatever the rigors of climate which may have been a factor in the extinction of the ruling reptiles, conditions during the late Paleocene and the succeeding Eocene had become comparatively mild. The tropical and temperate zones spread out much further toward the poles than they do today. Types of plants that now flourish only in such congenial climates as that of modern Malaya were found along the banks of the Thomas in England and in many of the present cold temperate zones of the United States Even regions within the Arctic Circle supported a rich flora.

Under such attractive conditions and with no competition from the giant reptiles, the mammals began to spread over the face of the earth. By the dawn of the Eocene all the main orders of placental land mammals, including insect eaters, rodents, hoofed mammals, carnivores and early primates, were flourishing. Alongside these lived primitive monotremes and marsupials, as well as aquatic placental mammals such as whales, which had become specialized to a nonterrestrial environment by the mid-Eocene.

Among the carnivores, a group known as creodonts (literally 'flesh teeth") foreshadowed the development of modern meat-eating forms. These had evolved from a line of

Fig. 7.4. Epigallus.

early insect eaters which were now acquiring varying adaptations to fit them as predators on larger game. Some were rather weasel-like in shape and habit, others more closely resembled wolves and lions. One of the largest, known as hyaenodon, could kill an animal as large as a modern rhinoceros.

These early carnivorous mammals preyed on a vast assemblage of herbivorous game. Among the most spectacular herbivores was a group known as the titanotheres, or "giant beasts." Superficially resembling modern rhinoceroses, these creatures were as much as 15 feet long and eight feet high at the shoulder and had massive, bony horns on their heads. Their relatives, the chalicotheres, were odd-toed forerunners of the ungulates, or hoofed mammals, with huge birdlike claws instead of hoofs. Other Eocene mammals were the uintatheres, named after the Uinta Mountains in Utah, where their fossil remains were first found. These creatures looked formidable enough, being also as big as a modern rhinoceros and wildling tusklike upper canines and three pairs of horny protective structures on the tops of their heads. They could live only on the softest vegetable foods, however, and their mental equipment was poor—fatal limitations in the struggle for survival which soon led to their extinction.

A much more progressive line of herbivorous mammals was that of the horses, which have survived right down to modern times. A little animal known as *Hyracotherium* or eohippus about the size of a fox terrier, was probably the ancestor of many later members of the group, including our familiar work horses and race horses. Instead of the characteristic hoof of modern horses, it had four toes on the front feet and three on the hind.

The elephants also trace back their pedigree to this remote epoch in the Age of Mammals. The ancestor of the modern elephant was a little swap-dwelling creature from Egypt known as *Moeritherium*.

Fig. 7.5. Paraceratherium.

About the size of a large pig, it had not yet developed the trunk which distinguishes its modern descendants, but had only the "blackish bulgy nose" which Rudyard Kipling described in the Just So Stories. *Moeritherium* was a successful little animal, nevertheless, for its trunkless snout was well adapted to cropping the vegetation along the fringes of the swamps where it made it home. Also inhabiting Egypt toward the very end of the epoch was a grotesque cousin of the elephant known as *Arsinoitherium*, with a massive tank-like body and a pair of enormous horns projecting forward above its nose.

Fig. 7.6. Syndyoceras.

The most dramatic water-dwelling mammals of the Eocene was an ancient whale known as *Zeuglodon*, which sometimes reached a length of between 60 and 70 feet. It was much slimmer than modern whales; in fact, its proportions remind us of descriptions of the legendary great sea serpent. Such ancestral whales were probably descendants of very early carnivorous creodonts which had abandoned evolutionary competition on dry land and explored the possibilities of the sea as a new environment for mammalian survival.

But the most potentially significant members of the mammalian cast in the Eocene were undoubtedly the primates. A wide variety of small, tree-dwelling mammals with exceptionally well-developed brains and flexible, grasping digits, they were already working out their evolutionary destiny in the treetops. They were not yet true monkeys, but were intermediate between these and their tree-dwelling insectivorous ancestors. Their fossil remains have been found in North America as well as in Europe and show that they must have been very similar to the living tarsiers and lemurs of today.

In the next epoch, the Oligocene, which lasted from 36 million to 25 million years ago, warm conditions remained widespread and mammals continued to flourish. Eohippus evolved into a somewhat larger type of horse, known as *Mesohippus*, which stood some two feet high. It has lost the fourth toe on each front foot, and adaptation to faster running. The elephants were also increasing in size and showed the beginnings of the trunk which is so typical of their modern descendants. Some now had four tusks, two above and two below, projecting a short distance beyond the jaws, and were probably moving away from the swamps to a life on the plains. Among the contemporaries of these creatures was a group known as the anthracotheres, meaning "coal-beasts," from the coal deposits in Italy where some of their fossils were found. These were probably the animals to which the modern hippopotamuses can trace back their pidigree.

Soon after the Oligocene epoch gave place to the Miocene some 25 years ago, the rich cast of the Age of Mammals was assembled in full splendor. The Miocene was the mammals' "golden age," a time when large-scale geological movements were buckling and twisting the earth's crust, leading to the creation of the Alpine and Himalayan chains and to a great shifting in the boundaries of land and sea. Intense volcanic activity also contributed to the changes taking place and the new mountain ranges had a profound effect on the circulation of the atmosphere. Tropical and subtropical regions were reduced and temperate conditions became more widespread. The modern world was beginning to take shape and living things began to approximate more closely the familiar types we known today.

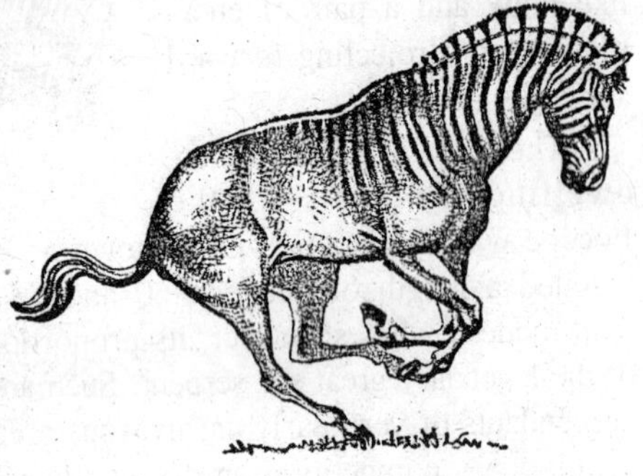

Fig. 7.7. The Quagga.

Vast herds of herbivorous mammals thronged the Miocene plain and forests. In addition to several different kinds of horses, these were rhinoceroses, giant pigs, the first antlered deer, camels, llamas and ancestral giraffes, among many others. Elephants included a number of very oddly specialized types: one with a pair of tusks in its lower jaw which curved back toward its chest, several others with their lower jaws modified into long trough-shaped shovels which they used for digging up water plants from the floors of swamps.

Preying on the herbivores was a varied assortment of carnivores, each adapted to hunting a different kind of quarry. These included civet cats and ancestral dogs and bears, but the most spectacular of all the carnivores were the so called "saber-toothed tigers." Now known much more accurately as "stabbing cats," from the enormously elongated upper canines which they probably used to penetrate the thick, tough hides of young mastodons, they were a remarkably successful group throughout the Age of Mammals, but it was in the Miocene epoch that they enjoyed their heyday.

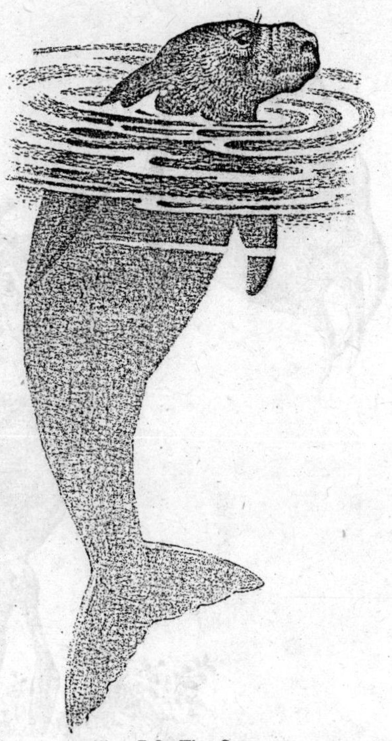

Fig. 7.8. The Sea cow.

After the late Miocene, mammals in general seem to have begun a period of decline which has continued to the present day. The primates, however, continued a successful evolutionary course despite the wavering fortunes of their mammalian contemporaries. Gibbonlike apes were already swinging through the branches of the Asiatic and European forests some 20 million years ago, and the ancestors of the various kinds of Old and New World monkeys were likewise flourishing. In East Africa an exceptionally interesting great ape of that time known as *Proconsul*, appears in the fossil record. This creature may well have been very close to the ancestry of the chimpanzee and even man.

In spite of the fact that mammals as a whole seem to have passed their peak in the late Miocene, the succeeding Pliocene and Pleistocene epochs still had a rich mammalian fauna. In the early Pliocene, giraffes were exceptionally numerous and included short-necked forms as well as the typical long-necked species we know today. A grotesque deerlike beast, *Synthetoceras*, had a huge Y-forked horn behind its nostrils, as well as conventional horns sprouting from its forehead. Sough America was inhabited by several giant creatures whose living cousins are today of only small or moderate size. A huge ground sloth, for instance,

Fig. 7.9: The Irish elk.

called *Megatherium* (literally "great beast") was as large as a modern elephant. There were also enormous armadillolike creatures known as glyptodonts (literally "sculptured tooth," from their characteristic dentition), some of which were armed with tails tipped with a cluster of vicious-looking spikes. In other parts of the world similar giant forms evolved, not only among the advanced placentals, but also among the marsupials. In Australia an extinct relation of the living wombat was as big as rhinoceros and there were giant kangaroos between 8 and 10 feet high.

The Pleistocene epoch was periodically characterized by very rigorous conditions of climate. From about a million years ago onward, arctic conditions on four separate occasions imposed themselves on much of the present temperate zones. During this great ice age, vast sheets of ice advanced from the polar regions over much of the earth, driving practically all life before them. The last withdrawal started as recently as 10,000 years ago and still continues today.

The mammals of the great ice age in many cases made very effective adaptations to the harsh conditions which challenged them. Thus the Pleistocene cousins of several creatures that today have almost hairless skins were characterized by thick, shaggy coats, which helped to retain their body heat. The so called "woolly mammoth" is a famous example of an ice age mammal that made this particular adaptation to cold. What is not so well known is that this creature also developed a thick layer of fat beneath its skin, especially on the shoulders, which probably served as a food reservoir when the frozen ground made its vegetarian diet hard to get. A woolly rhinoceros living in the same regions was protected by a similar thick covering of hair. Although hair is not normally preserved in fossil form, we know of its existence in these mammals because remains of their carcasses have survived, shaggy coat and all, in the permanently frozen ground of Siberia.

Many other species of mammals, however, were killed off in the great ice age, either through the climatic challenge of the times or by direct competition with man. This poses an interesting question: Why did some species of mammals win through while others failed? The answer lies in some of the laws controlling biological survival—and two major mammalian groups, the horses and the elephants, show in illuminating fashion how these laws work.

When the environment changes, all living things must either move to a new environment that resembles the old one or else subject themselves to the law "adapt or perish." Eohippus, the little "dawn horse," had teeth adapted only for dealing with fairy soft food, a fact which suggests that it was probably a forest dweller, living on soft shoots rather than the tough grasses of the plains. But as time went on the descendants of eohippus began to move into new environments. This occurred partly because the reduction of the forested areas of the earth led to increased competition for food and partly because of the evolution of new types of predatory carnivores with a liking for horseflesh. The descendants of eohippus thus gradually left the forest for the plains, and as a result, natural selection developed the plains horses, which were usually larger than their forest-dwelling ancestors. These gradually lost their full complement of toes and eventually supported themselves only on the center toe of each foot—the characteristics "hoof" of living horses. These larger horses could move more quickly than their tiny forebears, and their higher eye level gave them a better chance of spotting enemies on the open plains. The raising of the body on the forepart on the central toe, or hoof, gave them an exceptionally good turn of speed to outrun the plains carnivores that hunted them. Correlated with these changes came modifications in the teeth, which became adapted to grazing coarse grasses instead of nibbling the soft forest shoots. It was by such wonderful adaptations, all aimed at the single goal of survival, that the early horses became differentiated and spread across the face of the earth.

Elephant evolution took a different course, but obeyed the same evolutionary principles. The early, small swamp-dwelling elephants were forced by a changing climate to develop types suited to life in drier conditions. In these new circumstances increased size became an advantage to them, for it helped them to resist the attacks of predators in open country. Also, instead of concentrating on a speedy escape, they relied on the defensive advantages of tusks and an exceptionally thick skin. But in order to support the heavy weight of tusks, it was mechanically necessary for their necks to remain short instead of becoming elongated like those of the horses. This meant that they could not reach the ground to graze, while the tusks made browsing almost equally difficult. The answer was the development of a long prehensile upper lip, or "trunk" which could both reach the ground and extend beyond the tusks. Here again we see how, through the working of evolution, living things may find answers to the problems of survival in a constantly changing world.

The horses and elephants were comparatively successful in their adaptations. But many other mammals, such as the great titanotheres and chalicotheres, the gigantic *Paraceratherium* (formerly called *Baluchitherium*) and even the seemingly successful stabbing cats, have left no progeny. The hard truth of the matter, it must be re-emphasized, is that organisms must "adapt or perish," for nature does not appear to be interested in the survival of individuals, or even of different animal groups, but only in the continuation of life itself. Individuals play their part, but in themselves they are not the main object of the process.

The exceptional rigors of the ice age not only killed off many species but forced still others into extreme specializations which, in the long run, proved more disastrous than helpful. But there was another factor too: the emergence of man. Man is the most ruthless and efficient predator that evolution has yet produced. He hastened the extinction of many of the creatures of the Age of Mammals by hunting them for food and clothing. He domesticated others and harnessed them to his own use. Still others he destroyed inadvertently by drastic alterations of their environment. These processes continue today, as the spread of man's society over the face of the earth tends to eliminate more and more of the animals from whose ranks sprang.

Whether this immense destructiveness can be justified is a question that can best be judged when, in the final chapter of this book, the role of man as a mammals has been more fully described.

8

Moving on Four Legs

In the ability to move about from one place to another, mammals are by far the most efficient and versatile group of terrestrial vertebrates. They are not only pre-eminent on land in this respect, but have also invaded the air and reinvaded the water, where they can successfully compete with creatures who have occupied these environments for a much longer time.

The great majority of mammals may be technically classified from the point of view of locomotion as tetrapods—i.e., four-footed animals. Every order of mammals, with the exception of whales and sirenians, is mainly represented by tetrapod forms. This is a logical outcome of the evolutionary process, for the ancestors of all living terrestrial vertebrates are descended from fishes with four fins on the underside of the body, which became gradually modified for four-legged progression on land. Even the whales and sirenians were originally land-dwelling tetrapods. It was only when they reinvaded their ancestral habitat, the sea, that they lost their hind limbs, and their fore limbs became superficially finlike once more.

The movements most characteristic of land-dwelling tetrapods are walking and running. Mammals as different as the mouse, the hippopotamus, the chamois, the giraffe, the tapir, the dog and the buffalo all share the ability to walk and to run on four legs. But the story of mammalian locomotion is not altogether as simple as this. The whales and sirenians have limbs so highly specialized for aquatic life that nay form of terrestrial progression, let alone walking or running, is impossible. Even among seals it is extremely difficult. Bats, on the other hand, have become specialized for life in the air. Still other mammals, the kangaroos and the little rodents known as jerboas, sometimes mammals, the kangaroos and the little rodents known as jerboas, sometimes wall on all fours, but mainly they propel themselves on their hind limbs, using their forelimbs for other functions, such as grasping food. Several species among the primates, by contrast, progress mainly by the use of their *forelimbs*, which are much better adapted to swinging through the treetops than to making efficient movements on the ground. Finally, there is man, whose very dominance today is based partly on his ability to walk on his hind limbs alone, and thereby adapt his forelimbs for the manipulation of tools.

The horse provides a good example of how a typical tetrapod moves. When walking, if the right forefoot is carried forward first, then the left hind foot follows. The left forefoot goes forward next, and then the right hind foot. This completes the cycle, which is immediately repeated with another advance of the right forefoot, and so on.

In the walk the center of gravity is always contained within the triangle created by three limbs placed stably on the ground. The animals can therefore stop moving at any time without falling over. Such a slow form of locomotion is not of much use, however, in the more challenging crises of life. Walking would be unlikely to bring success, for instance, to a horse attempting to evade a lion. When the tempo is quickened to a trot, each foot is lifted *before* its predecessor in the cycle has touched the ground. There is, therefore, a brief period when the animal is supported by only two of its four limbs. This makes for greater momentum but less stability.

In horses, cats, dogs and many other mammals the two limbs which are raised in this faster gait are usually diagonally opposite—the left fore and right hind foot, and vice versa. But with certain other mammals a somewhat different procedure is followed. The camel and elephant lift the limbs *on the same side*, a lateral gait known as pacing, which imparts a peculiar and quite distinctive rolling motion. This is the reason why the sensation of riding on an elephant or camel is so different from that of riding on a horse. Pacing is sometimes practiced by young or old members of species that normally adopt a diagonal gait—colts, for instance, go through a pacing phase before adopting the diagonal rhythm of the adult horse.

The fastest gait characteristic of tetrapod mammals is the gallop. Here only one foot, or even no foot at all, is on the ground at certain points in the cycle. This type of movement is also well exemplified

Fig. 8.1. The advantages of a supple spine in mammals.

by the horse, but is practiced by such very different types of mammals as the hare and the weasel. In the gallop, the horse lifts its limbs in a different order from when it is walking, the sequence going right front, left front, right hind, left hind; or alternatively, left front, right front, left hind, right hind. At times the whole animal is flying through the air without any support whatever, and the imprints of the same hoof may be as much as 25 feet apart, some four times the length of the animals's body.

The horse's back remains comparatively rigid even at the fastest gallop, for the propulsive power is provided almost entirely by the muscles of the legs. But when such creatures as the weasel and greyhound gallop, they arch their backs at the moment of forward spring, and this greatly increases the thrust. The weasel, incidentally, shows the gallop in its simplest form Unlike the horse, which moves its four limbs individually, it gallops in a series of bounds, moving the fore and hind limbs in pairs.

Variations between the tow types of gallop practices by the horse and the weasel also occur. The hare, for instance, bounds on paired hind feet like the weasel, but puts down its forefeet one after the other at the end of the leap.

However they gallop, many mammals can attain quite exceptional speeds. The cheetah is the champion sprinter, overtaking its prey of swift gazelles and antelopes in half-mile bursts. French naturalist Francois Bourliere reports that a cheetah was clocked at 20 seconds over a distance of more than 700 yards —averaging just over 71 miles an hour, a record for mammals. Still more remarkable are its powers of acceleration, which will take it from a standing start to 45 miles per hour in two seconds—a performance that cannot be approached by even the fastest racing car. At full speed, the African wildebeest, the springbok and Thomson's and Grant's gazelles can do up to 50 miles per hour. The mighty Cape buffalo can charge at 35 miles per hour—nearly twice the average charging speed of an elephant.

Speed, of course, is very largely dependent on limb structure. Such rather slow mammals as bears and primates have feet with large, flat bearing surfaces extending form toes to heel, the whole of which are normally in contact with the ground during part of the walk. This stance is known as "plantigrade," derived from the Latin *planta*, meaning "the sole of the foot," and *gradior*, "to walk." The faster dog and cat, on the other hand, are "digitigrade"—they stand and walk on their "fingers," with the heel, or hock, being permanently raised. A third category of mammals, including the horses, antelopes and gazelles, is "unguligrade," progressing on the very tips of the "fingernails," which have become protected by the enlarged nails we know as hoofs.

These different structures, and their modifications, have a very strong bearing on the efficiency of their owners' movements under varying conditions. Thus the long, slender limbs of the horse are much better adapted for running swiftly over flat, but not slippery, surfaces than the shorter limbs of the flat-floated bears. Conversely, on ice a bear will do much better, for its broad sole gives it a much better frictional grip and it will not like the horse, be in constant danger of falling.

The sure-footed mountain sheep and goats have feet that are especially adapted for travel in rugged mountainous terrain. Their hoofs have sharp edges and the undersides are concave, enabling them to adhere somewhat like suction cups as these agile creatures make their way up and down seemingly impassable cliffs. Caribou have huge rounded hoofs to support their weight on snow or spongy tundra, and the edges of the hoofs are sharp to prevent skidding on ice. The camel's cloven hoofs are very broad and cushioned with thick soles—adaptations for travel over yielding desert ground. But Bactrian camels are animals of cold climates, which suggests that this hoof structure may have originally been an adaptation to snow. A number of arctic mammals as different as the polar bear, the Canada lynx and the snowshoe hare have broad, flat feet with furred soles that act as snowshoes.

Fig. 8.2. Trotting and pacing are economical gaits that allow animals to maintain speed over long distances.

Man, by nature a plantigrade primate, a unique among mammals in the extreme specializations he has made to a bipedal stance, but there are several members of other mammalian orders which have tended in this direction. The locomotion of such creatures is particularly well illustrated by the kangaroos and the jerboas. A typical large kangaroo, such as the great red kangaroo of Australia, has one of the oldest gaits in the animal kingdom. When moving slowly it uses not for supporters but five, for the thick muscular tail functions as a fifth supporting limb. With all four feet as well as its tail on the ground, the kangaroo raises its two hind limbs together, its body weight carried by the short forelimbs and the trial; as part of the some movement the hind limbs swing forward to a new position. The forelimbs are then likewise raised and moved forward, and the cycle repeats itself.

As the pace quickens, however, kangaroos adopt a quite different action—the familiar leaping gait for which they are famous. In this, only the hind limbs come into contact with the ground, which they do simultaneously and side by side. The body is carried well forward and is counterbalanced by the tail, which stretches out behind. The exceptionally long and powerful hind legs propel the animal forward

Fig. 8.3. Wallaby.

in a series of jumps, which can be maintained at a steady pace up upwards of 20 miles per hour. In short bursts a speed of about 30 miles per hour can be achieved. The leaps themselves are normally some five times as long as the kangaroos's body length—that is to say, some 25 feet—but leaps of 40 feet have been recorded. From a standing start the maximum height of a kangaroo's jump is about eight feet.

The Old World jerboas and the New World kangaroo rats and jumping mice move very much like miniature kangaroos. Their hind limbs are even longer in proportion to their forelimbs than are those of kangaroos, and their tails are likewise used for balance.

Such gaits as walking, running and leaping are practiced mainly by mammals that habitually live on the ground. A few mammals, such as moles and pocket gophers, spend practically their entire lives underground. Both have short, powerful forelimbs equipped with strong claws for digging, and moles have their entire forelimbs broadened and flattened into efficient shovels.

A much larger number of mammals have wholly or partly abandoned life on the stable earth for a new kind of existence in the leafy world of the treetops. In this comparatively unstable environment, the tails

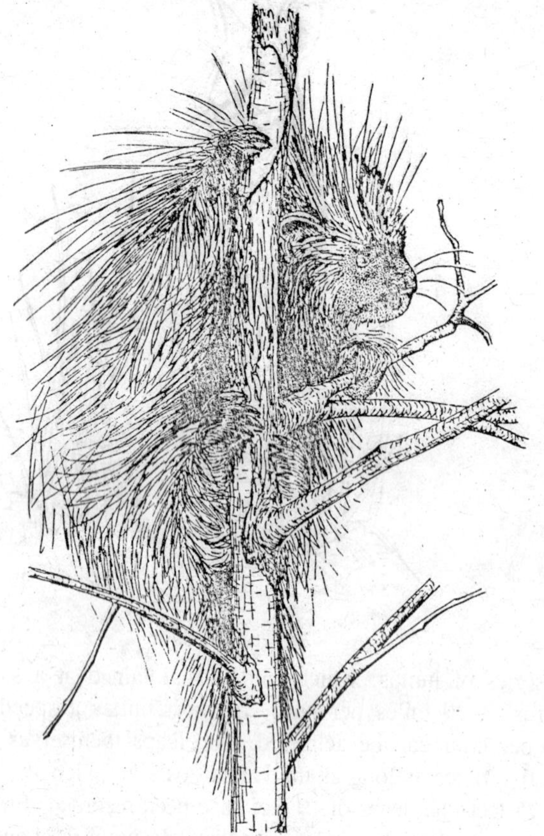

Fig. 8.4. Porcupine.

is mostly used as a balancing organ, but in some tree dwellers it has become modified to the point where it actually functions as a fifth grasping limb—a most useful adaptation. Many different mammals have tails that are more or less prehensile: the African long-tailed pangolin, the binturong of southeast Asia and the Papuan arboreal mouse, for instance. South America has numerous examples, among them monkeys, opossums, the tree porcupine, the silky anteater and the kinkajou. Such South American primates as the spider monkeys, woolly monkeys and howlers have such powerful prehensile tails that they can support the whole of their body weight by this organ alone.

Another even more important adaptation to movement in the trees is in the structure of the limbs themselves. There is a tendency, particularly noticeable in the arboreal primates, for the hind limbs to become stronger and heavier than the forelimbs. This is because the hind limbs are generally used for supporting the body weight while the forelimbs reach about in search of new holds. This tendency has had an extremely important effect on the evolution of man.

A number of characteristic modifications also occur in the digits of many tree-dwelling mammals. These are usually long and flexible and are well suited to grasping the branches through which the animals move. But a special advantage is provided by the thumb, which to a greater or lesser extend is separated from the other digits and can be opposed to them to grasp objects or to complete the encirclement of the limb of the tree. This opposable thumb, found in various stages of development in opossums and primates as well as other mammals, has reached a particularly high state of refinement in man—a fact that confirms other evidence of our descent from tree-dwelling ancestors.

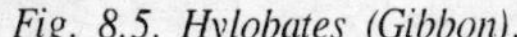

Fig. 8.5. Hylobates (Gibbon).

Fig. 8.6. Pongo satyrus (Orang utan).

As we move down the scale from the primates to less advanced mammals we find that the adaptations of the limbs to arboreal life are simpler, though these may still be very effective. Nails or claws are developed to varying degrees as aids to clinging to branches or bark. One very specialized adaptation is found in he tarsier, which in addition to flexible digits has expanded pads on the toes with cross striations which offer an exceptionally high degree of frictional resistance. They operate, in fact, on the same principle as the treated of a tire or a ridged rubber mat.

Various methods of climbing vertical tree trunks are used by different mammals. Monkeys usually ascend by grasping first with one hand and then with the diagonal, or opposite, foot. Squirrels may gallop up the trunk, taking alternate holds with their front and hind limbs. Bears hug the trunk with their forelimbs, taking successive holds with the sharp, curved claws of all four feet. The slow and deliberate Canadian porcupine is helped in its climbing by granular nonskid pads on the soles of its feet, as well as by modified spines growing at the base of the underside of the tail—a climbing aid found in similar form in the African spiny-tailed squirrel. When descending trees, most mammals come down tail first, but squirrels usually scuttle down head first, as does the fisher.

In certain primates, movement through the trees is achieved by the arm-swinging process known as brachiation. This is a kind of inversion of the bipedal walk, in which the forelimbs instead of the hind limbs are used alternately to carry to body along. Gibbons and spider monkeys, with exceptionally long forelimbs and hook like hands, are typical brachiators. The long limbs increase the momentum with which the animal can project itself forward. The fingers and the palm of the hand are exceptionally elongated, giving a greater surface for contacting a bough after a leap. A marked curvature of the bones of the hand still further increases its likeness to a hood or grapple. The thumb is correspondingly reduced so as not to impede rapid release in swinging from bough to bough, and the muscles which flex the digits are shortened so that the hand in a position of rest is curled inward. Swinging along hand over hand through the trees at great speed, gibbons often throw themselves 20 feet or more through the air between alternate holds.

Several groups of arboreal mammals are accomplished gliders. The North American flying squirrels are beautiful little nocturnal beasts which have membranes of fur-covered skin connecting the wrists of

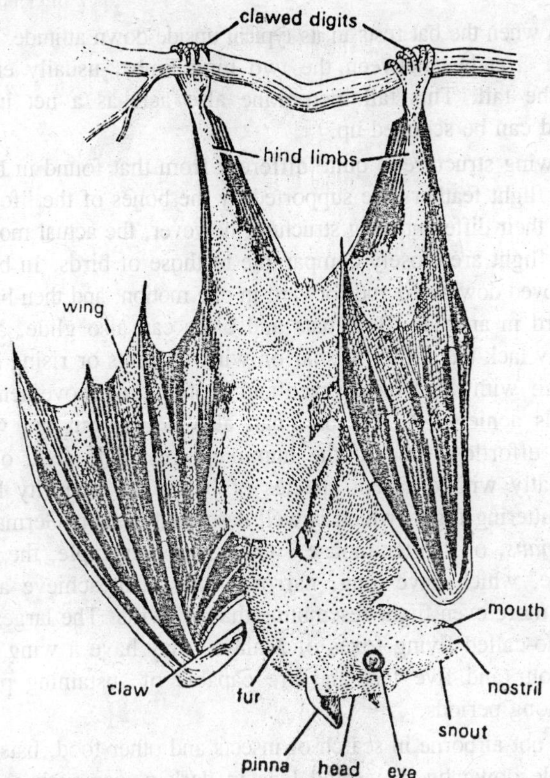

Fig. 8.7. Pteropus medius.

the forefeet and the ankles of the hind feet. Leaping outward into space, they spread their limbs and gliding membranes, and sail—sometimes 150 feet or more—to another and lower landing place. The Australian sugar gliders, which resemble the North American flying squirrels in appearance, glide in the same way. The flying lemurs of the Malay archipelago and the Philippines have similar gliding membranes between the neck, the limbs and the tail, but they extend through the fingers, giving them a webbed appearance.

The mammals which have achieved true flight are represented solely by the bats, whose greatly elongated forelimbs and fingers serve as supports, or struts, for the thin membranes forming the flying surfaces of their wings. Except for the short, clawed thumbs, which are free for grasping, the elongated fingers are all enclosed by these membranes. Extending back along the sides, the membranes are usually attached to the hind limbs at the ankles. The hind toes are free and are used for holding on when the bat rests in its typical upside-down attitude. Another membrane stretches between the two hind limbs, usually enclosing most of the tail. This tail membrane also acts as a net in which insect food can be scooped up.

This wing structure is quite different from that found in birds, in which the flight feathers are supported by the bones of the "forearm." In spite of their different wing structure, however, the actual movements of bats in flight are closely comparable to those of birds. In both, the wing is moved downward and forward in one motion, and then backward and upward in another. Both bats and birds can also glide, although bats largely lack the ability to soar on wind currents or rising currents of warm air with the typical grace and economy of movement which many birds achieve. A few bats can also hover, but not with the seemingly effortless ease of the hummingbirds. The style of flying varies greatly with different species of bats. The majority have an uneven, fluttering flight, which is well described in their German name of *Fledermaus*, or "flitter mouse." Others—for example, the tropical Molossidae, which have long, narrow wings—can achieve a faster, somewhat more even flight resembling that of swifts. The largest living bats, the so-called flying foxes of Malaya, may have a wing span of between four and five feet and are capable of sustaining powerful flight for long periods.

When not airborne in search of insects and other food, bats usually hand upside down by their hind legs in dark places such as hollow trees and caves. They do most of their flying at night, and zoologists were long intrigued by the mystery of how they avoided obstacles in total darkness. Then some years ago, two American scientists, Donald R. Griffin and Robert Galambos, applying modern techniques to the 18th Century experiments of the Italian Lazzaro Spallanzani and the Swiss Charles Jurine, conducted some ingenious research at Harvard University to find out just how bats and did this. Blinding the bats by sealing their eyes shut had no effect, for the bats avoided obstacles as skillfully as ever. But if their ears were plugged or their jaws tied shut and their mouths sealed, the bats blundered into wires. Other senses than sight were obviously being used. Hearing had something to do with it, and so did the bat's voice. Research eventually showed that the typical abilities of bats in the dark depend on a system of echo location.

What happens is this. Bats, when not actually asleep, give out a number of sounds, varying from an audible click to a cry so shrill that it cannot be heard by the human ear, although it can be detected by instruments. Even when the animal is stationary before taking flight, some half dozen or more of these ultrasonic cries may be given out every second. In normal flight the rate is stepped up to about 20 or 30 a second; when some obstacle is approached it is increased still more. The sound waves of these ultrasonic cries bounce off solid objects and the echoes travel back to the bat's ears, thus keeping the bat informed of the presence and distance of the obstacles. How the bat discriminates between various types of echoes—for instance, between an obstacle to be avoided and an insect to be caught—we do not yet know. This particular adaptation to a highly specialized way of living, now often termed "bat sonar," is one of the most fascinating and beautiful to be found in nature.

Finally we come to the swimming mammals, which represent another extreme form of adaptation to a special environment. All land-dwelling vertebrates probably developed originally from aquatic ancestors, the lob-finned fishes; and mammals themselves are the end product of an evolutionary line which had passed through an earlier amphibian and reptilian phase. Between 50 and 60 million years ago, however, certain mammals reinvaded the water, the once-familiar environment of their remote ancestors. The range of adaptations they made to aquatic life is considerable, and the creatures possessing them can be divided for purposes of simple description into three main groups.

Fig. 8.8. Phocaena (Porpoise)

First there are those mammals which, although normally land dwellers, can take to the water quite successfully in special circumstances. Thus, unlikely as it may seem such creatures as hares, hedgehogs, mice, moles, weasels, cats and even elephants are excellent swimmers if the need arises. The efficiency of these animals in water, however, is really based on their capacity to use physical equipment

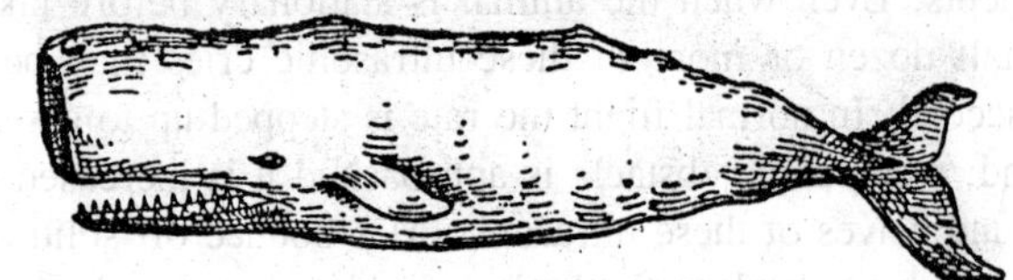

Fig. 8.9. Physeter (Sperm whale).

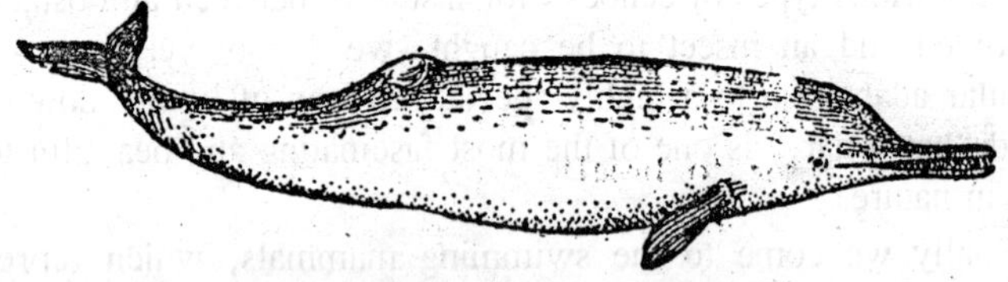

Fig. 8.10. Platanista gangetica (Dolphin).

primarily designed for land life. To take only one amusing example, swimming hamsters inflate their cheek pouches as an aid to buoyancy.

In the second group we can list a number of mammals with structures which are primarily adapted for aquatic use. The webbed feet of such mammals as the otter and the platypus, and the rudderlike tail of the beaver are obviously designed especially for aquatic locomotion. The otter swims mainly by fishlike undulations of the body, but the comparatively broad surface of the webbed types, which offers much greater resistance to the water than would individual digits, imparts an additional thrust and also allows the animal to make exceptionally rapid twists and turns. In the platypus, the webs of the front feet extend some distance beyond the toes, giving more thrusting surface in the water. The yapock, or water opossum, of tropical America uses its webbed hind feet while pursuing fish and other prey in streams. It is the only aquatic marsupial. An aquatic insectivore, the Himalayan water shrew has stiff hairs growing between its toes which serve the same function as webs. The beaver, which uses its horizontally flattened, blade-shaped tail for both sculling and steering in the water, shows how swimming ability can be achieved with an entirely different anatomical device. The muskrat uses its vertically compressed tail in much the same way.

The third group contains the most highly specialized water-dwelling mammals of all: the seals and their allies, the sirenians, or sea cows, and the whales and dolphins. Although their ancestors were four-footed land mammals, these creatures have long since become so efficiently adapted to water-dwelling life that many of them have entirely lost the ability to move about on land. The earless seals and the walrus may use all four webbed feet while swimming although seals often propel themselves—sometimes belly up—with their hind limbs alone. The sea elephants also swim with their hind limbs and usually keep their front flippers pressed to their sides. But eared seals such as the Alaskan fur seal and the California sea lion propel themselves principally with their forelimbs, while the hind limb are employed for steering.

The sea cows are so adapted to a watery existence that they have lost their hind limbs entirely, retaining only their paddle-shaped front limbs. Somewhat resembling the legendary mermaid in general shape (but not in appearance), sea cows have often been regarded as the source of the majority of mermaid "sightings" described by credulous voyagers through the centuries. One close-up look at a sea cow's wrinkled face and bristling cheeks, however, would change the mind of any believer. The largest of the sea cows, Steller's sea cow of North Pacific waters, measured from 25 to 30 feet long and weighed about three and a half tons. It was completely exterminated in the mid-18th Century by the depredations of fur hunters who slaughtered these harmless animals for food. The living sea cows, known as the manatee and the dugong, have also been much reduced in numbers by excessive hunting, even though they are now strictly protected in many areas.

Whales and dolphins represent a still more extreme stage of specialization. In fact, they have reacquired so many structures characteristic of their aquatic ancestors, such as fins, fluked tails and torpedo-shaped bodies, that they are still often mistaken for fish. The fin of a whale, however, differs as much in structure from the fin of a fish as the wing of a bat from the wing of a bird. As with sea cows, the hind limbs have been lost, though many species still have vestigial hind-limb bones buried deep in their flesh. Whales propel themselves with powerful up and down strokes of their tails, which have two flukes and, unlike the vertical tails of fishes, are flattened in a horizontal plane. The fins are used mainly for steering.

These are just a few of the varied structures that mammals have acquired as an aid to efficient movement in their different environments. Beautifully adapted to the functions they are called upon to fulfill, they form a vitally important part of each animals' equipment in the constant struggle for survival.

Mammalian Locomotion

The marked ability of mammals to move quickly and strongly is one of their outstanding characteristics. Generally more agile than the reptiles from which they are descended, and possessed of considerably more endurance, the mammals have developed a bewildering diversity of limbs, feet and toes, which enable them to get about effectively in the great variety of environments they inhabit.

Adaptation of the Skeleton

The different ways by which different mammals move about have been determined largely by their ability to respond to the evolutionary challenges of getting food and escaping their enemies. Some of the fastest-running mammals have evolved in a flat, dry habitat where feeding must be done over wide areas and out in the open. Among the slowest are the sloths, which have for their habitat trees of the American tropics, where they eat the foliage. They have developed extremely long forelimbs for moving upside down along branches, but are almost helpless on the ground. Among the greatest leapers for their size are the jerboas and kangaroo rats. Jumping about in search of food, they would be vulnerable targets were it not for their ability to dodge predators by leaping away from them. Their specializations for a bipedal mode of locomotion include enormously long hind legs, a long tail for balance and a chunky body which is easy to catapult.

Though different, the skeletons of the sloth and jerboa, like those of all other placental mammals, may derive from those of small ancient insectivores that scampered around on all fours. These archaic skeletons are believed to have been not unlike that of the hero shrew's below, which, despite its unusual backbone, retains many primitive features still fond among the insectivorous shrews.

Feet—and Their Many Uses

The foot shape of a mammal says a good deal about how its owner gets its food. The tree-dwelling tarsier leaps after insects and lizards and possesses hands and feet which provide a maximum of friction surface for a secure grip on branches. The platypus is a water-dwelling animal with a ravenous appetite. It paddles with its forefeet, searching the muddy bottoms of creeks and rivers for the hundreds of shrimps, tadpoles, worms, crayfish and larvae it can consume in a night. Its forefeet, one of which is seen at left, are generously webbed, and the swimming membrane even extends beyond the claws to make a large, efficient paddles. Moles, another group of active animals,

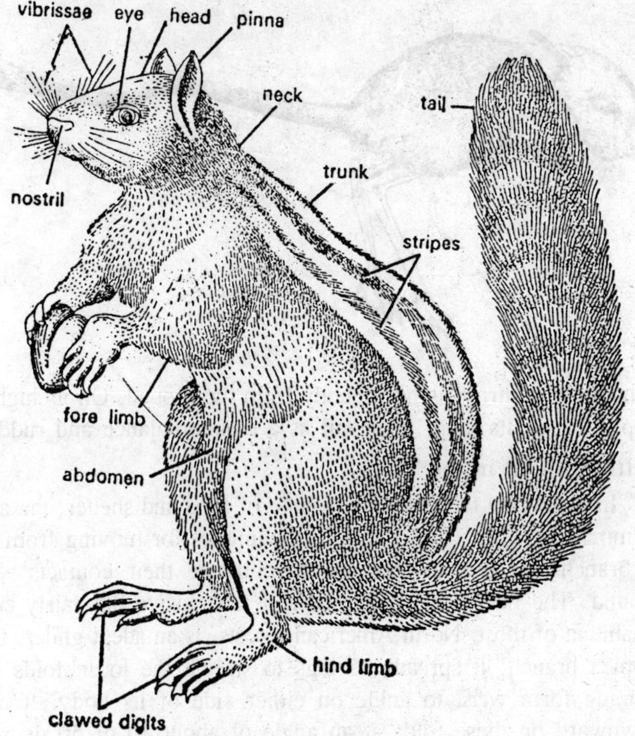

Fig. 8.11. Funambulus.

also have specialized forefeet with which they burrow after worms, insects and larvae. The walrus, though lacking feet in the strict sense of the word, has flippers, which are the modified limbs of its terrestrial ancestors. Feeding on shellfish in arctic and subarctic seas, it need not be a swift swimmer like some of the fish-pursuing seals, but must be a powerful diver, and its heavy frame and large rear flippers help it to be one.

Mammals in High Gear

Many of the fastest mammals are found among the large herbivores, and it is easy to understand why: they must be able to run from the large carnivores. The galloping giraffes shown on the opposite page and the impala below display several specializations of the hoofed animals for speed—long legs for a long stride, light shank muscles for quick action and heavy thigh muscles for power. The kangaroo, by contrast, has only one pair of long legs, and these are well muscled

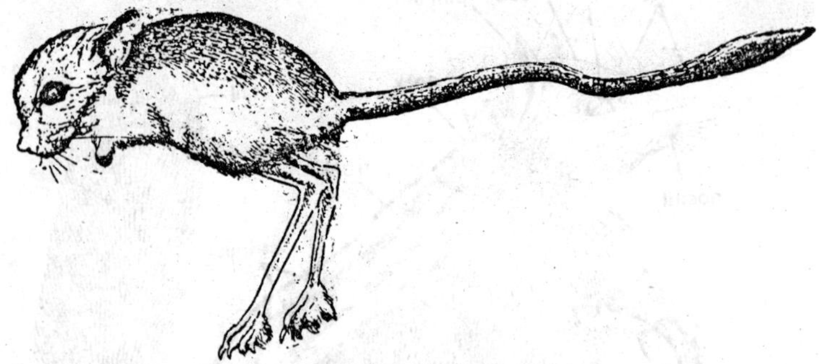

Fig. 8.12. Zerboa.

throughout to thrust its heavy body from the ground. On its high-flying leaps, it uses its long, thick tail as a counterbalance and rudder.

Getting About in the Trees

In exploiting the trees as a source of food and shelter, the arboreal mammals have developed various techniques for moving from branch to branch and tree to tree which minimize their contacts with the ground. The nocturnal flying squirrel, an elusive but fairly common inhabitant of most. North American forests, is an adept glider. Leaping from a branch, it spreads its legs to stretch the loose folds of skin running form wrist to ankle on either side of its body. It can sail downward on these folds at an angle of about 40 or 50 degrees for distances of 150 feet—or more, if the ground below slopes. It can even direct its course by raising, lowering or extending its legs, and balancing with its tail.

The sloth, lacking the flying squirrel's agility and swiftness is nevertheless well adapted to its jungle habitat. A highly specialized animal, it depends almost exclusively on the leaves of one type of tree for its food and an unobtrusiveness for its protection. It forages by inching along a branch in an upside-down position, using its claws as grappling hooks. Thus suspended, it is difficult to attack—and even more difficult to see, for green algae growing in its fur make it blend with the background.

9

Aquatic Adaptations in Mammals

Adaptation to environment is one of the most obvious and at the same time remarkable qualities of living organisms; in fact, it sums up nearly the whole result of evolution and the real cause of it all lies in the germplasm itself, although it is manifested only in the somatoplasm of the animal. Adaptation can also be defined as 'any modification in a part or in whole organism that makes it better far its existence in its present environment or enable it to live in a different environment so that it can secure food, protect itself and rear its young ones'.

A striking feature among various animals is their adaptation to their environment which is of great importance, because this adaptation helps them to survive in the struggle for existence. In aquatic mammals these adaptations are exhibited in change of body, locomotion, limbs, external covering etc.

The aquatic adaptations can be distinguished into two types. The one, which includes primitive gill breathing vertebrates. Their adaptation to the watery medium is perfect. Second includes those animals, which are secondarily adapted to the water medium due to their inability to breath in water. Fishes are the primary aquatic forms which have never had a terrestrial life, but have evolved from more primitive aquatic forms. On the other hand, secondary adaptations are studied in lung breathers which due to stressed circumstances including scarcity of food and great competitions for existence, returned once more to the primary home to their remote ancestors. As an effect of change in their life, they have some handicap in their life and the most important one is of the breathing. They breath by lungs but for other modifications they are well adapted for aquatic environment.

There are some examples leading amphibious life while some of them lead only aquatic life.

S.No.	*Order*	*Name of mammal*	*Ambhibious*	*Aquatic*
1.	Monotremeta	*Ornithorhynchus*	+	
2.	Marsupialia	*Chironectus* (aquatic opossum)	+	
3.	Insectivora	*Desmia* (aquatic-Shrews)	+	
4.	Rodentia	Musk rat, Water rat (Beaver)	+	
5.	Carnivora			
	(i) Fissipedia	*Latax* (Sea otter) *Lutra* (Aq. otter)	+	
	(ii) Pinnipedia	Walrus, Sea lion (*Otonia*)	+	
6.	Cetacea	Whale, Dolphin		+
7.	Sirenia	Dugong (sea cow) Manatee.		+
8.	Ungulata	Hippopotamous	+	

Modifications

Body Contour

The body contour becomes spindle shaped, the neck constriction disappears, the tail is enlarged, loss of external ears, external nostrils or nares move towards the apex of head are the other features. This assumption of the fish-like forms is best seen in the fully aquatic orders Cetacea, Sirenia, Pinnipedia and to a lesser extent in several other groups.

Skull

Shortening of craniun takes place which becomes proportionately higher and wider. The fascial portion of the skull tends to elongate. The zygomatic or temporal arch in Cetacca is reduced almost to a vestige.

Neck

Neck is reduced without affecting the number of cervical vertebrae. In whales the number of vertebrae is the standard mammalian seven, but they may be fused into a solid compressed mass of bones. In Manatee (Sirenia) there are six cervical vertebrae.

Vertebrae

Vertebrae tend to simplify secondarily because the body is equally supported throughout. Vertebrae are often completely fused into a continuous bony mass, or the atlas alone may be separated from the rest. Sometimes, however, all the vertebrae are complete and separate. Sacrum is more or less reduced because there is the loss of supporting or propelling functions on the part of appendages. Thus in cetaceans and sirenians the sacrum can not be distinguished from the other vertebrae. The zygapophyses are poorly developed, and are absent in the posterior portion of the trunk.

Chest

Chest tends to become cylindrical. It is also modified in such a way as to bring the internal cavity higher towards the back. This ensures greater stability of floatation and increases lung capacity, and is accompanied by the ribs tending to become highly arched dorsally and then to move upwards from their points of attachment. In the Pennipedia the position is transitional whereas in the whales the articulation of ribs and vertebrae is extremely loose probably to allow greater mobility of the chest for the rapid breathing necessary after prolonged submergence in water. In sirenians and cetaceans the diaphragin has become horizontal in position instead of being vertical as in most quadrupedal mammals.

Bones

Bones tend to become light and spongy. In whale the bones remain filled with oil. Sirenian is an exception to this because it feeds on submarine vegetation and hence remains at the bottom that is why the bones are comparatively heavy and massive as in manatee.

External Ears, Nostrils and Eyes

External pinna is absent in most of aquatic forms, however, it is present in Hippopotamus. The nostrils are shifted to the tip of snout and are often capable of being closed. Eyes tend to shift higher on the face as in Hippopotamus whose nostrils; eyes and ears can appear above the surface of water while the rest of head and body remain submerged.

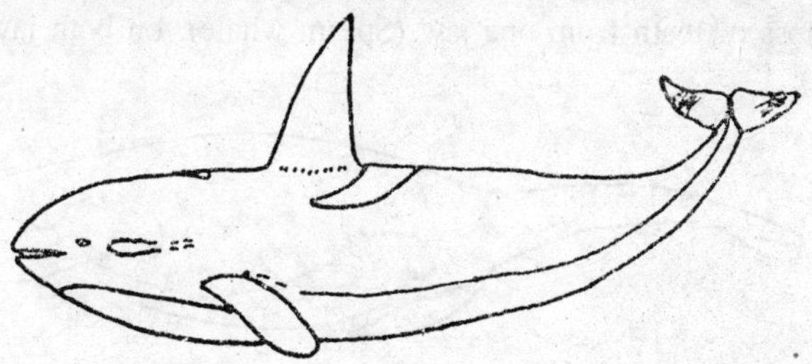

Fig. 9.1. Killer whale, Orcinu rectipinna.

Limbs

Webbed feet are the first adaptation. Extreme aquatic adaptation employs the development of a paddle in which there is a loss of mobility of the various joints, the entire skeleton being enclosed by the skin as a single mass showing no external division into fingers and toes.

Locomotive Mechanism

Fleshy fins-like expansion of body wall occur in the wale. These fins may be dorsal and caudal. The dorsal fin is a triangular structure. It is very high and narrow in the killer whale and is absent in white whale whereas in sulphur whale, the fin is small and situated well upon the tail. The development of this fin must be entirely a response to mechanical needs correlated with a certain bodily form and peculiarity of locornotion. In the Sirenia, the same is true, the Manatee having a round tail while Dugong has notched tail like that of a wale.

Integument

The modification of integument include the reduction of armouring of hair, absence of skin glands, and external ears. The whales and sirenian have lost all traces of hair except perhaps a few bristles around the mouth but most of them are well covered with foetal hair before birth. In white whale even the foetus has lost its hair thus showing extreme specialization.

As a compensation for the loss of hair, the aquatic mammals have developed layers of fat in the form of blubber beneath the skin and this serves admirely for the retention of the body heat which would otherwise radiate out very rapidly into the surrounding water.

Mouth Armourment

Except in the sea cow and Walrus, the jaws being no longer used for mastication but only for the prehensation of prey. Teeth modifications usually in the direction of simplification, the increased number or total loss of teeth from one jaw (Sperm whale), on both jaws (Beleen whale).

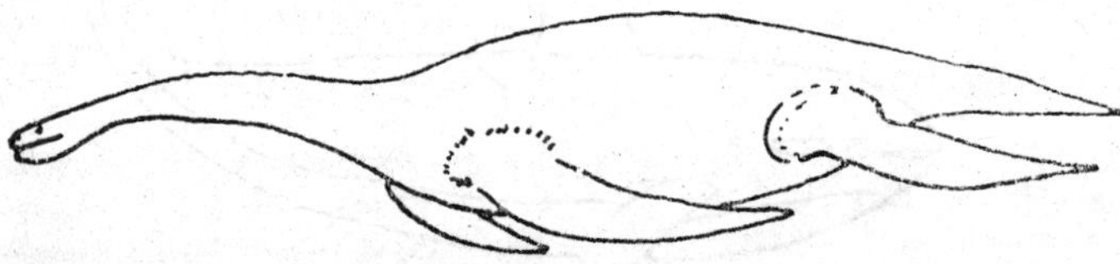

Fig. 9.2. Plesiosaur, Cryptocleidus.

Adaptive Peculiarities in Cetaceans

The group Cetacea includes whales and dolphin. This assemblage of large aquatic mammals is profoundly modified for marine life. They are unquestionably the most highly specialized structurally of all mammals, although certain of their characters are persistently primitive. Whales are the largest fish-like mammals and also largest in the animal kingdom. The return of whales to the sea is a fine example of convergence in evolution.

The cetaceans are widely distributed in almost all oceans of the world and also in large fresh water rivers. They show following peculiarities.

External Characters

Whales are fish like vertebrates, highly streamlined for efficient swimming. The body is torpedo shaped, and there is no visible neck distinct from the trunk. The head is continuous with the body. There is no hair on the body of the modern whales. The smooth skin makes a sleek surface that offers little resistance to the water as the animal moves forward. Eyes are strikingly small and with immovable upper and lower lids. Nictitating membrane is absent. External nostrils are

Fig. 9.3. Cetaceans.

shifted to the top and the back of the skull. The nostrils can be closed by valves. The anterior limbs are very much modified and are represented by a short externally unjointed structure known as *flipper*. Numerous phalanges are present to support paddle. Pelvis and hind limbs are reduced to mere vestiges.

The whales are unique among marine vertebrates in that the tail terminates in a horizontal fin that moves up and down, rather than in a vertical fin that moves from side to side. The horizontal tail fin of the whale is generally called *flukes*. It is stiff and strong and contains no bones. There is a fleshy dorsal fin that acts as stabilizer and prevents the animal from rolling.

Peculiarities of Skeleton

Skull bones are spongy in texture. The weight of body is reduced due to the less specific gravity of the oil. Brain case is large though small as compared to large facial part of the head. Bones of cranium are separated by sutures and åre loosely connected, this helps in the absorption of shocks. Temporal bones are hard and dense and loosely united with the squamosal. Maxilla is very large and extends as far as the anterior end of the snout. The premaxilla is devoid of teeth except in *Squalodon* and *Zeuglodon*. Snout is composed of maxilla, premaxilla, vomer and mesethmoid. Pterygoids meet one another to form hard palate and this helps in the formation of nasal passage. Mandibles have small coracoid processes. Hyoid is broad plate-like.

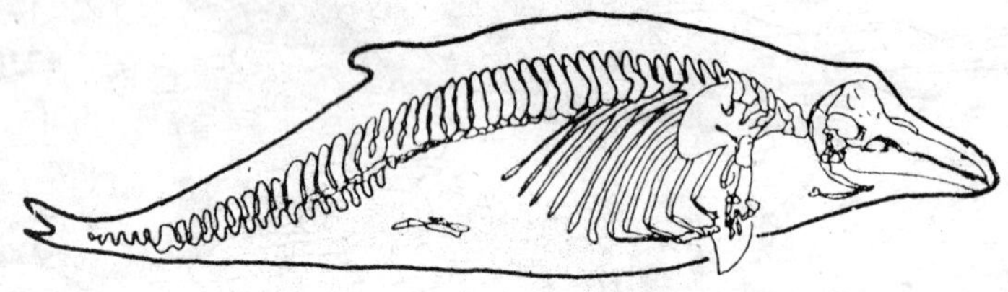

Fig. 9.4. Skeleton of Porpoise.

Dentition

Due to the restricted nature of the food available in the sea, the teeth except *Squalodon* and *Zeuglodon* are numerous and monophyodont. Their crows are compressed conical in shape. The lower jaw of sperm whale is long and narrow and is armed with from 40 to 48 conical teeth which fit into the toothless groove of the upper jaw.

Vertebral Column

The usual vertebral formula is C_7 T_{12} L_{21} C_{32}. The cervical vertebrae are shortened, compressed and fused together to a greater or lesser extent. (In *Balana*, for example, there is complete fusion; in *Balanopleura*, complete separation). The thoracic vertebrae are well developed with normal features of the vertebrae like neural spine, neural canal, transverse processes etc. In addition to these parts the thoracic vertebrae have the metapophysis showing some changes in the lumbar vertebrae. The lumbar and caudal vertebrae are numerous. Thick fibrous intervertebral discs are present between the vertebrae.

Sternum is short and ribs have become reduced in number and remain fused with the thoracic vertebrae.

Girdles

In pectoral girdle, the clavicles are absent. Scapula is large. In Dolphin, the scapula is very broad and flat and also fan shaped. The scapular fossa is extremely reduced. The acromian is long.

Pelvic girdle is greatly reduced due to the absence of hind limbs. It is represented by ischium which is longitudinal cylindrical in structure and placed parallel to vertebral column.

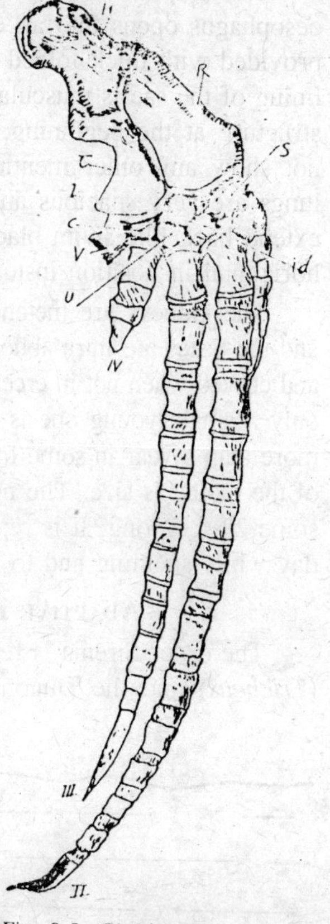

Fig. 9.5. Globicephala. Limb modification.

Internal Anatomy

Cetacea have redeveloped piscine characteristics, e.g., in body form, the almost total lack of bristles and absence of external ears; the long snout, lack of neck, and the broad powerful tail. These characters mimic the typical fish form and makes for smooth and unimpeded progress through water.

Even more remarkable adaptations peculiar to the group, have developed internally. The so called 'or' as 'blow' was traditionally explained as a jet of water, and is generally held to be the condensation of warm air expired from the lungs through the dorsal blow hole when the whale surfaces. A recent suggestion is that the 'blow' is composed of nitrogen-charged foam. Foam developed from a finely divided emulsion of fat droplets and gas.

Stomach is complicated and divided into three chambers. The oesophagus opens into an elongated blind sac like structure which is provided with thick folded epithelium and serves as store house. The lining of the sac is muscular and soft. The duodenum is also a dilated structure at the beginning. Rest of the parts of alimentary canal do not show any other mentionable modification. Trachea is short and lungs are very spacious and are not divided further into lobes. They extend back like swim bladder of fishes. The diaphragm has become horizontal in position instead of being vertical.

The kidneys are metanephric and lobulated. Whales lack scrotum, and the testes are intra-abdominal. Seminal vesicle absent. Penis is large and curved when not in erection. There is a bicornuate uterus, but usually only a single young one is produced at a time. The gestation period is more than a year in some forms and the new born calf may be one-third of the mother's size. The neonatus weights more than two tons and in some 23 feet long. It is reported to grow at a rate of about 200 lb per day while suckling and to reach puberty at about four years of age.

Adaptive Peculiarties in Sirenians

The order Sirenia is represented by two living forms, the Manatee (*Tricheus*) and the Dugong (*Halicone*). They are fully aquatic and

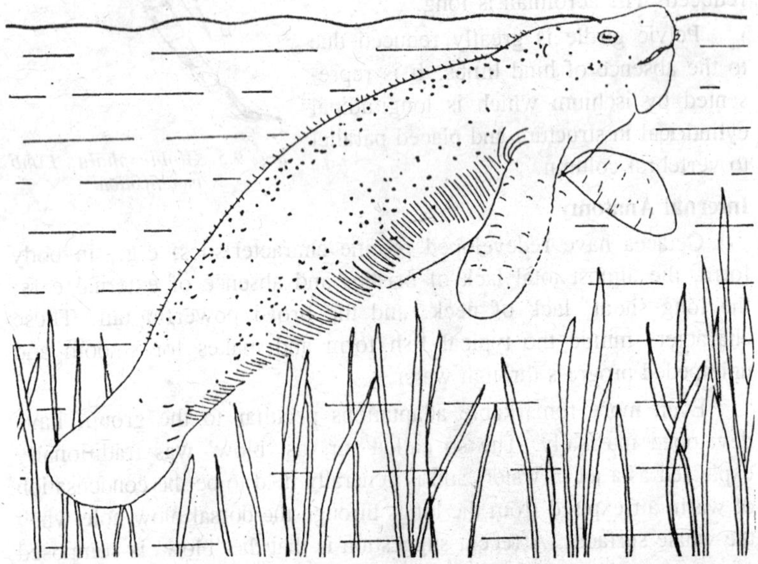

Fig. 9.6. The Sirenia.

present a number of structural features convergent with those of the Cetacea. The adaptive peculiarities of sirenians are given below.

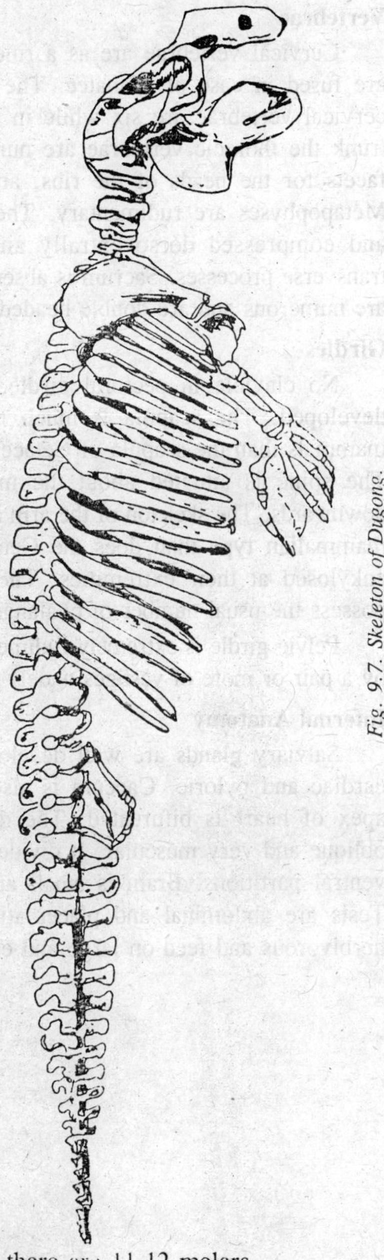

Fig. 9.7. Skeleton of Dugong.

External Features

Body is fish-like in appearance. Thick and almost hairless skin with an underlying layer of blubber. Head as compared with the size of entire body is larger in both the animals. No external pinna. Eyes are small with nictitating membrane. Fore limbs paddle like and extremely movable to all sides. Digits are enclosed in a common cutaneous covering and the number of phalanges is normal. The dorsal fin absent. The tail is horizontally expanded forming the caudal fin.

Skeletal Peculiarities

The skull is characterized by its extreme hardness. The cranial cavity is rather long and narrow as compared with that of the cetacean. The external nares are very wide, but they are situated relatively farther forwards. The nasals are vestigeal. The tympanic and periotic are readily separable from the other bones. The premaxillae are enormous in the Dugongs. The mandible has a well developed ascending ramus and coronoid process. Dentition is variable. Dugong has 5-6 molars while is Manatee there are 11-12 molars.

Vertebrae

Cervical vertebrae are as a rule free except 2nd and 3rd which are fused in case of Manatee. The centra are thin. In Manatee the cervical vertebrae are six while in Dugong the number is seven. In trunk the thoracic vertebrae are numerous. All have well developed facets for the heads of the ribs, and well developed zygapophyses. Metapophyses are rudimentary. The caudal vertebrae are numerous and compressed dorsoventrally and are also provided with wide transverse processes. Sacrum is absent. Sternum is a broad bone. Ribs are numerous and are double headed.

Girdles

No clavicle in pectoral girdle. Scapula and coracoid are well developed. The scapula is much more like that of the terrestrial mammals than the scapula of Cetacea, and is nearer that of the Seals. The spine is situated about the middle; the acromian is directed downwards. The skeleton of the arm also departs less from the ordinary mammalian type than does the Cetaceans. The radius and uina are ankylosed at their extremities. There are five digits, all of which possess the usual number of phalanges.

Pelvic girdle is extremely rudimentary and if present is represented by a pair or more of vestiges widely separated from the spinal column.

Internal Anatomy

Salviary glands are well developed. The stomach has two parts cardiac and pyloric. Caecum is also present in large intestine. The apex of heart is bifurcated. The diaphragm, like the Cetacean, is oblique and very muscular. It divides the body cavity into dorsal and ventral partitions. Brain is small and convolution is merely absent. Tests are abdominal and placentation is of zonary type. They are herbivorous and feed on large and estuarine plants.

10

DENTITION IN MAMMALS

Teeth are the first hard structures of the body to put in an appearance during vertebrate development, even before any part of the body skeleton. The structure, kind, number and arrangement of teeth are collectively known as *dentition*. The initial function allotted to the dentition, in vertebrates, is to hold the prey or to prevent the escape of prey from mouth. In mammals the teeth perform many functions like cutting, grinding, tearing and sometimes locomotion. They are also used in defence and offence. Teeth are the derivatives of stomodaeal region of the alimentary canal. Thus these integumentary derivatives are homologous with the placoid scales of Elasmobranches. These are the exoskeletal derivatives.

In lower vertebrates including fishes and amphibians, the teeth are of *acrodont* type because they are fixed on the mucous membrane of the jaws and whenever they are worn out they don't leave any gap or socket there. In crocodiles, *Archaeopteryx*, and mammals the condition of dentition is *thecodont*. In this condition the teeth are fixed in alveolar sockets in jaws. In reptiles, a somewhat intermediate stage has been observed. Here the teeth are partly fixed on the jaw bones and partly fixed in sockets where they are placed obliquely. This stage is called the *pleurodont*.

Adaptive Radiation

The most primitive eutherians-the insectivores, feed upon invertebrates specially the insects, and from these evolved other mammals like carnivores, ant-eaters, sanguivores, omnivores etc. The modern classification of eutherian mammals is based upon the trophic adaptations of their teeth. Dentition provides clue to the diet of the mammal and it also help in deciding the pedigree or ancestory of certain mammals.

Toothless Mammals

Teeth are present in most of mammal but, however, some mammals have no teeth. This category includes, the *Tachyglossus* (spiny anteater), and the giant anteater *Myrmecophaga*, a relative of the Armadillo. In *Ornithorhynchus* (duck-billed platypus) embryonic teeth are replaced by horny epidermal plates in adults. Similarly in whalebone whales foetal teeth are replaced before or soon after birth by sheets of a fibrous, stiff, horn-like epidermal derivative known as *baleen* which extends downward from the upper jaw. Baleen whales are filter feeders, straining small organisms known collectively as plankton, from the water by use of their baleen sieves.

Structure of a Tooth

Typically a mammalian tooth is embedded in a socket in the jaw bone, - the dental alveolus, lined by a vascular membrane, known as *alveolar membrane*. A tooth is differentiated into three parts *crown*, *neck* and *root*.

Crown

The part of the tooth which projects out of the jaw is termed as the *crown*. It is white glistening part of the tooth.

Root

The basal part of tooth that remains embedded in the alveolar socket. Its terminal part is called *root canal* that may remain either open or closed. In open rooted tooth the growth is unlimited as in the incisors of rodents, lagomorphs and elephants. And in close rooted tooth, the growth is restricted as in the majority of cases in mammals.

Neck

The part of tooth between crown and the root is the neck. It remains embeed in the gum.

The major part of a tooth is composed of hard dermal bony substance, the *dentine*. It encloses the *pulp cavity* that remains filled with *pulp* made up of connective tissues, blood vessels, nerves and mesenchyme cells or *odontoblasts*. In the crown, dentine is covered externally by a thin, very hard and shining layer of *enamel* which is the hardest part of the tooth. The dentine has fine net-work of *canaliculi* to get nourishment from blood. The root of tooth is surrounded and fixed in jaw bone by a layer of *cement* and a vascular *periodontal membrane*.

DEVELOPMENT OF A TOOTH

The embryonic origin of mammalian tooth is two fold. *Enamel* is secreted by stomodaeal ectoderm, and the rest of the tooth is formed from mesoderm. The *dentine* or *vitrodentine* is entirely of mesodermal origin but is formed under the organizing influence of an enamel organ which comes from stomodaeal ectodermal epithelium.

The first step in this direction of development is the appearance of ectodermal thickening or the *dental primordium* by a process of

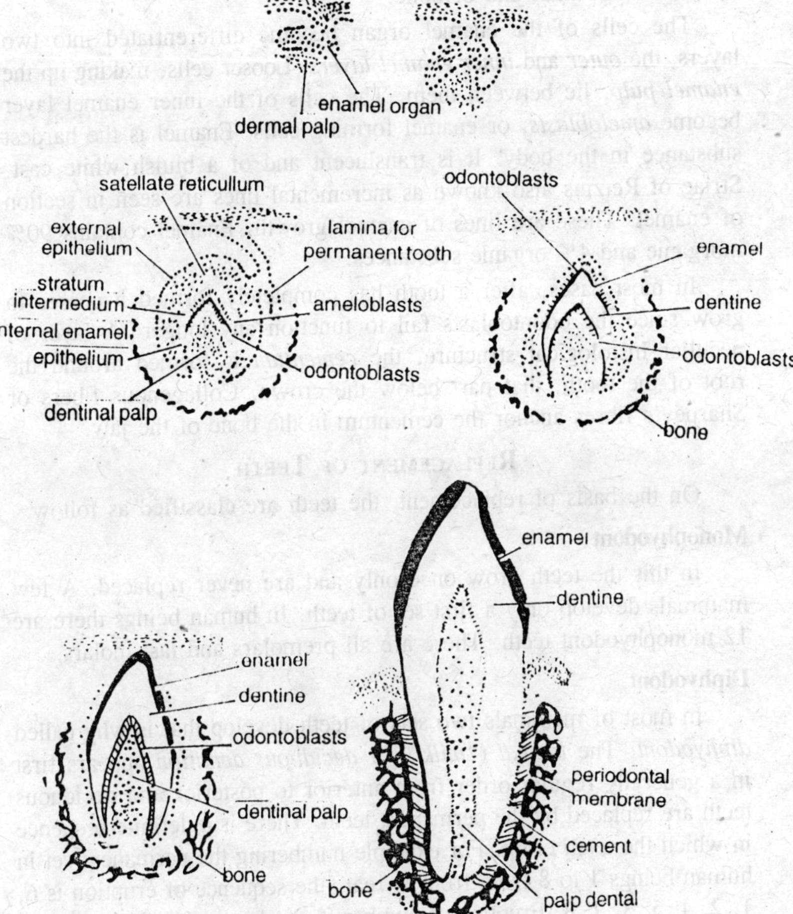

Fig. 10.1. Stages in the development of a tooth.

invagination into the surrounding mesenchyme to become the shelf like *dental lamina*. From this, at intervals, small epithelial buds proliferate, each of which develops into a *tooth gum* by multiplication of the ectodermal cells. The mesenchymal cells under the ectoderm also increase in number and push against the latter so that an inverted, cup shaped structure, the *enamel organ* takes form. This will give to the tooth that appears later on in life. The mesenchyme within the enamel organ is known as the *dental papilla*. Those cells of the dental papilla next to the enamel organ become differentiated into *odontoblasts*, which deposite a layer of dentine just under the ectodermal cells of the enamel organ. Dentine is harder than bone. Its organic and collagen content is less than that of bone.

The cells of the enamel organ become differentiated into two layers, the *outer* and *inner enamel layers*. Looser cells, making up the *enamel pulp*, lie between them. The cells of the inner enamel layer become *ameloblasts*, or enamel forming cells. Enamel is the hardest substance in the body. It is translucent and of a bluish white cast. Striae of Retzius also known as incremental lines are seen in section of enamel. These are lines of enamel growth. Enamel contains 90% inorganic and 4% organic substances.

In most cases, after a tooth has completely formed it ceases to grow since the odontoblasts fail to function any longer. A layer of peculiar histological structure, the *cementum* is formed around the root of the tooth, that part below the crown. Collagenous fibers or Sharpey's fibres anchor the cementum in the bone of the jaw.

Replacement of Teeth

On the basis of replacement, the teeth are classified as follow:

Monophyodont

In this the teeth grow once only and are never replaced. A few mammals develop only a first set of teeth. In human beings there are 12 monophyodont teeth. These are all premolars and last molars.

Diphyodont

In most of mammals two sets of teeth develop that is why called *diphyodont*. The *lacteal* ("*milk*") or *deciduous dentition* appears first in a generally regular order from anterior to posterior and deciduous teeth are replaced by the permanent teeth. There is a definite sequence in which the teeth erupt. For example numbering the permanent set in human beings 1 to 8 from front to rear, the sequence of eruption is 6, 1, 2, 4, 5, 3, 7, 8. Eruption of number 8, the last molar, is delayed in higher primates, and this "*wisdom tooth*" is sometimes imperfect, unerupted or is missing.

Polyphyodont

Most vertebrates through reptiles have a succession of teeth, and the number of replacement during a life time is indefinite but numerous. This stage is called the *polyphyodont*. It has been estimated that an elderly crocodile may have replaced its front tooth 50 times. Polyphyodont condition does not occur in mammals.

Morphological Differentiation of Teeth

On the basis of shapes the teeth can be distinguished as:

Homodont

When all the teeth are of a similar form or shape, the dentition is said to be *homodont*. The toothed whales are homodont mammals. Non-mammalian vertebrates are usually homodont. The initial function allotted to homodont dentition is to prevent the escape of prey from the mouth cavity.

Heterodont

When teeth are arranged in groups which differ in shape and function; the term *heterodont* is applied to this dentition. Most vertebrates below mammals that possess teeth have homodont teeth. However, there are several kinds of fishes like Jackson sharks, *Heterodontus*, where heterodont condition is seen.

Types of Mammalian Teeth

The teeth of a generalized mammal may be designated from anterior to posterior, on each side of each jaw, as; *incisors* (usually two to five), *canines* (never more than one), *premolars* (generally two to four), and *molars* (variable but often three).

Incisors

Incisors, located anteriorly, have one horizontal cutting edge and a single root. They are best developed in herbivorous mammals, which use them for holding, cropping, or grazing. There are two pairs of incisors in most of mammals lodged in alveolar sockets in premaxilla. The *anterior incisors* are long and curved almost into a semicircle and the *posterior incisors* are much smaller and less curved. Incisors may be totally absent as in sloths, or lacking on the upper jaw, as in bovine. Elephant tusks are modified incisor *teeth*. In rodents and lagomorphs, incisors are open rooted and grow throughout life. In lemurs, incisors are denticulated like a comb, serving for cleaning fur.

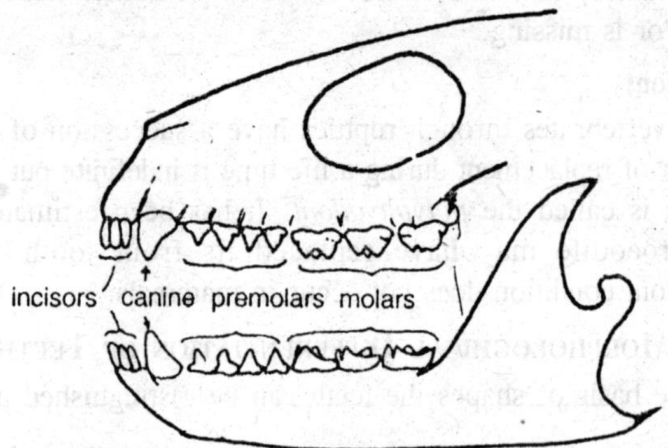

Fig. 10.2. Dentition of a hypothetical generalized eutherian mammal.

Canines

Canines lie immediately behind the incisors. Caninas are generally elongated, single rooted and with a conical sharp monocuspid crown. They are meant for piercing, tearing, and offence and defence. In carnivores the canines are spear-like and used for piercing flesh. They are also the tusk of the walrus and are used in digging molluses and also in locomotion on ice. Canines are absent in rodents and lagomorphs and so there is a toothless interval, a *diastema*, between the last incisor and the first cheek tooth.

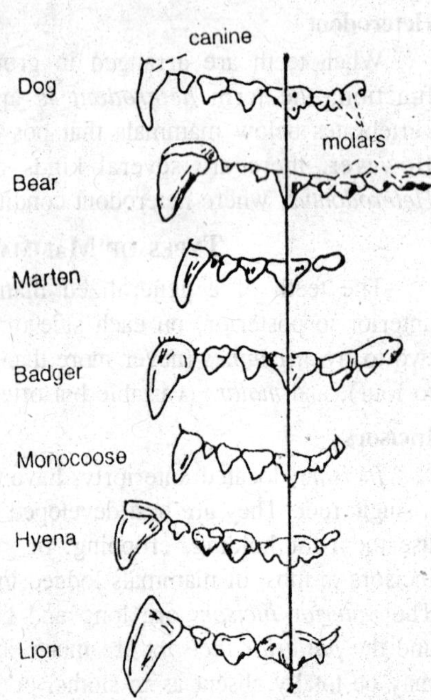

Fig. 10.3. Teeth of the upper left jaw of various carnivores. The vertical line passes through the fourth premolars, or carnassial teeth.

Cheek Teeth

Premolars and *molars* are the *cheek teeth*. These are modified to masticate food stuffs. During mastication, the food is repeatedly worked between interlocking *occlusion surfaces* of the upper and lower dentition by the action of the mobile, sensitive tongue and a characteristic mammalian facial structure, the *cheeks*. The food thus becomes a loose pulp that readily absorbs gastric juice.

Premolars and *molars* of carnivores and herbivores are very different, those of carnivores being specialized for shearing flesh and those of herbivores for grinding. The occlusal or grinding surface is not smooth, but is worked by one to several sharp *cusps*, which interlock when the jaws close, or occlude. This cusps pattern is referred to as *tubercular* and permits both piercing and crushing of the food. On the basis of these cusps the mammalian cheek teeth are categorized into following types:

Bunodont

The cheek teeth of some mammals lack sharp ridges and pointed cusps and have, instead, low rounded cusps completely covered with enamel so that no dentine reaches the surface. These bunodont teeth wear down evenly. These are found in rhinos, some hogs, some primitive ruminants, some rodents such as the white footed mouse, and human being. These teeth function for both animal and plant materials.

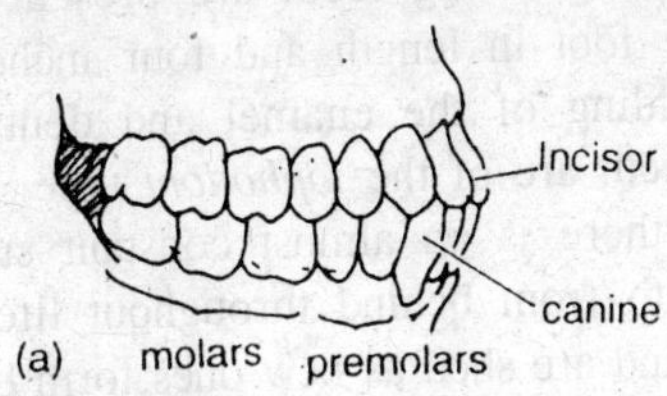

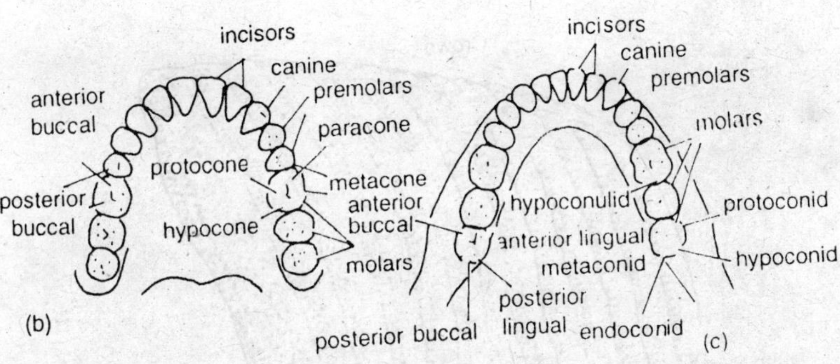

Fig. 10.4. Diagrams of the dentition of man (a) in protile, (b) upper jaw, (c) lower jaw. In (b) and (c) the cusps are labelled with the nomenclature of human anatomists on the right.

Secodont

Carnivores must be able to shear or cut away chunks of flesh from a large carcass. Hence, the cusps are sharp and knife like and present little grinding surface. In living carnivores *secondont* teeth reach their height of development in a shearing, *carnassial apparatus* of the fourth upper premolar and the first lower molar, which slide past one another like scissor blades to cut flesh.

Selenodont

Tougher plant materials such as leaves, twigs, and stems require more specialized dentition. The crowns of selenodont teeth are broad, and the enamel is disposed in vertical, crescent- shaped columns separated by dentine. Thus the crescent shaped folds of hard enamel enclose softer areas of dentine. Chewing causes the softer dentine to wear down more quickly than the enamel, there by providing sharp, crescentic, rasping ridges with a wide variety of configuration and that macerate the food during the complex side to side and forward-backword movements of the artiodactyl jaw.

Lophodont

In elephant, enormous grinding teeth are present which may measure as much as one foot in length and four inches in width. There is an intricate folding of the enamel and dentine to form transverse ridges. Such teeth are of the *lophodont* type. In elephants as well as the *Manatee* there is an almost constant succession or replacement of molar teeth from behind throughout life. The front ones are pushed forward and are shed as new ones form behind. Only one fully developed molar tooth is present at a time on each side of each jaw in elephants.

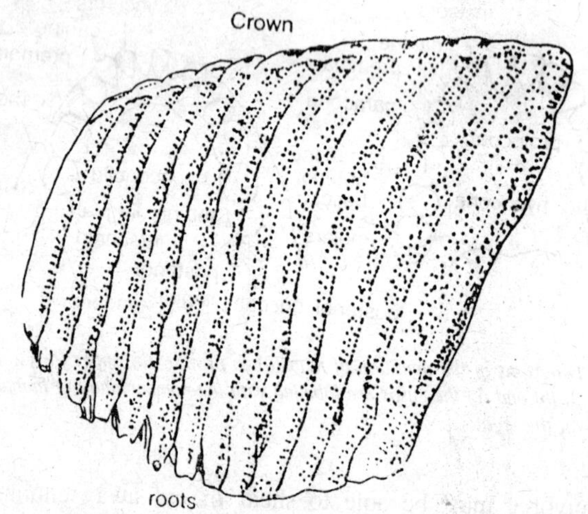

Fig. 10.5. Unerupted molar of an elephant to show formation by union of a number of separate units, each with its own pulp and root.

Brachydont and Hypsodont

Rodents, the largest mammalian order and with the largest variety of diets, exhibit the largest variety of teeth, some of them are low crowned with long roots (*brachydont*), as in squirrels, some high crowned and with short roots (*hypsodont*), as in wood rats. The cheek teeth of man are *brachydont*, and those of horses are *hypsodont*. Tusks are hypsodont.

Triconodont and Trituberculate

Cheek teeth of early placentals were triconodont, having three cone like prominences arranged in a straight line. *Cynognathus* had triconodont teeth. Later on these cones of *triconodont* dentition became arranged in a triangle, resulting in *trituberculate* secodont teeth. Still later, other parts may have developed from these three original tubercles so as to form additional cusps, ridges, and folds, and thus are finally arrived at the many and varied types of mammalian cheek teeth that exist today.

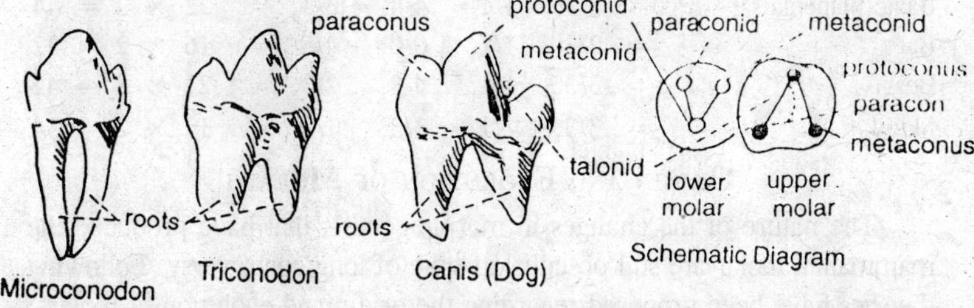

Fig. 10.6. Plan of molar teeth. Anterior is to left in each case.

DENTAL FORMULA

Only in mammals there is a fixed number of teeth characteristic of each species. The maximum number of teeth in placental mammals with the heterodont dentition is 44. The number of teeth in any particular species is surprisingly constant. For purposes of classification, a so called *dental formula* has been devised. Since the two halves of the jaw correspond, only the number on one side are recorded. Those of the upper and lower jaws are separated by a horizontal line, so that they appear in the formula as numerator and denominator. The kind of tooth is represented by the initial letter i,c,pm,m, indicating incisor, canine, premolar and molar, respectively. The number of teeth shown by a dental formula is multiplied by 2 to get the total number of teeth. Thus, the dental formula of man is i. 2/2, c. 1/1, pm 2/2, m. 3/3. In order to simplify these dental formulas, the initial letter is customarily omitted, so that for man it appears as 2/2, 1/1, 2/2, 3/3. When a certain type of tooth is lacking, a zero is used to indicate this fact. The dental formulas of some mammals are given below:

Beaver	=	1/1,	0/0,	1/1,	3/3.	=	10 × 2 = 20
Cow, Sheep and Goat	=	1/3,	0/1,	3/3,	3/3.	=	16 × 2 = 32
Cat	=	3/3,	1/1,	3/2,	1/1.	=	15 × 2 = 30
Dog	=	3/3,	1/1,	4/4,	2/3.	=	21 × 2 = 42
Mole	=	3/3,	1/1,	4/4,	3/3.	=	22 × 2 = 44
Squirrel	=	1/1,	0/0,	2/1,	3/3.	=	11 × 2 = 22
Rat	=	1/1,	0/0,	0/0,	3/3.	=	8 × 2 = 16
Horse and pig	=	3/3,	1/1,	4/4,	3/3.	=	22 × 2 = 44
Elephant	=	1/0,	0/0,	0/0,	3/3.	=	7 × 2 = 14
Kangaroo	=	3/1,	1/0,	2/2,	4/4.	=	17 × 2 = 34
Rabbit	=	2/1,	0/0,	3/2,	3/3.	=	14 × 2 = 28
Opossum	=	5/4,	1/1,	3/3,	4/4.	=	25 × 2 = 50
Lemur	=	2/2,	1/1,	3/3,	3/3.	=	18 × 2 = 36
Tarsius	=	2/1,	1/1,	3/3,	3/3.	=	17 × 2 = 34
Basic placentals	=	3/3,	1/1,	4/4,	3/3.	=	22 × 2 = 44
Bat	=	2/3,	1/1,	0/0,	4/5.	=	16 × 2 = 32
Bear	=	3/3,	1/1,	4/4,	2/3.	=	21 × 2 = 42
Skunk	=	3/3,	1/1,	3/3,	1/2.	=	17 × 2 = 34

Origin and Evolution of Molars

The nature of the changes in morphogenesis that have produced the mammalian molar are still obscure, in spite of long controversy. Following theories have been proposed regarding the origin and evolution of molars:

Concrescence Theory

This theory was proposed by *Rose*. According to this theory, the modern molars are the products of the fusion of a number of primitive cone shaped reptilian teeth. The prosterior teeth in the jaw of *Sphenodon* offer evidence in support of this point of view. Some compound molars, those of elephant, for instance are formed by the union during development of many, at first separate cones.

Dimer Theory

The credit for this theory goes to *Bolk* that is why it is also called the "*Theory of Bolk*".

According to this theory, each ancestral reptilian tooth carried a row of three cusps, one behind the other. Such teeth are termed *triconodont* teeth. *Bolk* suggested that each mammalian molar was formed by the fusion of two such teeth that came to lie side by side.

He believed that there was evidence of this double or *dimeric* nature of the teeth in a septum. Sometimes seen in an early stage of the enamel organ. The fundamental pattern of the tooth would thus show six cusps, three on the labial and three on the buccal side. Since the remains of the molars of many of the earliest mammals show three cusps and therefore do not support Bolk's theory which requires that the most primitive molars should have six cusps.

Differential Theory or Theory of Trituberculy

It was given by *Cope* and *Osborn*. This theory is more widely accepted. The theory postulates the budding out and growth of additional contact surfaces, or *cusps*, upon the crown of an originally conical tooth. This theory is based largely upon evidence presented by the ancestral teeth of fossil mammals.

The addition of two such cusps give rise to the *tritubercular tooth* which is the typical molar of mammals generally from the earliest representatives down to *Eocene* times. Even today the forms like *Chryochloris* (mole), *Didelphys* (opossum) and some lemurs exhibit this ancestral tritubercular type.

The three cusps of a tritubercular molar are arranged in the form of a triangle. Molars of the lower jaw have a more lateral cusp, the *protoconid*, medial to which are the two secondary cusps, the *paraconid* in the anterior position and the more posterior *metaconid*. The corresponding cusps on molars of the upper jaw are *protoconus* (a medial) and two lateral *paraconus* (anterior) and a *metaconus* (posterior). The addition of extra cusps on more highly developed molars may bring the total number to a maximum of six. Molars of the lower jaw elongate some what through the development of a posterior extension, the *talonid* or heel, which may bear two or three extra cusps. The above cusps can be recognized in the molars of the most mammals and others may be added. An internal posterior cusp on the upper molar, called the *hypocone*, makes the tooth *quadrilateral. Protoconule* and *metaconule* added between the main upper cusps then produce a six cusped molar, from which many modern types can be derived. Further ridges along the outer side of the tooth are known as *styles*. In the lower molar *entoconid* and *entoconulid* added to the heel also make a six cusped tooth.

Unusual Teeth

Certain unusual teeth found in mammals deserve special mention. The *tusks* of elephant are incisor teeth with open root canals. They grow continuously throughout life. The largest known tooth is the tusk

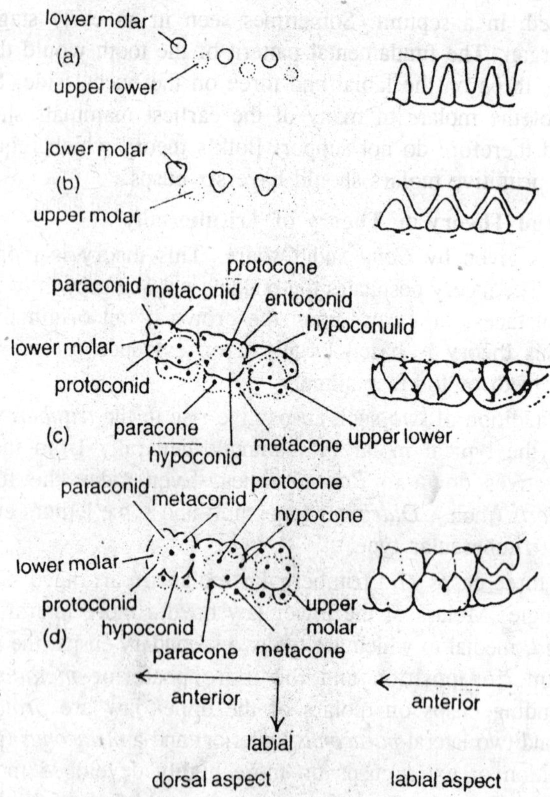

Fig. 10.7. Diagrams to show stages of arrangement of teeth and cusps.

of an extinct mammoth, *Archidiskodon*. It weigh over 250 pounds and is more than sixteen feet in length. In wild boar, the tusks are formed from lower canines, while the male dugong, *Halicone*, a sea cow develops tusks from upper incisor. In both sexes of *Phacochoerus*, there are four upward curving tusks, which are the transformed canines of both jaws.

The male norwhal, *Monodon* has lost all its adult teeth except an upper left one which is prolonged enormously into a formidable twisted pikestaff that may reach seven to nine feet in length.

Embryonic monotremes, being the only mammals to develop within an egg shell outside the body of the mother, are the only mammals which have a *horny egg tooth* at the end of the snout. This serves the young in escaping from the egg shell.

11

FEEDING AND DIGESTION

INTAKE OF FOOD : MOUTH AND TEETH

Supply of Raw Materials

The matter within the system of any animal needs continual renewal; the body is in a steady state, not an equilibrium, and life depends upon the availability of fresh supplies. Moreover the homeostatic actions of the body depend upon the expenditure of energy for moving substances against concentration gradients and for many similar purposes. In a mammal, energy-giving foods are required to maintain the high temperature and to make possible all the many special actions that ensure homeostasis. The whole system of life of a mammal is thus based upon the expenditure of large amounts of energy, the information available through the nervous system ensuring that such expenditure brings the body into conditions in which life can be maintained.

A large part of the activity of every mammal, therefore, is devoted to obtaining supplies of material for building and rebuilding the body and for the provision of energy. Mammals have acquired the power to gain the necessary nourishment from all sorts of odd corners of the varied world of land, sea, and air. Yet the basic pattern of their digestive system shows few special features and is essentially that common to all vertebrates. It is surprising that such a wide variety of diets can be digested by means of the familiar apparatus of a stomach with its acid and pepsin, and an intestine with enzymes acting in alkaline solution.

The differences between alimentary systems are mainly in the apparatus for obtaining the food from the environment, the mouth and teeth, and in the presence of special chambers of the gut for the cultivation of symbiotic bacteria to assist in breaking down cellulose. The placental mammals of the Cretaceous Period, more than 70 million years ago, were insectivorous; from these all the specialized carnivores and herbivores have been evolved and many types, like man, have become omnivorous. Indeed most mammals show considerable flexibility in the way they obtain the materials necessary for their maintenance, as would be expected from animals with such an adaptable nervous organization.

Whatever their specializations, all mammals take some proteins, carbohydrates, and fats, together with water and certain special molecules, inorganic salts, and vitamins. Proteins, besides providing the necessary nitrogenous materials, can be used as fuel for the provision of energy and warmth, which are otherwise supplied by carbohydrates and fats.

Selection of Food

The activities by which the animal seeks, selects, and eats its food are an important part of its behaviour and the types of food chosen depend largely on the nervous organization. In this, as in other matters, the mammals show exceptional powers. They are provided even more fully than other animals with a system that provides a 'drive' to seek food, and they use great ingenuity until they find it. We know little of the source of its drive or how it is related to the needs of the organism. Whatever the mechanism may be it arranges that the precise amount required is taken. Thus an adult man eats about 12000 kg of food in 25 years and may remain of the same weight to within 2 kg or less.

The basic regulation is almost certainly from the hypothalamus, where there are centres controlling both eating and drinking. The feeding centres are partly controlled by the level of the sugar in the blood, but this is not the only factor, since diabetics, with a high blood-sugar level, may be often hungry. There is a ventro-medial hypothalamic satiety centre, stimulation of which, in a rat with implanted electrodes, produces reduced food intake, while destruction leads to obesity. A lateral feeding centre operates in the opposite direction.

There is some evidence that these centres operate by stabilizing the fat stores. When one of a parabiotic pair of rats was given a hypothalamic lesion it became fat and the other rat then reduced food intake and lost weight.

It was shown long ago that gastric motility, measured by a swallowed balloon, is increased during hunger pains. Such contractions

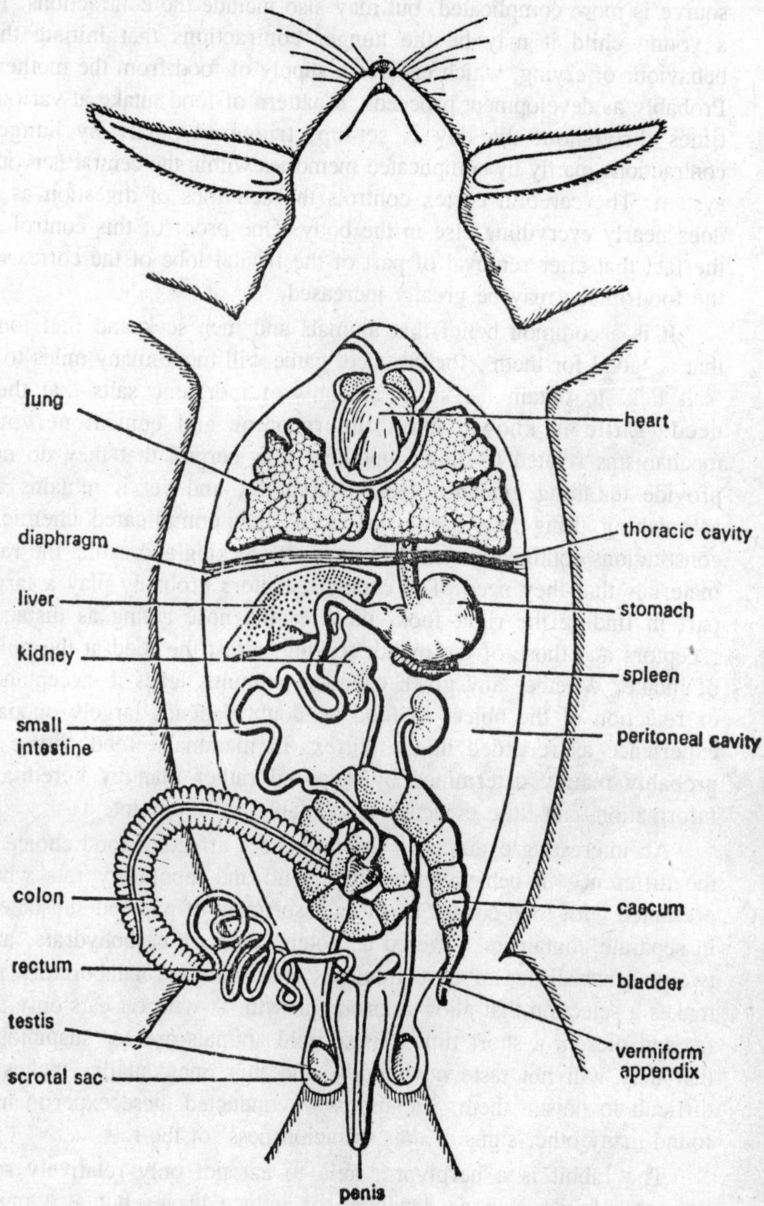

Fig. 11.1. Coelom and viscera.

are often supposed to be the source of hunger, but they seem not always to be present, and hunger persists after denervation of the stomach. Its source is more complicated, but may also include the contractions. In a young child it may be the hunger contractions that initiate the behaviour of crying, which ensures a supply of food from the mother. Probably as development proceeds, a pattern of food intake at various times throughout the day is set up, triggered partly by hunger contractions, partly by complicated memories within the central nervous system. The cerebral cortex controls the activities of digestion as it does nearly everything else in the body. One proof of this control is the fact that after removal of part of the frontal lobe of the cortex of the food intake may be greatly increased.

It is a common belief that animals and men seek and find food that is 'good for them', for instance, game will move many miles to a 'salt lick' to obtain the small amounts of inorganic salts that they need. Little is known about the receptor and central nervous mechanisms related to these appetites. It is certain that they do not provide unfailing guidance for the creature, and yet it remains an astonishing thing that organisms with such complicated chemical constitutions continually renew themselves, seeking and eating the raw materials that they need. The chemo-receptors probably play a large part in finding the right food, those of the nose acting as distance receptors and those of the mouth serving to test the food at the point of intake. Whether any given chemical stimulus leads to acceptance or rejection of the object as food no doubt depends largely on past experience as recorded in the cortex. In mammals, food choice is probably mainly determined by acquired rather than by hereditary information, but little evidence is available on this point.

An interesting example of the influences affecting food choice is the difference in behaviour between wild and laboratory rats when presented with food consisting of an assortment of purified substances in separate containers. Offered a protein, a fat, a carbohydrate, and twelve other dishes of pure minerals and vitamins, a laboratory rat makes a selection that allows normal growth. A wild rat eats only the fat and dies in a short time. These wild animals are so 'suspicious' that they will not taste new foods, and this, incidentally, makes it difficult to poison them. Richter, who conducted these experiments, found many other signs of the 'suspiciousness' of the rats.

The rabbit is a herbivore, able to eat not only relatively soft vegetable foods such as dandelion or lettuce leaves but also much harder roots and grasses. For this activity it possesses sharp, continually growing incisors to crop the grass, large grinding molars, and an enormous intestinal sac, the caecum, in which bacteria make the cellulose of the grass available to the rabbit. Such unpromising food is not easily digested and rabbits have the habit of passing it through the gut several times by eating their faecal pellets and thus ensuring maximum use of the material available. Many mammals have specializations allowing cooperative digestion by means of symbionts. In artiodactyls there are elaborate stomachs. In perissodactyls (horses) there is a special fauna in the intestine. Leaf-eating monkeys have an enlarged stomach.

The dentition and gut of man, like his tastes, are those of an omnivore, able to obtain his nourishment from either plant or animal food or from a mixture of both. Neither the teeth nor gut of man are able to deal with the hardest plant foods.

Dissolution of Materials

The process of breaking up the food is begun in the mouth by the teeth, tongue, and by ptyalin in the saliva. The rest of the alimentary canal is concerned with completing the physical break-up of the tissues that have been eaten and the absorption from them of the molecules needed by the mammal.

The dissolution of the food is accomplished chemically by the combined effects of enzymes secreted by the gut and its glands and the action of symbiotic bacteria living within the gut. The high temperature of the body assists the action of the enzymes, and the gut becomes a great vat in which the macerated food is quickly acted upon by the enzymes and bacteria. The most striking special developments of the mammalian plan of digestion are indeed not regions of special enzyme production but large chambers harbouring symbionts, such as the elaborate 'stomach' of sheep, cows, and other ruminants, or the caeca of the large intestine of rabbits and horses. The movement of food along the gut and the secretion of the glands are so regulated as to keep the mass in the right physical and chemical conditions for these processes. The movement of peristalsis by which the food is propelled operates rhythmically and continually, rather as does the heart-beat. It is regulated by the operation of receptors in the wall of the gut, which signal whether conditions are favourable for digestion, setting up reflexes that speed or slow *peristalsis*. Chemical signaling by hormones is also employed.

Thus the regulation of digestion is ensured mainly by a series of mechanisms laid down by heredity. Like the food intake it is influenced also by the operation of higher nervous centres in the medulla oblongata, hypothalamus, and cerebral cortex. Electrical stimulation of the hypothalamus, or of parts of the frontal lobes of the cerebral cortex, produces changes in gastric and intestinal motility. Following removal of parts of the frontal lobes from monkeys it was found that the stomach becomes more active and excessively large amounts of food are consumed; similar symptoms may appear in man after injury or operations on the frontal lobes.

The processes of digestion thus also come under the influence of the higher cerebral centres, whose actions are influenced by memories set up during the lifetime of the individual. In man digestion may be profoundly influenced by learning processes. When habits of food consumption or digestion are interfered with, the balance of the organism may become severely disturbed.

Alimentary Tract

The alimentary tract is a long hollow muscular tube starting at the mouth and ending at the anus. Along its length are well defined regions, buccal cavity, pharynx, oesophagus, stomach, duodenum, ileum, caecum, appendix, colon and rectum. Opening into the alimentary tract are the ducts of several glands which produce secretions concerned with the digestion of food; these glands include the salivary glands, pancreas and liver. The gut is suspended from the dorsal wall of the coelomic cavity by a double layer of the peritoneum which lines the coelom, and between the two layers of periotoneum pass the blood vessels, nerves and lymphatics which supply the gut.

The basic structure of the gut is shown in transverse section. The innermost layer consists of the mucosa. In the buccal cavity and anal canal this consists of a layer of squamous stratified epithelium since these regions were formed in the embryo by intuckings of ectoderm. The remaining portions of the gut (except the oesophagus) are lined by a simple columnar epithelium. In some regions there are intuckings of this epithelium into the deeper layers of the wall of the gut to form glandular structures producing digestive secretions. In the small intestine the mucosal layer is highly folded to form the intestinal villi, which greatly increases the surface area available for the absorption of digested food-stuffs. Lying beneath the mucosa is a loose connective tissue, containing blood vessels and lymphatics, called the submucosa. In most regions of the gut this submucosa contains a thin layer of muscle, the muscularis mucosa.

Outside the mucosa and submucosa are the muscle layers of the gut, an outer longitudinal and an inner circular layer. These layers of

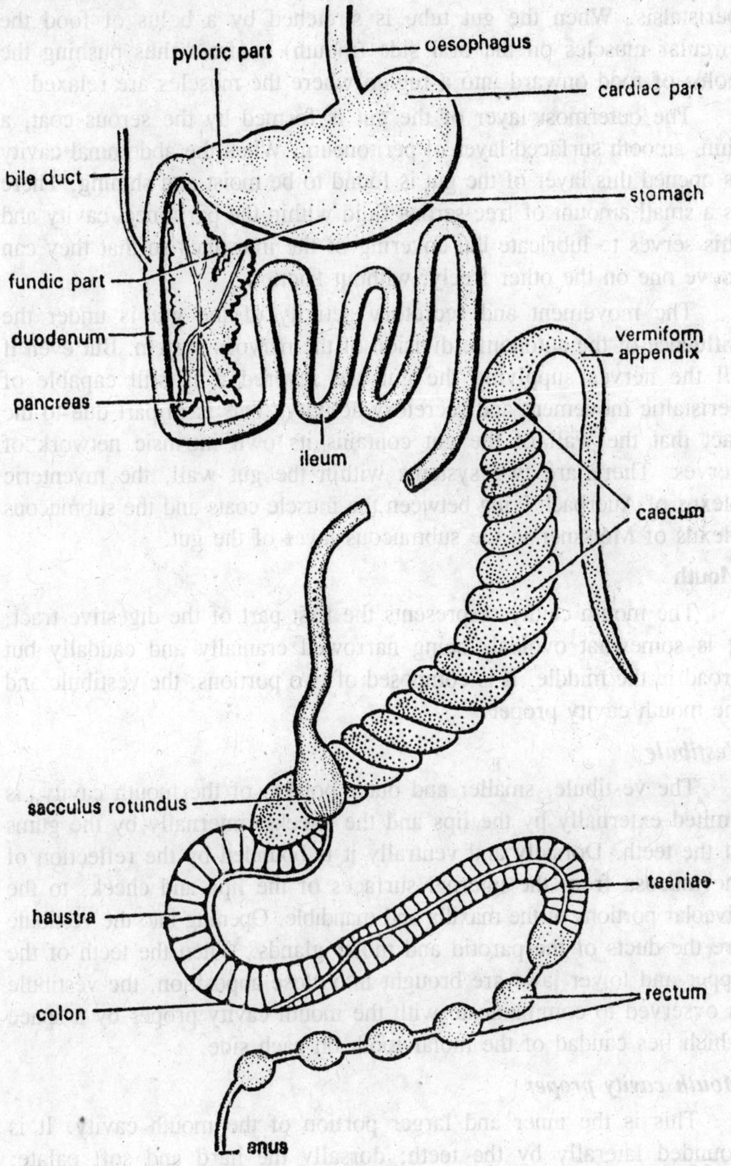

Fig. 11.2. Alimentary canal.

muscle consist of smooth muscle fibers except in the upper part of the oesophagus where striated muscle is found. The muscle coats are responsible for propelling the food along the gut in a movement called peristalsis. When the gut tube is stretched by a bolus of food the circular muscles on the oral side (mouth) contract thus pushing the bolus of food onward into a region where the muscles are relaxed.

The outermost layer of the gut is formed by the serous coat, a thin, smooth surfaced layer of peritoneum. When the abdominal cavity is opened this layer of the gut is found to be moist and shining. There is a small amount of free serous fluid within the peritoneal cavity and this serves to lubricate the covering of the intestines so that they can move one on the other freely, without friction.

The movement and secretory activity of the gut is under the influence of the autonomic division of the nervous system. But even if all the nerves supplying the gut are severed it is still capable of peristaltic movement and secretory activity. This is in part due to the fact that the wall of the gut contains its own intrinsic network of nerves. There are two systems within the gut wall, the myenteric plexus of Auerbach lying between the muscle coats and the submucous plexus of Meissner in the submucous layer of the gut.

Mouth

The mouth cavity respresents the first part of the digestive tract. It is somewhat ovoidal, being narrowed cranially and caudally but broad in the middle. It is composed of two portions; the vestibule and the mouth cavity proper.

Vestibule

The vestibule, smaller and outer portion of the mouth cavity, is limited externally by the lips and the cheeks; internally by the gums of the teeth. Dorsally and ventrally it is bounded by the reflection of the mucosa from the internal surfaces of the lips and cheeks to the alveolar portions of the maxilla and mandible. Opening into the vestibule are the ducts of the parotid and molar glands. When the teeth of the upper and lower jaws are brought into close apposition, the vestibule is ovserved to communicate with the mouth cavity proper by a space which lies caudad of the molar tooth on each side.

Mouth cavity proper

This is the inner and larger portion of the mouth cavity. It is bounded laterally by the teeth; dorsally the hard and soft palate; ventrally by the tongue and the reflection of the mucosa from the lateral and ventral surfaces of the tongue to the gum on the internal surface of the mandible. The mouth cavity receives at the cranial end of its floor the secretions of both the sublingual and submaxillary glands.

Lips

The lips are two thickened folds surrounding the orifice of the mouth and forming the cranial wall of the mouth cavity. Each lip is formed externally of skin covered with hair and internally of mucosa. The upper lip is characterize in the midline by a broove which extends dorsad onto the septum of the noise. The internal surface of each lip is attached in the midline to the corresponding gum by a fold of mucous membrane, termed the frenulum. Laterad of each side of the frenulum of the upper lip, the mucosa bears a fairly large number of conspicuous papillae. In addition, the lateral portions of the lower lip are connected by rather thick fold of mucous membrane to the diastematic area of the lower jaw between the canine and first premolar teeth. The upper and lower lip are continuous with each other at the angle of the mouth and with the cheeks canded.

Cheeks

The cheeks are rather small in the cat and only include that area from the angle of the mouth to the ramus of the mandible. Externally, the cheek consists of skin covered with hair; internally, of mucosa continuous with that of the lips. The ducts of the parotid and molar glands drain into the vestibule of mouth cavity on the internal surface of the cheek.

Palate

The palate constitutes the roof or dorsal boundary of the mouth cavity proper. It is composed of two portions, the hard palate lying craniad and soft palate caudad.

Hard palate

The hard palate is formed by the palatine processes of the premaxillae and maxillae together with the horizontal portion of the palatine bones; Cranially, it is continuous with the alveolar borders of the premaxillae; laterally, with the alveolar borders of the maxillae; and caudally, with the soft palate. The hard palate, thus constituted, separates the nasal cavity from the mouth cavity proper. The oral surface of the hard palate is covered with a thick mucosa, which is corrugated into transverse palatine ridges with their convexities directed craniad. The cranial and caudal surfaces of each palatine ridge are papillated. Immediately caudad of the incisior teeth there is situated in the middle of the cranial end of the hard palate's mucosa, a papilla, on each side of which is the opening of the small incisive duct. The latter duct passes ventrad through the incisive foramen form the vomeronasal organ on the floor of the nasal cavity.

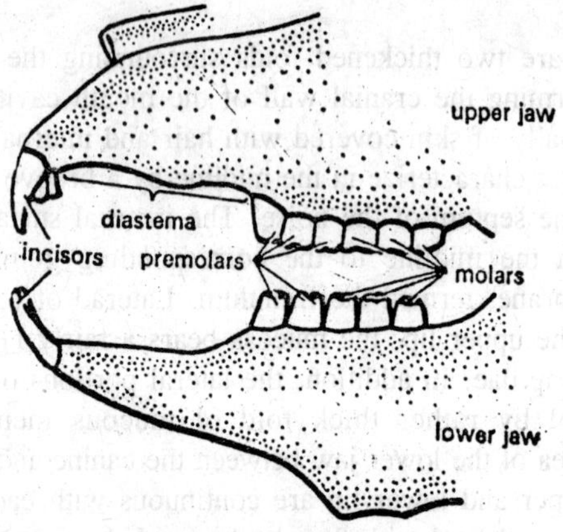

Fig. 11.3. Disposition of teeth on the jaws.

Soft palate

The soft palate is the caudal portion of the palate and, in its relaxed position, incompletely separates the caudal portion of the mouth cavity proper from the nasopharynx. It is essentially a movable fold, about three centimeters in length and extending caudad from the hard palate to the level of the epiglottis. Its cranial end is attached to the horizontal portion of the palatine bone; craniolaterad its nasal surface attaches to the ventral borders of the vertical plates of the palatine bone and to the internal pterygoid processes and their hamuli. Caudally, the soft palate presents a free concave margin, each side of which arches laterally into two folds, continues ventrad into the side of the base of the tongue and is termed the glossopalatine arch; the caudal palatine arch, because it passes ventrocaudad to blend into the pharyngeal wall, is called the pharyngopalatine arch. Between the palatine arches lies the palatine tonsil, an aggregation of lymphoid tissue with the greater part of its longitudinal axis obscured from view and directed craniad. The narrow aperture of communication between the mouth cavity proper and the pharynx – bounded dorsally by the free caudal margin of the soft plate, laterally by the glossopalatine arch, and ventrally by the dorsum of the tongue – is known as the faucial isthmus, or simply as the fauces.

Structurally, the soft palate is a fold of mucous membrane which, between its oral and nasal surfaces, encloses muscles, an aponeurosis, nerves, and blood vessels.

The important musculature of the soft palate follows:

(a) *Tensor veli Palatini*. This is a small triangular muscle with its apex just craniad of the tympanic bulla and its base diverging craniomediad over the hamulus of the internal pterygoid process. *Origion:* scaphoid fossa on the ventral surface of sphenoid just caudad of the foramen ovale. Insertion: aponeurosis in the soft palate with similar muscle from the opposite side. Action: tensor of the soft palate. Innervation: branch of the pterygoideus nerve from the mandibular division of the trigeminal. The pterygoideus nerve, after leaving the mandibular division of the trigeminal nerve, sends a branch to the pterygoideus muscles; then passes caudad to the tympanic bulla, and in so doing sends a branch to cranial border of tensor veli palatine. The pterygodeus nerve then continues into the tympanic bulla to innervate the tensor tympani muscle.

(b) *Levator veli palatine*. A flat muscle lying mediad of the tensor veli palatine and turning caudad over the dorsal surface of the constrictor pharynges superior muscle to enter the soft palate. *Origins :* caphoid fosssa of he sphenoid bone and styliform process of the bulla tympani, Insertion: soft palate, with some fusion of fibers from the opposite muscle in the midline. Action: levator of the soft palate. Innervation: pharyngeal plexus.

Teeth

The cat is supplied with two sets of teeth, each of which appears at a different period of its life. The first set develops during the suckling period and is termed the milk or deciduous dentition. Around the age of four months, the permanent teeth begin to erupt and gradually push out and replace the deciduous teeth.

The deciduous teeth number twenty-six: twelve incisors, four canines, and ten molars (six in the upper jaw and four in the lower).

The permanent teeth aggregate thirty in all: twelve incisors, four canines, ten premolars, and four molars.

Mammalian heterodonty is generally represented in a dental formula, which indicates at a glance the number of each type of teeth in each half-jaw from the anterior midline to the posterior end of the jaw. The numerator in the formula represents the upper teeth; the denominator, the lower teeth.

General characteristics

Each tooth consists of four portions: the crow, the portion projecting beyond the gum and covered with enamel; the cingulum, the ridge of enamel that obtains at the base of the crown so as to protect the surrounding gum; the root, which may be single or divisible into component fangs, is the part embedded in the alveolus of the jaw and covered with a bone-like substance termed cementum; the neck, the faint constriction that occurs where the root meets the crown. The peridental membrane anchors the tooth in the alveolus.

The surface of the tooth directed toward the lip or cheek is called the labial or buccal surface. The surface which is directed toward the tongue is called the lingual surface.

Permanent teeth

The different types of permanent teeth are modified in shape of crown and root structure according to their position and use. Accordingly, a more detailed study of the different types is presented below:

Incisors

The incisors are the six small teeth embedded in the alveoli of the premaxilla, with a similar number of the same type of teeth contained in the alveolar border of the mandible below. The crowns of the incisor teeth are characteristically chisel-shaped and are each notched into three minute cusps. The lingual surface of the crown of each upper incisor is transversely grooved while the similar surface of the crown of each lower incisor is not. The root of the incisor in either jaw is longer than the crown; it is single, laterally compressed, and directed backward.

Canines

The upper canines, two in number and contained in the alveoli of the maxilla caudad of the incisors, are long, curved, pointed teeth. Their buccal surfaces are convex and ridged; the cingula are not too distinct. The root of each upper canine is single and larger than the crown. The lower canines, also caudad of the incisors on the mandible, possess a more curved crown than the upper canines; the buccal surface of each lower canine is also convex and ridged. The root of the lower canine is similar to that of the upper canine. The upper canine may be readily distinguished from the lower canine by its relatively straighter and longer crown.

Premolars

These are teeth with laterally compressed crowns. The premolars are so-called because they lie craniad of the ture permanent molars, which have no precursors in deciduous dentition. The upper and lower premolars are considered individually as follows:

(a) *First upper premolar*. The first upper premolar is small and possesses a characteristically triangular crown. Its root is generally single. Between the upper canine and first premolar there is a diastematic interval.

(b) *Second upper premolar*. The crown possesses a large principal cusp at the caudal base of which is a smaller cusp, called the caudal basal cusp. The caudal area of the cingulum is modified into a small cusp, the talon. The root is double, with the caudal root being characteristically prismatic and larger than the cranial root.

(c) *Third upper premolar*. The crown has a large principal cusp at the cranial base of which are two smaller cusps, called the cranial basal cusp. Caudad of the principal cusp and separated there form on its lingual surface by a vertical cleft is the large caudal basal cusp. The root is divisible into three fangs; two are cranial and support the cranial basal cusp; the third one is caudal, triangular and flattened, and supports the principal and caudal basal cusps. The ciingulum of the tooth in rather well developed.

Because of the disposition of its principal and caudal basal cusps, the crown of this tooth presents a blade-like or cutting edge and, for that reason, is referred to as a carnassial or upper sectorial tooth. Together with the mandibular molar, which is a sectorial tooth, this premolar embodies an efficient scissor-like shearing mechanism for cutting food.

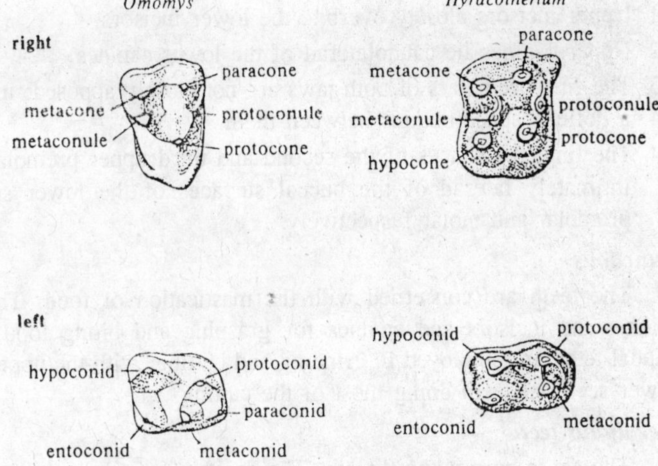

Fig. 11.4. Molar teeth of primitive placentals.

(d) *Upper molar.* The molar is rather small and lies caudomediad of the last premolar. Its crown is tuberculate rather than cuspidate. Its root is divisible into two fangs, the outer fang being generally larger than the inner one.

(e) *First lower premolar.* The first lower premolar lies caudad of the lower incisor and is separated from the latter by a distematic interval. The crown possesses a large principal cusp, at the cranil and caudad bases of which are situated respectively a cranial basal cusp and a caudal basal cusp. The cingulum is rather well developed and is elevated into at alon a just caudad of the caudal basal cusp. Its root is diviusible into a cranial and a caudal fang.

(f) *Second lower premolar.* The second lower premolar is somewhat longer than the first lower premolar but is otherwise morphologically similar to it.

(g) *Lower molar.* The lower molar is the largest tooth in the lower jaw. Its crown is bicuspidate and adapted to from a cutting edge. For this reason the lower molar is often referred to as the lower sectirial tooth. Its relation with the upper third premolar in embodiment of a shearing mechanism has already been indicted. The root of the lower molar is divisible into two fangs; the cranial fang is larger and supports the greater part of the crown.

Relations of the permanent teeth

When the jaws are closed, as when at rest, the teeth are disposed into the following relations:

1. Upper incisors closely overbite the lower incisors.
2. Upper canines lie caudolaterad of the lower canines.
3. The first premolars of both jaws are not closely apposed; in fact, a definite space obtains between them.
4. The lingual surfaces of the second and third upper premolars lie intimately laterad of the buccal surfaces of the lower second premolar and molar respectively.

Functions

The teeth are concerned with the mastication of food. The cat utilizes the incisors and canines for grasping and biting food. The caudal teeth are employed in grinding and cutting, with the upper and lower sectorial teeth doing most of the cutting.

Deciduous teeth

The cat is born without teeth. The deciduous teeth begin to erupt about two weeks after birth. Normally within sixty days after birth, deciduous dentition is complete. The number and types of teeth have already been enumerated in the dental formula of deciduous dentition. A brief tabulation of the resemblances and differences that obtain between the deciduous teeth and their permanent successes is presented below:

1. Deciduous incisors and canines are to great degree, similar to the permanent incisors and canines.
2. The first deciduous upper molar resembles its permanent successor, the first upper premolar.
3. The second upper deciduous molar, in being a sectorial tooth, does not resemble the second permanent premolar, which replaces it. In fact, the upper deciduous molar is more comparable to the upper sectorial or third premolar.
4. The third deciduous molar resembles the permanent molar rather than its permanent replacement, the third upper premolar.
5. The first lower deciduous molar, in addition to being separated from the lower deciduous canine by a spacious diastema, resembles the first lower premolar, which succeeds it.
6. The second lower deciduous molar differs from its permanent replacement, the second lower premolar, in being a sectorial tooth. Like the second upper deciduous molar, it is comparable to the lower sectorial or true molar rather than to its permanent successor, the second lower premolar.
7. The permanent molar in both jaws, as has already been indicated, have no precursors in deciduous dentition.

Structure of a typical tooth

Enamel

The structure of a typical mammalian tooth. That part of the tooth which sticks out above the gum is called the crown. It is capped with a substance called enamel, which is the hardest material of the body. It consists of cylinders of very hard inorganic material which are aligned perpendicularly to the surface. The enamel consists of 96% mineral matter and is the only part of the tooth formed from the epidermis, all the rest of the tooth being mesodermal in origin. The enamel is completely formed before the tooth erupts from the gum.

Dentine

This material forms the bulk of the tooth and consists of an organic fibrillar network in which mineral salts, mainly of calcium, are deposited. About 70% of dentine is mineral matter. It is a hard

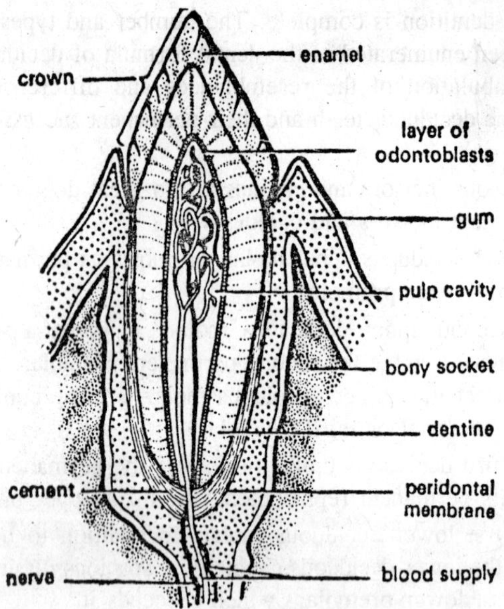

Fig. 11.5. L.V.S. of a tooth.

substance, harder than bone and is commonly called ivory. The dentine is perforated by canals, which run at right angles to the surface, in which lie the processes of the odontoblasts, the mesodermal cells which produce the dentine. The odontoblasts line the inner edge of the dentine layer and continue to lay down dentine as long as they are adequately nourished. They are supplied by the blood vessels of the pulp cavity in which the cells lie. In most teeth the base of the pulp cavity, at the root of the tooth, becomes constricted soon after the teeth have reached their full size, so reducing the blood supply to the odontoblasts. Dentine production now ceases and the tooth stops growing. Some teeth such as the incisors of the rabbit or the molars of a sheep continue to grow throughout life as they are worn away by the friction with the food material and with the opposing teeth. They can do this because the base of the pulp cavity never closes and thus the odontoblasts receive a rich blood supply. Teeth which have a non-occluded pulp cavity are said to have open roots.

Cement

Cement is a modified type of bone which surrounds the roots of the teeth and serves to bind them to neighbouring structures. In some herbivourses it extends between folds of the tooth so as to form part of the grinding surface. It is less hard than dentine. Between the cement and the surrounding bone of the jaw is a system of collagenous fibers, the *periodontal ligament* (or membrane) by which the tooth is attached.

In most teeth the tip of the root narrows considerably, but remains open allowing blood vessels and nerves to enter the pulp cavity. In teeth that grow continually, such as the incisors of the rodents, the pulp is widely open below and these teeth are said to be 'without roots' and to have 'persistent pulps', they are worn away continually.

Innervation of the teeth

The periodontal ligament is well supplied with nerve fibers, which are probably mechanoreceptors. There are brush-like nerve ending in the dental pulp. These extend among the odontoblasts and into the inner dentine. Electron microscopy has shown that nerve fibers enter one in every few hundred dentinal tubules. They lose their schwann cells and come into direct contact with peripheral fibrocytes and odontoblasts. This gives 50 nerve fibers per square millimeter, which is high density of innervation. These provide the basis for the notorious pain sensitivity, which is the only sensation elicited from a tooth. These nerve fibers do not reach to the amelo-dentinal junction which is nevertheless highly sensitive. Some effect therefore must be transmitted along the tubules.

Development of teeth

Teeth are formed from what are called tooth buds which arise in the developing jaw. The tooth bud consists of a cup like enamel organ which fits over a group of mesodermal cells called the dental papilla. The cells of the papilla produce the dentine forming the bulk of the tooth. The enamel is produced by the cells of the enamel organ.

The first sign of the formation of teeth is the development of a ridge of thickened epithelium along the line of the jaw, called the dental lamina. At regular intervals along the dental lamina groups of epithelial cells proliferate into the mesodermal tissues of the jaw and give rise to the enamel organs. The enamel organs consists of a two layered cup and within the cavity of the cup are the mesodermal cells of the dental papilla. When there are two sets of teeth, the milk teeth and the permanent teeth, the dental lamina has two rows of tooth buds, the upper most being the milk tooth buds, the lower row the permanent teeth, which may develop several years later pushing the milk teeth out of the jaw as they grow upwards.

The cells of the inner layer of the enamel organ produce the enamel and induce the cells of the outer edge of the dental papilla to form the dentine. The remainder of the papilla is eventually invaded by nerves and blood vessels to form the tissues of the pulp cavity. The stages in the development of the tooth.

The jaw bone eventually grows round the developing tooth to form the socket or alveolus, to which the tooth eventually becomes firmly attached by way of the periodontal membrane.

Variations in dentition

Food is complex organic material which we have seen has been synthesized in the first place by green plants. This food may be transformed into the protoplasm of animals when the plants are eaten. The protoplasm of plants is surrounded by a cellulose cell wall which during the period known as secondary thickening in the life of the plant, may become chemically altered or impregnated with various substances like lignin. Walls of plant cells are often woody or in the case of grasses they have a high silica content which makes them very hard. Plant material is much harder to chew than the soft juicy flesh of animals.

During the evolution of mammals from reptiles one of the significant trends has been the formation of the false palate in the roof of the mouth, separating the breathing passage from the mouth. The maxillae and the palatine bones have grown to form this shelf. The significance of the palate is that its possessors have the ability to breathe whilst the mouth is full of food. This development may well be associated with the evolution of warm bloodedness and the need for constant breathing to supply energy in respiration. As soon as food can be retained in the mouth without preventing breathing there is virtue in having a battery of teeth which can deal with this food. In all classes of vertebrates below the mammals the teeth are only of use to stop the prey escaping from the mouth and all the teeth are of similar shape – a simple pointed cusp. With the development of the palate in the mammal is associated the development of teeth of different kinds suited to the special kind of food which the animal eats. Special kinds of teeth are absolutely essential to deal with food such as grass, so much so, that one of the reasons given for the sudden eclipse of the giant dinosaurs at the end of the Mesozoic era is that the grasses were becoming the dominant vegetation in the cretaceous period (100 million years ago) and the dinosaurs' dentition was not capable of dealing with such hard siliceous material. In order to understand the adaptive radiation which has occurred in mammals let us look more closely at a few different types of mammalian dentition. It is interesting to comment here that in evolution once a great step forward has been taken e.g. when the amphibians came onto land, or when the mammals learned how to maintain a constant body temperature there frequently follows a period of great evolutionary experimentation. Different combinations of genes produce different characters and those which are successful are retained as their possessors live long enough to reproduce and pass on these gene combinations to their offspring. When the palate was formed there seems to have been an opportunity to try several methods of dealing with food. The attempts to exploit this new opportunity for efficiency constitutes a good example of what we call adaptive radiation.

Dentition of the sheep

The teeth of mammals are characterized by being heterodont and diphyodont. These two technical terms mean that there are teeth of different kinds and that there are two sets of teeth, a set of 'milk' teeth followed by the permanent set. There are four kinds of teeth in the typical mammal and they are called incisors, canines, premolars and molars. The numbers of these teeth present can be expressed in the dental formula e.g. I.3/3, C.1/1, P.M.4/4, M.3/3 or more simply 3.1.4.3/3.1.4.3. This formula is derived in the following way. The letters represent the different types of teeth, and the top number after each letter indicates the number of teeth of that kind found in half the top jaw. The denominator of the fraction indicates the number of teeth of this kind found in half of the lower jaw. Thus the typical mammal with the above quoted formula has twelve incisors four canines, sixteen premolars and twelve molars, that is forty four teeth in its permanent dentition.

The formula for the sheep is 0.0.3.3./3.1.3.3, but this does not tell us enough about the dentition so we must draw and describe it. The incisors of the lower jaw bite against a horny pad on the top jaw. In the upper jaw the canines are absent and a space called the diastema separates the horny pad from the premolars. This space is characteristic of herbivores. The premolars look similar to the molars and together they constitute an effective grinding battery. The teeth in this grinding battery have open roots i.e. they continue to grow throughout life. As they wear away they do so very unevenly because they are made of three different substances. The dentine, enamel, and cement, wear at different rates and leave a crescentic pattern on the surface of the tooth. There are sharp ridges of the hard enamel and slightly softer dentine passing from front to back of the tooth surface and also from side to side of the tooth surface, so that whichever way the bottom jaw moves, grinding is sure to occur. If we look at the jaw joint we see that it is very flat allowing the lower jaw to move in a circular path. If you watch a sheep chewing you will see that this is in fact how the lower jaw works thus exploiting the grinding ridges on the tooth surface. The sheep's jaw is not a strong one and examination of a sheep's skull will show how very easy it would be to dislocate the jaw, but this does not matter to the sheep for the grass does not struggle violently when it is bitten. If you look carefully at the molars and premolars of the sheep you will see that the sides of these teeth are strengthened, especially at the corners, by buttresses of cement. These pillars of cement serve to prevent the edges of the tooth being chipped off by the siliceous food.

Dentition of the dog

The appearance of the dog's indicates at once that it is a flesh eater although many animals which are very closely related e.g. the fox are known to have a very mixed diet often including insects. The formula is 3.1.4.2./3.1.4.3. The dog belongs to the genus Canis and it is not for nothing that the canine is so called. These canine teeth are well developed in the dog and are used as a weapon of defence and attack. They are used to spear the prey. If you watch a dog chewing meat or better still, a bone, you will see how he turns his head on one side and gets the food to the angle of his jaw. This is where the carnassial teeth are. They are specially designed teeth for cutting flesh and are developed from the last upper premolar and the first lower molar, although in other carnivores different teeth in the molar battery may be involved. The carnassial teeth of the dog; they have vertical surfaces which act like a pair of scissor blades because the jaw joint in the dog does not allow side to side movement but only movement in the vertical plane. The lower jaw is inserted into the skull by the long transversely running condyle of the lower jaw, which is housed in a deep transversely disposed groove in the skull. The deep jaw joint is essential in order to prevent dislocation of the joint when the prey struggles. The emphasis in the carnivore is on the canine and carnassial teeth, on attack and on chopping up the meat into chunks, which are then quickly swallowed.

Dentition of the rabbit

If the sheep can be called a grinder and the dog a chopper-up of its food, then the word for the rabbit is a nibbler. The formula is 2.0.3.3./1.0.2.3. and the emphasis is undoubtedly on the incisors, which have open roots and grow continually throughout life so much so that if you keep any rodents as pets you should be sure to provide them with something hard to chew like a piece of wood so that they can keep their incisors sharp and prevent them from becoming overgrown. The front surfaces of rodents' incisors have a very thick layer of enamel and as the curved tooth wears it leaves this sharp edge of enamel exposed, because the enamel is more resistant to wear than the rest of the tooth. The rabbit has the typical large diastema of the herbivore. There is no superficial difference between the molars and premolars, and together they form a very efficient grinding battery. The ridges on the surfaces of the molars are transverse and well designed, when it is realized that the jaw moves to and fro in the vertical plane only. The movement of the jaw is restricted by the method of insertion of the dentary (lower jaw bone). The articulating surface is the top edge of the end of the dentary and if fits into a longitudinal groove in the squamosal bone of the skull, whilst any side to side movement is obviously restricted by the quadratojugal arch of the cheek bones.

We have seen in the sheep, dog and rabbit three very different sets of teeth, each designed with a jaw joint adapted to the special type of dentition. Each of these three animals is a very specialized feeder and by comparison man has a very unsepecialized set of teeth. This may be related to the fact that there is little selective value for man in having specialized teeth, since he has the ability to use tools to help him in his feeding.

Tongue

The tongue is a mobile muscular organ ensheathed by mucous membrane and situated in the floor of the mouth cavity proper. It is divisible into a root and a body.

The root is the slightly narrowed caudal area of the tongue lying in the region of the faucial isthmus. Laterad, the root of the tongue is continuous with the region of the palatine tonsil, and caudal it is connected to the cranial surface of the epiglottis by a small fold, the frenulum glossoepiglottica.

The body of the tongue is the relatively larger part of the tongue lying cranisd of the faucial isthmus and forming the floor of the mouth cavity proper. Ventrad, the mucous membrane of the body of the tongue is firmed into a vertical fold, the frenulum linguae, which connects the tongue to the floor of the mouth and lingual surfaces of the gums. The cranial border of the frenulum is modified on each side into a

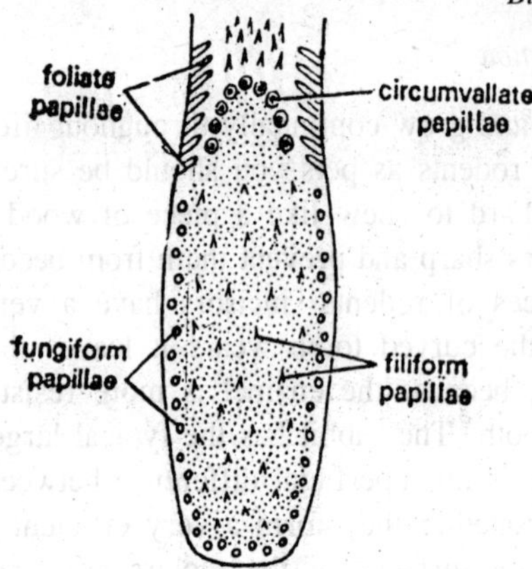

Fig. 11.6. Tongue.

prominent papilla, the apex of which bears, on its lateral and medial surfaces, the openings of the ducts of the submaxillary and sublingual glands.

The dorsal surface of the tongue, or dorsum linguae, is separated from the ventral surface by the lateral margins which meet craniad at the tip or apex of the tongue. The dorsum is marked by a very shallow and imperfectly formed furrow, the median sulcus; and, in the addition, presents a rather rough appearance because of numerous small projections called the lingual papillae.

The lingual papillae are elevations of the subepithelium tissue of the tongue and are covered with stratified squamous epithelium. The are distributed over covered with stratified squamous epithelium. They are distributed over practically the entire dorsum of the cat's tongue and may be classified into three types, which are described as follows:

Filiform papillae

These are spinous in from and are distributed over the greater part of the dorsum. They are characteristically presented on the cranial third of the dorsum where they are cornified into small tusk-like spines, which are directed caudad. These papillae give this part of the tongue its typically file-like or speculated aspect. Microscopically each papilla consists of a core of connective tissue capped by a pointed mass of cornified, str squamous epithelium.

Fungiform papillae

There are broad and rounded at their free ends, and narrow at point of attachment to the tongue. They occur typically along the lateral edge of the middle third of the dorsum and are scattered in the vicinity

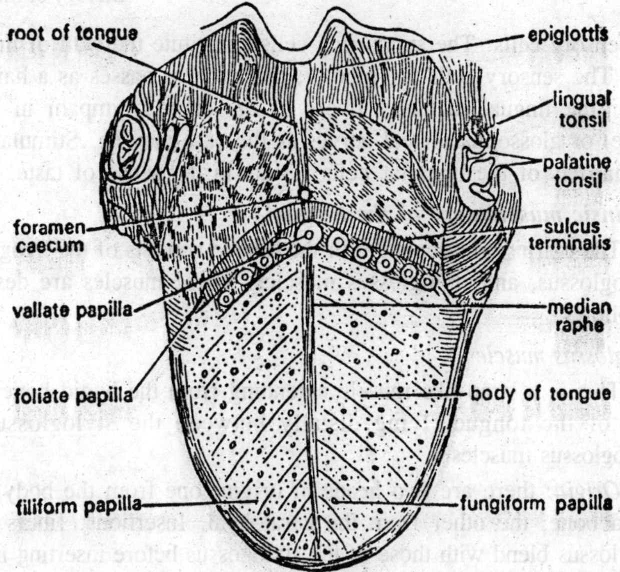

Fig. 11.7. Dorsal view of human tongue.

of the median sulcus just caudad of the filiform papillae. Microsopically they are elevations of the subepithelial connective tissue that broaden out to form rounded eminences above the surface of the tongue; they possess taste buds in their epithelium.

Circumvallate papillae

There present a knob-like, flattened summit, and lie contained within a deep trench or fossa. They occur in rows of two or three on each side of the midline of the caudal part of the body of the tongue and are disposed into a V-shaped pattern which coverges toward the root or base of the tongue. Microscopically they differ from the fungi from papillae in having a wide base and in being surrounded by a deep trench and wall. In the walls of the circumvallate papillae are contained numerous taste buds.

Taste buds

Taste buds are barrel-shaped structures which occur in the epithelia of the fungi form and circumvallate papillae. The board base of each taste bud rests on the connective tissue of the tongue while the narrow, tapering end terminates in a pore, termed the gustatory pore, which opens on the free surface of the epithelium. The taste bud itself embodies two types of cells, the sustentacular or supporting cells and the sensory cells. The supporting cells constitute the wall of the taste bud. The sensory cells are bipolar; one process passes as a hairlet of either the lingual (actually fibers of the chorda tympani in lingual nerve) or glossopharyngeal nerves as the case may be. Stimulation of the hairlets of the bippolar cells results in the sense of taste.

Extrinsic musculature of the tongue

The exirinsic musculature of the tongue consists of the hyoglossus, genioglossus, and styloglossus muscles. These muscles are described as follows.

Hyoglossus muscle

This is a strap-like muscle extending from the hyoid bone to the side of the tongue, lying distally between the styloglossus and genioglossus muscles.

Origin: there are two heads of origin; one from the body of the hyoid bone; the other from the ceratohyal. Insertions: fibers of the hyoglossus blend with those of the stylogossus before inserting into the dorsum of the tongue. Action: retraction and depression of the tongue. Innervation: branch of the hypoglossal nerve.

Genioglossus

A fan-shaped muscle, the apex of which diverges out from the mandible to contain finally its broad base within the frenulum of the tongue. The cranial portion of the base radiates into the apex of the tongue; the caudal portion arches into the root. The muscle lies dorsad of the geniohyoid.

Origin: just mediad of the symphysis of the mandible. Insertion: Cranial portion inserts into the apex of the tongue; caudal fibes fan into the root of the tongue. Action: contraction of the cranial portion effects the withdrawal of the tongue's apex; contraction of the caudal portion results in pulling the base of the tongue forward. Innervation: branch of the hypoglossal nerve.

Styloglossus muscle

A relatively long muscle, somewhat strap-like in from. Initially it passes ventrocraniad between the internal pterygoid and digastric muscles; then it turns mediad over the distal end of the hyoglossus and, finally, craniad to the apex of the tongue.

Origin: mastoid process of the temporal bone and cartilaginous portion of the cranial cornu of the hyoid. Insertion: apex of the tongue. Action: retraction and elevation of the tongue. Innervation: branch of the hypoglossal nerve.

Intrinsic musculature of the tonuge

The intrinsic musculature of the tongue is that which is completely contained within the substance of the tongue. The intrinsic muscles are generally disposed into bundles which run longitudinally, vertically, and transversely; thus they cross each other at right angles. The bundles are separated from each other by fibroelastic and adipose tissue. The intrinsic musculature may be profitably studied in a microscopic section of the tongue.

Functions of the tongue

In the cat, it functions as a ladling organ by which liquid is conveyed to the mouth. It is an organ of taste by virtue of its taste buds. It aids in mastication by its disposition of the food to the proper dental mechanisms. It assists in deglutition by driving the bolus of food to the back of the mouth cavity.

Pharynx

Caudally both the nasal and mouth cavities communicate with a common passageway, the pharynx, according to its relations, is divisible into a nasal pharynx, and oral pharynx.

The nasal pharynx is that area of the pharynx lying dorsad of the soft palate and caudad of the nasal cavity. Its roof is constituted of the sphenoid bone and basilar portion of the occipital bone, together with the longus capitis muscle along its dorsocaudal aspect. Its lateral walls are formed by the internal pterygoid processes and pterygoid musculature. Its floor is the soft palate. Craniad, the nasopharynx communicates with the nasal cavity by way of the choanae; caudally, it is continuous with the oral pharynx when the soft palate is relaxed; laterally, it communicates with the middle ears by means of the Eustachian tubes. The openings of the Eustachian tubes occur as slit-like apertures, one of which is situated on each wall of the nasophraynx just caudad of the posterior border of the internal pterygoid process of the sphenoid bone.

The oral pharynx is smaller than the nasopharynx. It extends from the faucial isthmus to the opening of the esophagus. The floor of the oral pharynx consists of the caudal portion of the base of the tongue and the epiglottis. Its lateral wall sand roof are formed, for the most part, by the middle constrictor muscle of the pharynx. Dorsad the roof of the oral pharynxisx separated from the basilar portion of the occipital bone and atlas by the longus capitis muscle. Craniad the oral pharynx communicates with nasopharynx when the soft palate is relaxed, and with the mouth cavity by way of the faucial isthmus; caudad it communicates with esophagus dorsocaudally and with the larynx presents, between the mucosa of each lateral wall of the oral pharynx presents, between the palatine arches, a fossa in which the caudal end of the palatine tonsil is contained.

Muscles of the pharynx

The muscles which are related to the functioning of the pharynx are presented as follows:

Glossopharyngeus muscle

This muscle consists initially of a small sheet of muscle, the fibers of which arise from the lateral surface of the genioglossus and that area of the styogolossus luing a short distance craniad of the cranial cornu of the hyoid bone. The fibers of the muscle converage from a flat sheet into a narrower band as they pass, over the external surface of the cranial cornu of the hyoid bone, to continue dorsad for insertion on the dorsal surface of the pharyunx.

Origin: lateral surface of the genioglossus and ventral surface of the styloglossus a short distance craniad of the cranial cornu of the hyoid bone. Insertion: median dorsal raphe of the pharynx. Action: constrictor of the pharynx. Innervation: glosopharyngeal and vagus nerves by way of the pharyngeal plexus.

Salivary Glands and Saliva

In man there are three pairs of salivary glands, the parotid, submandibular and sublingual glands. These discharge their secretions by ducts which open into the oral cavity. The openings of the submandibular glands can be seen as two papillae under the anterior end of the tongue where the fold of mucous membrane called the fraenum is attached to the floor of the mouth. If the taste buds have been suitably stimulated e.g. by some lemon juice placed on the tongue, then saliva will be seen to be pouring out of these papillae. The opening of the parotid ducts can be seen on the inner side of the cheek opposite the upper second molar tooth.

There are two ways in which the salivary glands may be stimulated. First, by means of a reflex from the taste buds on the tongue. There are receptors on the tongue which enable us to distinguish bitterness, sweetness, sourness and saltness and the position of these receptors has been mapped out by placing different substances at different sites on the tongue which has been prepared for the experiment by removal of excessive saliva. In this way maps have been made showing the sites of the different sense organs. Saliva is also secreted in response to such stimuli as the sight or smell, sometimes even the thought, of food.

Saliva contains water, mineral salts, mucin and an enzyme called ptyalin. The exact composition varies according to the type of stimulus, thus the saliva contains more mucin and ptyalin following stimulation by bread and meat than by lemon juice. About 1.0-1.5 litres of saliva are produced each day in man. The water and mucin moisten and lubricate the food so facilitating swallowing. The enzyme ptyalin is an amylase and it begins the conversion of starch (amylase and amylopectin) and glycogen into maltose, with the production of a small amount of glucose. Ptyalin works best in the slightly alkaline conditions of the mouth, and the action of the enzyme is helped by the presence of chloride ions. Because food does not stay an appreciable length of time in the mouth the salivary amylase does not play a significant part in the digestion of carbohydrates. However, the action of ptyalin continues for a time in the centre of the bolus of food even when it has reached the stomach, until the hydrochloric acid of the gastric juice reaches the centre of the bolus and inactivates the ptyalin, by virtue of the low pH.

Mastication and Swallowing

The degree to which food is masticated by the teeth and tongue varies from species to species. Carnivores are notorious bolters of the food. In man the food is masticated to a variable degree to produce a bolus of food, by the action of the tongue against the hard palate. The bolus is then pushed to the back of the mouth by raising the front of the tongue. The bolus is then rapidly and reflexly ejected into the pharynx and the upper end of the oesophagus. During this time the mouth cavity is closed from the pharynx by the tongue pressing against the palate, and the naso-pharynx is closed off by elevation of the soft palate. Breathing movements cease and the larynx is lifted up under the base of the tongue so that the laryngeal opening is protected against the inhalation of food.

Oesophagus

The oesophagus is a thick walled muscular tube through which the food passes from the pharynx to the stomach. Most of the oesophagus has no outer peritoneal coat and its outer layer consists of a loose connective tissue, the adventitia. The muscular coats are very prominent particularly the outer longitudinal layer. At the upper end of the tube the muscle fibers are striated, giving way lower down to unstriped muscle fibers. There is a thick submucous layer in which many secretory glands occur, the oesophageal glands. These pour their secretions into the lumen of the oesophagus and so lubricate the epithelial surface to reduce the friction of the passage of food. The epithelial lining of the oesophagus consists of a stratified epithelium. In the resting organ the surface of the epithelium is folded and this permits the accommodation of food.

Liquids and soft food pass down the oesophagus at a fast rate, by the force of the initial act of swallowing and may reach the lower end of the oesophagus in 0.1 sec. Food of this consistency tends to pass directly through the cardiac sphincter into the stomach. More solid food passes down the oesophagus by peristaltic action of the oesophageal muscles. A peristaltic wave takes about five seconds to reach the cardiac sphincter which relaxes to allow the passage of the food into the stomach. There has been much argument as to whether there is a true sphincter at the junction of the oesophagus and the stomach, the cardiac sphincter. Although there may be no anatomical sphincter there seems to be a functional one.

Stomach

The stomach is a muscular bag whose main function is to store the food and convert it into a semi-liquid material called chime before it passes into the duodenum. Part of the stomach, indeed sometimes almost the entire organ, may be removed surgically, without interfering generally with life. But the patient has to eat small regular meals, emphasizing the normal function of the stomach – which is storage. The human stomach is a J-shaped organ as shown. Its size varies greatly, according to the food contents, but the average capacity is about two pints. Four main regions of the stomach are described:

1. The cardiac region which surrounds the region of the cardiac sphincter.
2. The fundus, an air-filled portion, lying adjacent to the cardiac region.
3. The body of the stomach, the main portion where food is stored and where the food becomes more fluid as the process of digestion begins.
4. The pyloric region, the distal part of the stomach which opens into the duodenum by way of the pyloric sphincter.

Histological structure of the stomach

The mucous membrane of the stomach is thrown into large folds which are taken up as food is accommodated; this is made possible by the loose submuscous tissue. In addition to these large folds there is a

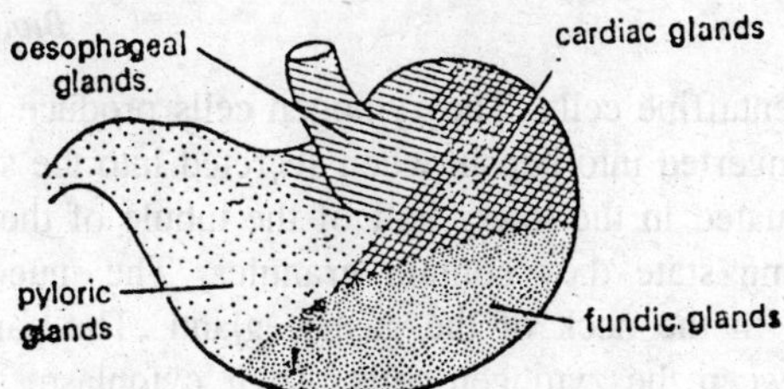

Fig. 11.8. Stomach showing distribution of gastric glands.

system of very much smaller grooves covering the entire surface of the mucosa. The surface of the mucosa is also pitted by small apertures which open into the lumen of the gastric glands which occupy the thickness of the mucosa. The glands vary in structure from one region of the stomach to another. The surface of the stomach is covered by a layer of columnar cells which secrete an alkaline mucus, and this layer of cells merges into the stratified epithelium of the oesophagus at the cardia and with the intestinal epithelium at the pylorus.

There are four types of cell in the gastric glands, the chief or zymogen cells, the parietal (or oxyntic) cells, the mucous neck cells

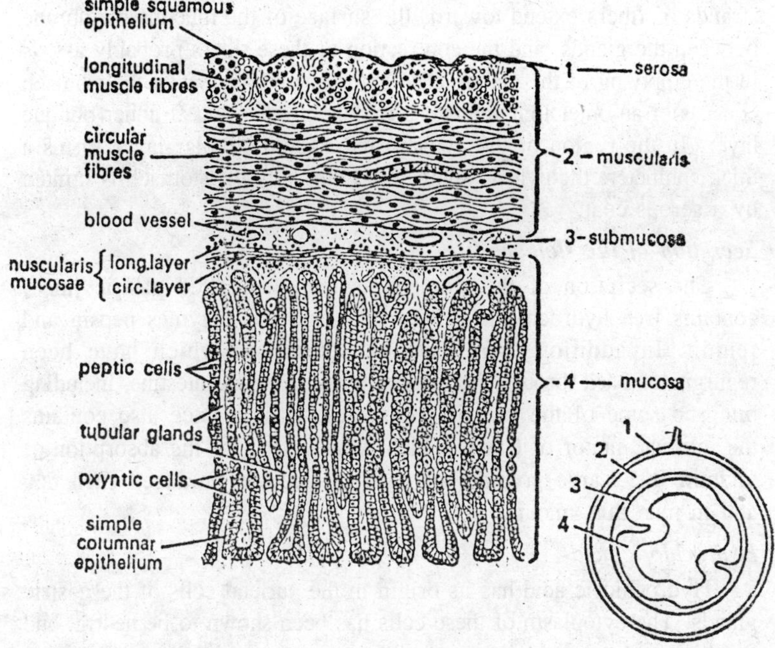

Fig. 11.9. T.S. of stomach.

and the argentaffine cells. The zymogen cells produce the pepsinogen which is converted into pepsin when secreted into the stomach. These cells are situated in the lower half of the tubule of the gastric gland. In the resting state they contain granules. The mucous neck cells are situated in the neck of the gastric gland. The parietal cells are situated between the zymogen cells. Their cytoplasm stains red with aniline dyes. They contain an intracellular canaliculus, where it is thought that the hydrochloric acid is produced, away from the cytoplasm of the cell. The argentaffine cells are thought to be concerned with the production of the intrinsic gastric factor (of Castle). These cells which is a section of the fundus of the human stomach. The glands of the pyloric region are rather different from those in the fundus and body of the stomach in that the gastric pit reaches much deeper in the mucous membrane. In addition the glands are more branched and coiled. The secretion of the pyloric glands contains mucus but no enzymes.

As in other regions of the gut the stomach wall contains a submucous layer and muscularis mucosa. From the muscularis mucosa strands of fibers extend towards the surface of the mucous membrane between the glands, and the contraction of these fibers probably assists in the emptying of the glands. The external muscle coat of the stomach consists of an outer longitudinal, a middle circular and an inner oblique layer. In the region of the pylorus and middle circular layer forms a thick sphincter, the pyloric sphincter. Externally the stomach is limited by a serous coat.

Secretion of the stomach

The secretion of the stomach, sometimes called gastric juice, contains free hydrochloric acid, mucus and the enzymes pepsin and rennin. In addition there may be substances which have been regurgitated back through the pylorus from the small intestine, including bile and some of the intestinal enzymes. Gastric juice also contains the intrinsic factor of Castle which is necessary for the absorption of vitamin B_{12}. Large volumes of gastric juice are produced each day and in man this amounts to 2-3 litres.

Hydrochloric acid

Hydrochloric acid has its origin in the parietal cells of the gastric glands. The cytoplasm of these cells has been shown to be neutral and the hydrochloric acid is presumably formed within the intracellular canaliculus. These cells contain the enzyme carbonic anhydrase which catalyses the formation of carbonic acid from carbon dioxide and water.

The hydrogen ions of hydrochoric acid are derived from dissociated carbonic acid, the chloride ions from sodium chloride.

The hydrochloric acid released from the parietal cells performs two main functions. Firstly it forms the acid medium in which the enzyme pepsin works under optimum conditions. The pepsin is secreted from the chief or zymogen cells in an inactive form pepsinogen, which is converted into the active form pepsin under the influence of the hydrochloric acid. In addition the acid medium of the stomach prevents the growth and multiplication of bacteria and it kills most of the organisms ingested with the food.

Control of gastric secretion

Three phases of gastric secretion are described, the nervous, gastric and intestinal phases. The first stimulus to the production of gastric juice is a nervous one. When we see, smell or taste food, impulses pass along the vagus nerve to the stomach stimulating the production of gastric juice, and this prepares the stomach to receive the food. This was first demonstrated by Pavlov in dogs. He operated on dogs in order to bring the oesophagus out to open in the neck so that food could not pass directly into the stomach. In the same dogs he prepared pouches of stomach which were brought to the surface of the body so that specimens of gastric juice could be easily obtained. When these animals ate, the food passed out from the opening in the neck but gastric juice was still produced from the stomach. When the vagus nerves to the stomach were cut the production of gastric juice ceased.

The next phase of gastric secretion is called the gastric phase and depends upon the fact that the contact of food with the gastric mucosa stimulates the production of gastric juice. This is not a direct effect but the food causes the mucosa of the pyloric region to produce a hormone called gastrin which circulates in the blood back to the rest of the stomach which responds by the production of gastric juice. The gastric juice produced in response to gastrin appears to be predominantly acid and is not rich in pepsin.

When semi-digested food reaches the small intestine there is a further reflex production of gastric juice, perhaps based upon the production of another hormone from the mucosa of the duodenum.

Digestion in the stomach

There are only two known enzymes produced in the stomach, pepsin and rennin. Pepsin is secreted in the inactive form pepsinogen and activated by hydrochloric acid. Once pepsin is formed it is capable of activating more pepsinogen. Pepsin belongs to a group of protein digesting enzymes called endopeptidases (which include the enzymes trypsin and chymotrypsin produced by the pancreas). These enzymes by acting upon the peptide bonds within the molecules of protein break up the large protein molecules into smaller fragments. These smaller fragments are then further broken down by a group of enzymes, produced by the small intestine, called exopeptidases. These exopeptidases act by breaking down terminal peptide bonds to liberate free amino acids. The differences between pepsin, trypsin and chymotrypsin lie in the fact that each enzyme is capable of breaking down only certain peptide bonds, depending upon the chemistry of the protein molecule adjacent to the peptide bond.

In the stomach then, protein digestion is begun by the action of pepsin, which, by acting as an endo-peptidase, breaks down the molecules of protein into smaller fragments. These smaller fragments, sometimes called polypeptides, are broken down further when they pass into the duodenum and small intestine and meet other enzymes.

There is no fat digesting enzyme produced in the stomach but the condition of the fat is altered by the warmth and churning action of the stomach. Further, globules of fat are liberated from animal tissues when these become softened and partially digested by the enzyme pepsin.

There is no enzyme produced which is capable of digesting carbohydrates, but the action of the enzyme ptyalin, which is mixed with the food, continues until the pH of the gastric juice falls sufficiently to inactivate this enzyme. Some disaccharides e.g. sucrose, are hydrolysed by the presence of dilute hydrochloric acid into monosaccharides. The special features of the stomach of some herbivores for the digestion of carbohydrates is described.

Rennin is another proteolytic enzyme and is characteristically found in the gastric juice of young mammals. It is secreted in an inactive form, pro-rennin, which is activated by the hydrochloric acid of the gastric juice. Rennin catalyses the conversion of the protein of milk, caseinogen, into paracasein which is precipitated inn the stomach as a calcium salt. The precipitated paracasein forms a firm curd in the stomach. This process ensures that milk stays for some time in the stomach so that it becomes exposed to the action of the proteolytic enzymes.

Stomach movements and emptying

The contents of the stomach are mixed by waves of peristaltic contraction which pass along the stomach from cardia to pylorus. The pyloric sphincter is open for most of the time but contracts as a wave of peristalsis reaches it. Food is free to leave the stomach for the duodenum as soon as the consistency is sufficiently fluid and as soon as the peristaltic waves are strong enough to force the chime out through the bottle neck which the pylorus represents. The rate at which the stomach empties depends upon the type of food contained. Naturally fluids tend to pass out quicker than solid foods. Fats have a characteristic effect in slowing the rate of stomach emptying and this explains why hunger does not return so quickly after a meal containing a good proportion of fat as after a light carbohydrate meal. There is also a possibility that a hormone, called enterogastrone, produced by the duodenum exerts controlling influence on gastric motility and emptying.

Small Intestine

In the stomach the food is churned up into a paste of mucus and enzyme known as *chime*, ready to be passed in a liquid condition into the duodenum.

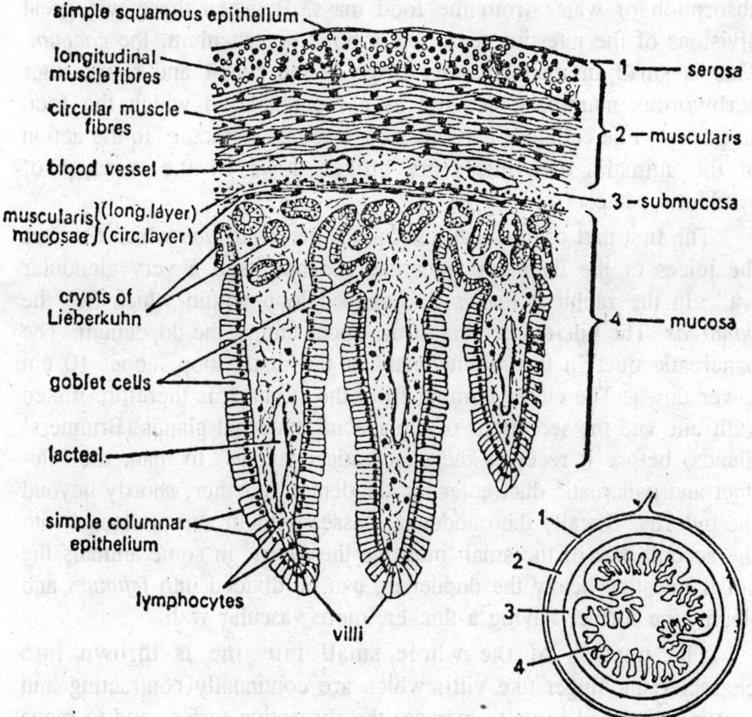

Fig. 11.10. T.S. of doudenum.

A sharp change takes place in the process of digestion as food passes the pyloric sphincter and enters the intestine. This corresponds to the point at which the mid-gut proper began in the lowest vertebrates, and here the breaking up of all the various types of food-stuff proceeds actively. The stomach, for all its importance, can be considered as a bag to contain the food that has been taken in. Until the food passes the pylorus, the only important enzymatic actions upon it are those of the ptyalin and pepsin, and little absorption has taken place. Beyond the pylorus the food is mixed with enzymes responsible for breaking down all the chief classes of foodstuff and absorption of the products into the blood-stream begins.

The gut distal to this point is essentially a tube with the basic plan already described, namely a mucous lining, connective tissue submucosa, two layers of muscles, and an outer covering of peritoneum. The intestine may be divided into two main sections: the small intestine, in which the breakdown of foodstuffs is completed and absorption takes place, and the large intestine, which is a region concerned with the absorption of water from the food mass. Between these two great divisions of the intestine there lies a blind diverticulum, the *caecum*. This is small in man but very large in the rabbit and some other herbivorous mammals, serving as a receptacle in which the food stagnates while cellulose and other constituents refractory to the action of the animal's own juices are broken down by the enzymes of symbiotic bacteria.

The first part of the small intestine is the *duodenum* which receives the juices of the liver and pancreas and has itself a very glandular wall. In the rabbit it makes a U-shaped loop within which lies the *pancreas*. The bile-duct enters at the upper end of the duodenum. The pancreatic duct in the rabbit enters at the distal loop, about 10 cm lower down. The chime coming from the stomach is therefore mixed with bile and the secretions of the special duodenal glands (Brunners' glands) before it receives the pancreatic enzymes. In man, the bile-duct and pancreatic duct enter the duodenum together, shortly beyond the pylorus. Distally the duodenum passes without abrupt change into the second part of the small intestine the *ileum*. In some animals the small intestine below the duodenum can be divided into *jejunum* and ileum, the former having a thicker, more vascular wall.

The mucosa of the whole small intestine is thrown into characteristic finger-like villi, which are continually contracting and re-expanding and serve to increase the absorptive surface and to move the liquid in contact with it. The surface of the extremities of the villi is covered by columnar cells, having on the face turned towards the intestinal contents a series of microvilli. These appear under the light microscope as lines running at right angles to the surface. The intake of food molecules is at least partly an active process. The outer region of the cell is rich in the enzyme phosphatase which is probably part of a system for making the necessary energy available. Mucus-secreting cells are also present on the villi and are known from their form as *goblet cells*.

The depressions between the villi are known as the *crypts of Lieberkuhn*, and they are lined by columnar and goblet cells, together with other cells also apparently secretory, the *argentaffin* cells, rich in serotonin, and the *Paneth cells*, which are probably responsible for the production of some intestinal enzymes. The cells at the bases of the crypts show numerous mitoses and are continually replacing the lining of the intestine, the epithelial cells being continually shed. The entire epithelial lining of the human duodenum is renewed every 2 days. In the duodenum there are further numerous *pyloric* or *Brunner's* glands, lying deeper within the submucosa and having a structure and section similar to those of the glands at the pyloric end of the stomach. This secretion is alkaline and contains mucus but little enzyme. The mucous membrane of the intestine obviously requires protection against foreign invaders and is richly supplied with lymphoid tissues. In the ileum this forms large masses, the *Peyer's patches*. Lymphocytes are frequently seen migrating through the intestinal epithelium and apparently entering the lumen.

As the chime enters the duodenum it is mixed with bile and the *succus entericus* from the intestinal wall and Brunner's glands. These render the chyme less acid and provide powerful enzymes, so that the whole becomes a copious liquid mass of rapidly changing composition. In spite of the alkaline juices added to it the reaction of the duodenal contents is usually acid, about pH 5.5 in man. The pancreatic enzymes therefore do not work at their optimum, which is alkaline. The lower regions of the small intestine are also acid, as a result of the presence of organic acids produced by bacteria. The contents of the large intestine are usually slightly acid or neutral.

Pancreas

The *pancreas*, occupying the space between the limbs of the duodenum, is a branching set of tubes ending in blind secretory sacs or *acini*. The cells of these secrete the pancreatic enzymes. They contain much basophil material in the cytoplasm, which the electron

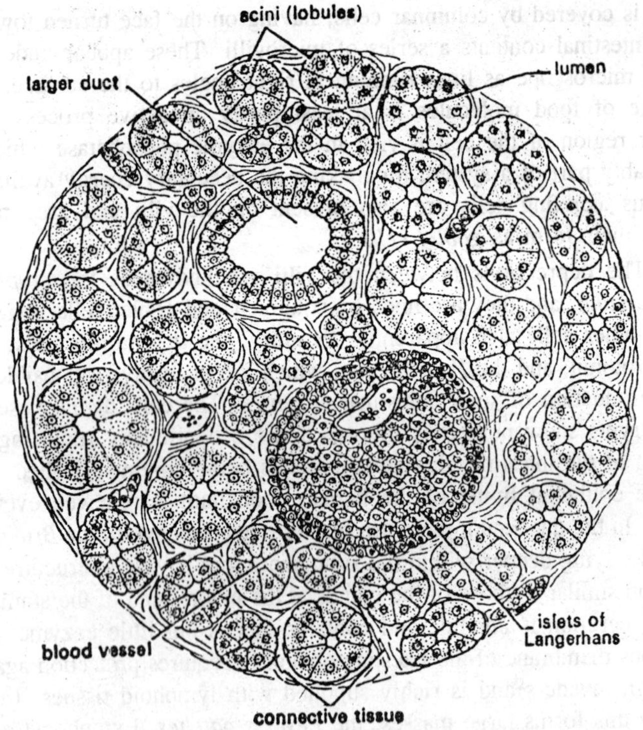

Fig. 11.11. T.S. of pancreas.

microscope shows as rough endoplasmic reticulum. This becomes labeled within 5 minutes of an injection of radioactive leucine. Later the activity appears in the Golgi region (smooth reticulum) lying more peripherally, and in 12 minutes is found in the granules of zymogen, which contain the precursors of the pancreatic enzymes and are discharged into the ducts. Among the alveoli lie a number of masses of cells, the *islets of Langerhans*, constituting an endocrine organ that is responsible for secreting into the blood the hormone *insulin*, whose presence is necessary to allow the tissue to utilize glucose.

Control of secretion of the pancreatic juice is both hormonal and nervous. In 1902 Bayliss and Starling showed that the presence of food in the duodenum excites the production of the hormone *secretin*, which circulates to the pancreas and excites its secretion. This was the first clearly demonstrated case of chemical action at a distance in the body and its discovery led to the introduction of the word 'hormone' and to the development of the concept of endocrine secretion. Secretin, like gastrin, is a polypeptide, with 28 amino-acid residues. Stimulation of the vagus also excites secretion of pancreatic juice in small amounts but especially viscous and rich in enzymes.

Digestion in the intestine

The pancreatic juice has a wide variety of actions, as its name implies. It is an alkaline liquid, containing the protease trypsin, also amylase, lipase, maltase, and the clotting enzyme rennin. The *trypsin* is secreted in an inactive form *trypsinogen*, which is readily activated in the intestine by an enzyme *enterokinase*, which removes a hexapeptide. Acting in alkaline solution trypsin splits proteins or polypeptides in the middle, being like pepsin an *endopeptidase*. The chymotrypsins are other endopeptidases, acting specifically on certain bonds. These actions produce fractions that contain two amino acids (dipeptides) or at most a few amino acids. These are then acted upon by a further set of enzymes the *exopeptidases*, which split off single amino acids, either that next to the COOH-(*carboxypeptidases*) or the NH_2-group (*aminopeptidases*). The cells shed by the walls of the intestine also contribute enzymes (cathepsins). The products of protein breakdown are absorbed by the intestinal cells, either as amino acids or small peptides, the latter being further broken down there by the contained peptidases, since only amino acids appear in the bloodstream. Absorption and transfer are active processes in the course of which amino acids are moved against concentration gradients. The *pancreatic amylase* is more powerful than the amylase in the saliva and rapidly breaks down either boiled or unboiled starch. The digestion of starch and other carbohydrates is completed by enzymes of the succus entericus, such as *mltase* and *sucrase* (= invertase); the end-products that are chiefly absorbed are mono-saccharides such as glucose.

For the digestion of fats, the pancreatic lipase is the most important agent; it splits them into glycerol and fatty acids, which are absorbed. Owing to the high surface tension, fat suspended in water tends to form large globules which would not be easily acted upon by the enzymes. The breaking up of these to make an emulsion of fine globules is largely the work of certain substances in the bile secreted by the liver, such as sodium glycocholate, whose parent acid is formed from the amino acid glycine, and cholic acid. These *bile salts* have great powers of reducing surface tension and thus of emulsifying the fats. In some forms of jaundice the bile-duct is blocked by infection or the liver otherwise disturbed; fat is not emulsified, and foods containing it therefore remain undigested.

The fatty acids absorbed mostly do not reach the blood-stream directly but pass either through or between the intestinal cells into the lymph spaces lying behind. In the cells fats are resynthesized and discharged into special lymph vessels, the *lacteals*, hence ultimately reaching the thoracic duct. If an animal is dissected soon after taking a fatty meal the mesentery will be seen crowed with rows of these beautifully white lacteal channels.

The secretions of the duodenal wall, pancreas, and liver therefore provide a number of enzymes able to split many types of ingested matter into products useful for the body. In addition to the enzymes mentioned there are also others in the succus entericus able to act on nucleotides and other materials, making them available for the body.

The food is passed from the stomach into the duodenum as fast as it can be acted upon in the latter. The pylorus is a thick ring of circular muscle that relaxes each time that a peristaltic wave arrives from the stomach, provided that the contents of the duodenum are not too acid. Probably there are receptors in the duodenum that respond to the presence of the acid, and their discharge reflexly inhibits the opening of the pylorus, thus ensuring sufficient neutralization of the stomach acid.

Food is passed along the intestine by waves of peristalsis, and it is mixed, shaken, and churned by the seginenting and pendular movements. During passage along the duodenum the processes of breakdown are completed and most of the absorption takes place in the jejunum and ileum.

Liver

The provide for the absorption of the food products, the intestine has a large blood supply from the branches of the *superior mesenteric artery*, which leaves the aorta close behind the celiac artery and sends many branches spreading out through the mesentery. From the intestine the blood does not pass directly back to the heart but to the fiver, by the *hepatic portal vein*. This enables the liver to perform those of its functions that are concerned with conversion and storage of the food taken in. Perhaps at an early stage of evolution the liver may have served directly as an organ for breaking down food. In *Amphioxus* the liver is represented by a hollow diverticulum, which secretes digestive enzymes. This strictly digestive function of the liver is still represented in mammals by the action of the bile in emulsifying fats. But the main action of the liver is to convert the foodstuffs into whatever products are needed by the body.

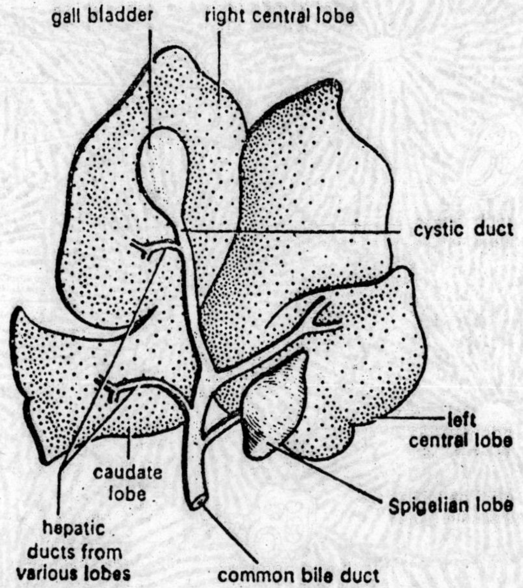

Fig. 11.12. Liver lobes and gall bladder.

The liver develops as an outgrowth of the gut, but comes to form a compact mass of gland cells, *hepatocytes*. These are usually described as being arranged as plates, mostly one cell thick and organized into lobules. Blood flows in sinusoids between the plates from branches of the portal vein around the outside (hence *interlobular veins*) in towards the *intralobular* veins at the centre, which are branches of the hepatic vein. The branches of the hepatic artery also run between the lobules and discharged into the sinusoids. The external secretion of the liver is discharged into minute *bile canaliculi* lying within the plates between the live cells. These drain towards the outside of the lobule into ductules, which unite to form the bile duct. Thus blood and bile run in opposite directions in the lobule. A complication is that there are holes in the plates, so that the liver sinusoids anastomose. Moreover, the vascular capillaries have a discontinuous endothelial lining, so that the blood plasma comes into direct contact with the cells. There is debate as to how this arrangement is arrived at from the original liver diverticulum.

The liver is an organ with many functions, which may be grouped under five headings: (a) those concerned with metabolism and storage of the food; (b) production of the bile; (c) activities concerned with the composition of the blood (such as production of plasma proteins

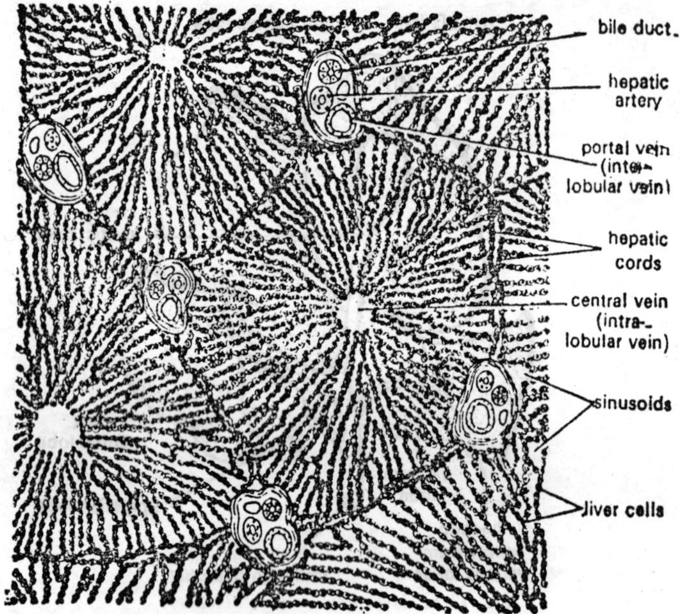

Fig. 11.13. T.S. of liver.

including those needed for clotting), regulation of blood sugar and other substances; (d) protective activities and detoxification; (e) formation and destruction of red cells. Under several of these headings we could recognize subdivisions, and it is clear that the liver occupies a very important place in vertebrate life; its cells, primarily digestive in function, have come to constitute a chemical workshop for the body. It has been said that the liver has over 100 functions but only two types of cell, hepatocytes and macrophages. However, the former, although they all look alike, probably include many different clones, each specialized to produce certain enzymes.

With all these activities the turnover of proteins is very rapid in the liver. The cells themselves last for about five months (in a rat) but the half-life of total liver protein is only 2-4 days. The plasma proteins produced by the liver turn over with a half-life in man of about 20 days. The level is very precisely regulated; greater protein degradation (outside the liver) is compensated by increased synthesis, but the mechanism of control is not well understood.

The actions of the cells of the liver on the food are varied and depend upon the condition of the animal at the time (Rouiller 1963). Thus if amino acids are needed for building the body they are passed on in the blood-stream, but in an adult more amino acids are usually taken than are needed for building and the excess is *de-aminated* in the liver, the nitrogen being converted into urea which is passed to the kidney for removal, while the remaining parts of the molecules are available for combustion to provide energy. Similarly carbohydrate may be either passed on or converted into glycogen, in which form it is stored in the liver.

The liver plays an essential part in providing a continual supply of glucose for the support of the metabolism of the muscle, brain, and other organs. This glucose is formed by the process of *glycogenolysis*. But there is only enough glycogen in the liver to last for a few hours, and when it is exhausted the liver makes glucose from amino acids or other sources (*gluconeogenis*). The level of blood sugar is accurately regulated largely by the action of the hormones adrenaline and glucagons (from the pancreas), which promote glycogenolysis; and insulin, which increases the storage of glycogen from blood glucose. A third set of activities in the liver is concerned with fat metabolism and the liver cells are commonly filled with fatty granules. Many other substances are also dealt with by the organ, for instance vitamin A is synthesized from its carotene precursor, hence the richness of fish liver oils in this vitamin. Iron and copper are metabolized and stored in the liver and many other processes go on there. The bile is secreted continuously and stored in the *gall-bladder*, from which it is repeatedly discharged when food enters the intestine. If not used, its water is e-absorbed. The bile is alkaline in man but acid in the cat and dog. It contains, besides the bile-salts already mentioned, the *bilep-pigments*, which are derived by breakdown of the haemoglobin of the blood, and also the polyhydric alcohol *cholesterol* and the phosphatide *lecithin*; the former of these is the main constituent of gall stones. Several functions have been suggested for the bile; only those concerned with the digestion of fats are certain but the bile probably also has an action on the movement of the gut.

The detoxicating functions of the liver are perhaps connected with those that are more strictly metabolic. Many substances that have a pronounced action when introduced into the body, such as alcohol or anaesthetics, are broken down in the liver and rendered ineffective. Many of the pollutants now normal in our environment are dealt with in the liver including toxic gases, dyes, preservatives, and drugs. These are excreted only very slowly by the kidney, especially if they are lipid-soluble, but the liver cells have non-specific drughydroxylase enzymes on their endoplasmic reticulum able to oxidize lipid-soluble compounds. These require the co-operation of reducing vitamins such as NADPH, which may be absent in poorly nourished people who then show marked toxic effects of drugs tolerated by other people.

The liver also has an important function in regulating the amount of steroid hormones circulating in the blood. Thus, in man, the whole of the aldosterone is removed by a single passage through the liver.

The sinusoids of the liver contain in their walls many cells of the reticulo-endothelial system (histiocytes), here known as the *cells of Kupffer*. These are responsible for removing from the blood-stream particles of foreign matter, dead bacteria, etc. The liver is also concerned in regulating the amount and composition of the blood, destroying old red cells and, in the embryo, forming new ones, as well as producing some of the protein components of the blood, the fibrinogen and prothrombin.

Large Intestine

The large intestine is shorter than the small intestine, being about thirty-five centimeters in length; but it is wider in diameter than the small intestine. It is divisible into a caecum, ascending colon, transverse colon, descending colon, and rectum. The course and relations of these segments of the large intestine are considered as follows:

Caecum

This is a pointed but blind sac-like pouch lying just cauded of the plane of entrance of the ileum into the ascending colon. Dorsally, it lies in relation with the transverse part of the duodenum; ventrally, with the coils of the small intestine; laterally, with the abdominal wall.

Ascending colon

The ascending colon extends craniad from the caecum to the level of the third lumbar vertebra where it turns sinistrad into the right colic flexure. Dorsally, the ascending colon lies in panacreas; ventrally, with the coils of the small intestine. Craniad it approximates the caudal surface of the cystic (right lateral) lobe of the liver, where it turns into the right colic flexure.

Transverse colon

The transverse colon extends transversely sinistrad from the right colic flexure to the point where it bends into the left colic flexure. Dorsally it lies in close relation with the body of the pancreas and ventrally with the dorsal surface of the pyloric portion of the stomach and coils of the small intestine.

Descending colon

The descending colon courses caudad from the level of the transverse process of the third lumbar vertebra to the level at the point of junction between the seventh lumbar and first sacral vertebrae. Dorsally, it lies in close approximation with the dorsal body wall and, dorsolaterally, with the medial surface of the left kidney. Caudad of the posterior extremity of the left kidney, the descending colon swerves caudomediad to lie in the relation with the dorsal body wall and coils of the small intestine.

Rectum

The rectum extends caudad from the level of the first sacral vertebra to the anal aperture. Dorsally, the rectum lies in relation with the flexor caudae longus muscle; laterally, with the caudorectalis muscle; ventrally, with the neck of the bladder in the female and with the urethra in the male. The rectum is about five centimeters in length.

Digestive Enzymes

Enzymes are highly specific, heat labile, organic catalytic substance manufactured by living cells, but which act independently of the cells. The enzymes which thus far have been isolated in the pure state are protein in nature. A catalyst is a substance which will initiate or alter the speed of a chemical reaction without itself being permanently changed by the reaction. The majority of chemical reaction occurring in the animal organism are enzymic reactions. Such physiological activities as digestion, cellular respiration, muscular contraction, and a myrid of others are dependent on enzymic action.

The enzymes which are concerned with the digestion of foodstuffs are termed hydrolases or hydrolytic enzymes. A hydrolase is an enzymes which brings about hydrolysis, that is, the splitting of the substrate by the addition of water. This group of enzymes is further classified according to the chemical nature of the substrate upon which the enzymes act into : (a) the carbohydrases which split the peptide linkage of proteins-these include the proteinases which are capable of splitting linkages not adjacent to a terminal group and the peptidases which attach terminal peptide linkage; (c) the esterases which hydrolyze nucleic acid. Pertinent facts concerning the hydrolases involved in the process of digestion are shown in tabular form.

Saliva

The saliva is colorless, odorless, watery fluid supplied to the mouth by the salivary glands. Ordinarily it is elaborated at a rate sufficient to keep the mouth moist, but the ingestion of food, or even the taste, the odor, the sight, or the thought of it will increase the production of saliva. The pH of human saliva usually ranges from 6.2 to 7.6, and its specific gravity varies from 1.002 to 1.008. The volume secreted per day varies anywhere from a few hundred cubic centimeters to several liters when the gland is stimulated by a drug, such as pilocarpine, or irritated by some substance, such as bichloride of mercury.

The digestive function of saliva is dependent on the presence of salivary amylase or ptyalin. This enzyme is absent in the saliva of the cat and dog and is present in the saliva of man, rabbit, rat, and mouse. The substrate for salivary amylase is starch, although glycogen is also hydrolyzed by this enzyme in omnivoral but not in herbivora. The hydrolysis of starch involves the following steps: Starch → erythrodextrin + maltose → alpha achroodextrin + maltose → beta achroodextrin + maltose → maltose. A salivary maltase which catalyzes the splitting of maltose to glucose and other enzymes are also present in traces, but their source (salivary glands, salivary bacteria, etc.) has not been definitely established. Since food remains in the mouth only for a short time, the chief site of salivary digestion is in the stomach.

Other constitutents of the saliva are water (over 99%), mucin, a little serum albumin and serum globulion, urea acid and various inorganic salts. It may also contain leucocytes, desquamated epithelial cells, and bacteria. The mucin and other proteins give the juice its lubricating quality which makes it an aid in swalloing.

The function of saliva are as follows:

(a) It has a digestive function in those animals where it contains an amylase.

(b) It moistens and lubricates and thus aids in deglutition.

(c) It dissolves soluble substances and renders them capable of being tasted.

(d) It plays a role in the regulation of the water balance of the body. The volume of saliva secreted is related to the amount of water in the body. When a sufficient amount of water is lost from the body, the secretion of salva is decreased or suppressed entirely. The dryness of the mouth and phyrynx which ensues results in the production of the sensation of thirst.

(e) It has a cleansing action on the mouth and teeth.

(f) It serves as a route for the excretion of a number of substances (glucose in pancreatic diabetes; mercury, lead, and other heavy metals are excreted in part by the salivary glands; various drugs etc.)

The secretion of saliva is under nervous control. This is necessarily the case since food remains in the mouth only for a short time and thus the need for saliva is immediate. The nervous control is a reflex one and consists of a salivary center located in the medulla oblongata, of afferent fibers to the center, and of efferent fibers from the center to the glands.

The afferent fibers take origin in the oral mucosa and are stimulated both mechanically and chemically. The salivary center also receives afferent impulses under certain conditions from other regions of the body. Irritation of the stomach or other portions of the gastrointestinal tract will set up reflex salivation. Impulses from the brain at the smell, sight, hearing, or thought of food can stimulate salivary secretion. Such salivation is called psychical secretion and involves the operation of a conditioned reflex. The secretion of saliva can also be inhibited as, for example, when one is frightened.

The efferent or motor fibers from the center to the glands arise from both the parasympathetic and sympathetic divisions of the autonomic nervous system and are true secretory fibers. The parasympathetic innervation to the submaxillary and sublingual glands is carried in the chorda tympani branch of the facial nerve. If a chorda tympani nerve in the cat is exposed and cut, and the distal end stimulated electrically, there is produced a profuse amount of a watery juice poor in solids together with vasodilatationand increased blood flow through the gland. After a few days, the glands deprived of their parasympathetic innervation exerts an inhibitory as well as an excitatory control over salivary secretion. The sympathetic innervation to the submaxillary and sublingual glands is derived from the superior cervical ganglion. Electrical stimulation of this innervation results in the production of a small volume of a t hick viscid salvia, vasoconstriction, and reduced blood flow. The vasomotor effects produced by electrical stimulation are due to the presence of vasomotor fibers in the efferent nerves to the glands.

Gastric Juice

The gastric juice is secreted by the gastric glands which are buried in the gastric mucosa. It is poured into the lumen of the stomach through the gastric crypts. The glands are of three different varieties:

(a) the cardiac glands which extend for a short distance beyond the esophageal communication

(b) the fundic or gastric glands which occupy about two-thirds of the mucosa; and

(c) the pyloric glands which are in the pyloric portion of the stomach.

The gastric juice in man is a clear, colorless, acid, watery fluid with a specific gravity of 1.003 to 1.006. It contains approximately 99 per cent water and 1 per cent solids. The solids are hydrochloric acid, enzymes, mucin, inorganic salts, and an intrinsic factor which acts on an extrinsic factor in food to produce an antianemic principle essential for the normal maturation of the red blood cell.

The enzymes present are pepsin, rennin, and possibly a gastric lipase. Pepsin splits native protein to proteoses and peptones, free amino acids being formed in relatively small amounts in the process. The action of rennin on the caseinogen of milk yields two soluble substances, paracasein and why protein. The paracasein combines with calcium to form a curdy precipitate, called calcium paracasein, which is then attacked by pepsin. The action of gastric lipase, if present, is slight. The lipolytic activity of gastric juice may be due to hydrochloric acid.

The hydrochloric acid is formed by the parietal cells located in the bodies of the fundic glands. The concentration of free acid in the juice is 0.4 to 0.5 per cent. The acid is also present in combination with protein. Some of the functions of hydrochloric acid are as follows:

(a) it provides a suitable hydrogen ion concentration for the action of pepsin;
(b) it activates pepsinogen (converts it to pepsin);
(c) it hydrolyzes some sugar and fat and forms acid metaprotein;
(d) it acts on prosecretin to convert it to secretion and the latter calls forth pancreatic and bile secretion;
(e) it has a bacteriocidal function;
(f) it may digest certain forms of cellulose;
(g) it may possibly play some role in cardiac and pyloric sphincter control.

Mucin in the juice has a protective function since it prevents the stomach from digesting itself. It has a high hydrochloric acid combining power, and it is also antipeptic. An antienzyme is also present which inactivates pepsin before it attacks the gastric mucous membrane.

Experimental studies indicate that there are three phases to gastric secretion. The first phase is called the cephalic or reflex phase and it involves both conditioned and unconditioned reflexes. Gastric secretion evoked by the taste, smell, sight, or thought of food is called psychic secretion (appetite juice) and is due to conditioned reflexes. These reflexes involve the cerebral cortex and are mediated over the vagi. If the vagi are cut, no appetite juice is secreted. Experimental electrical vagal stimulation results in the production of a highly acid juice rich in pepsin. Stimulation of the mucous membrane of the mouth by food also calls forth gastric secretion. This is due to the operation of an unconditioned or inborn reflex. The second or gastric phase is elicited by the presence of food in the stomach. Evidence indicates that it is due to chemical excitation of the gastric glands by gastrin. This hormone is, in reality, histamine and is liberated by partially digested food from the gastric mucosa. A third or intestinal phase, that is, gastric secretion due to the presence of food in the duodenum, is probably caused by a hormone formed in the upper intestine.

Pancreatic Juice

The acid chime which enters the small intestine from the stomach has only been partly digested. Digestion is completed in the intestine by the action of the pancreatic and intestinal juices and the bile. The pancreatic juice enters the duodenum through the ampulla of Vater (the main duct of the pancreas becomes confluent with the common bile duct to form this ampulla) and the accessory pancreatic duct.

Pancreatic juice is a colorless, distinctly alkaline, watery fluid with a specific gravity of 1.04. In man about 500-600 cc, are secreted daily. It contains various enzymes and inorganic salts. The pancreatic enzymes include pancreatic amylase or amylopsin, sucrase, trypsin, chymotrypsin, carboxypeptidase, steapsin or pancreatic lipase, polynucleotidase, protaminase, rennin, and others.

Trypsin is a powerful which breaks down protein, proteoses, and peptones to polypeptides and liberates some amino acids. The precursor of trypsin is trypsinogen and it is activated by enterokinase of the intestinal juice. Chymotrypsin in its action on protein is like trypsin, but it has a greater milk coagulating or rennitic effect, and it is activated by trypsin. The amylase exerts an action on starch and dextrins and the end product is maltose. The action of sucrase is on sucrose or cane sugar and it splits it into fructose and glucose. Carboxypeptidase is a protease of the peptidase type which acts on polypeptides, provided their carboxyl groups are free, and converts them into simple peptides and amino acids. Steapsin acts to split fats to fatty acids and glycerol. Polynucleotidase is a nuclease which splits the nucleic acid molecule into nucleotides. Protaminase assists in the hydrolysis of protamines. A rennin is said to be present, but since trypsin has a rennitic action under suitable conditions, it is doubtful whether a pancreatic rennin exists.

The secretion of the pancreatic juice is both under nervous and chemical control. Vagal stimulation after a longlatent period gives a viscid juice rich in enzymes, and it is possible that the nerve control is for the release of enzymes. Stimulation of the splanchnics also augments secretion, but this is probably due to vasomotor action. If the pancreas is denervated, it still continues to elaborate pancreatic juice. Although the exocrine secretion is regulated in part by nervous influences, the chief control is chemical or hormonal and consists of the secretion mechanism. Acid chyme, when it enters the duodenum, extracts secretion from the mucosa. The secretion is absorbed and is carried by the blood stream to the pancreas where it excites the secretion of a large quantity of a juice rich in alkali. Hence the volume of juice secreted and its alkali (largely bicarbonate) content are no doubt determined by secretion.

Bile

Bile is a bitter, transparent, slightly alkaline, viscid, yellowish green or golden brown, vital digestive juice formed by the liver and poured into the duodenum. It is secreted continuously at low pressure, but enters the intestine only during periods of digestion. The discontinuous entry of this fluid into the gut is made possible by the existence of the gall bladder, in which bile formed during the intervals between digestive periods can be stored. A chemical analysis of bile shows it to contain water (less in gall bladder bile due to water absorption by the gall bladder mucosa), the bile salts, the bile pigments, inorganic salts, mucin (formed by the epithelium of the gall bladder and larger bile ducts), lipids (cholesterol, lecithin, and neutral fat), and traces of urea and carbon dioxide. A lipase is said to be present in the bile.

The bile salts, sodium taurocholate and sodium glycocholate, are manufactured solely by the liver cells, and the most important physiologic functions of the bile are due to their presence. These salts undergo a circulation in the body. They are in part reabsorbed from the intestine and make their way back via the portal blood to the liver where they exert a cholagogue effect (augment the secretion of bile) and are then re-excreted. Some bile salts are destroyed by bacterial action in the intestine and some are lost from the body by way of the feces. An important role in fat digestion and fat absorption is played by these substances.

The bile pigments are bilirubin and biliverdin. Bilirubin comes from the breakdown of hemoglobin. This disintegration is carried out by the cells of the reticulo-endothelial system, especially of the liver, spleen, and red bone marrow. Biliverdin is formed by the oxidation of bilirubin. A reduction product of bilirubin due to bacterial action in the intestine is called stercobilogen or urobilogen. It imparts the yellow brown color to the feces.

Some of the functions of the bile are as follows:

(a) It is of great importance in the absorption of fat. In biliary obstruction much of the fat in the feces is fatty acid which means that the fat has undergone digestion. It is probable that fatty acids are absorbed after forming complexes with the bile salts.

(b) The bile salts, by lowering surface tension, aid in the emulsification of fats. By this process the fat is broken up into small particles so that it presents a greater surface area for the action of pancreatic lipase.

(c) The bile serves as a channel for the excretion of the bile pigments, cholesterol, various inorganic salts, toxins, drugs, and dyes.

(d) The alkalinity of the bile is a factor in neutralizing the acidity of the gastric juice.

(e) The bile salts hold cholesterol in solution.

(f) The bile salts activate pancreatic lipase.

(g) The bile salts are necessary for the absorption of vitamin K and the fat-soluble vitamins.

(h) The bile stimulates peristalsis and thus has a laxative effect. (i) The bile probably assists in the liberation of secretion from the intestinal mucosa.

(j) The bile precipitates metaproteins and proteoses and retards their passage through the intestine.

The secretion of bile is to a small extent subject to nervous influences. The vagus contains both excitatory and inhibitory fibers, and, hence when it is stimulated, the effects of the two types of fibers almost neutralize each other; but the inhibitory effects is usually slightly stronger and there is some diminution in the bile flow. If the vagus is cut high in the neck, the inhibitory fibers degenerate more rapidly than the excitatory fibers, and, then if the nerve is stimulated, there is an acceleration of secretion. Bile formation is also under chemical control. It is augmented by the action of secretion and by a number of digestive products, but the bile salts undoubtedly constitute the chief cholagogues.

INTESTINAL JUICE

The juice of the small intestine consists of the secretion of the glands of Brunner and the intestinal juice proper, or success entericus, which is formed by the crypts of Lieberkuhn. The glands of Brunner are found for the most part in the submucosa of the duodenum in the vicinity of the pyloric sphincter. They secrete a viscid, alkaline fluid containing a proteolytic enzyme which closely resembles pepsin. The enzyme is activated by hydrochloric acid.

The intestinal juice proper is a yellowish, alkaline liquid with a specific gravity of approximately 1.010. It contains a variety of enzymes, inorganic salts (chiefly sodium bicarbonate and sodium chloride), a trace of carbon dioxide, bacteria, desquamated epithehal cells, migrated leucocyte and clumps of mucus. In man about 3000 cc. are secreted daily. The number of enzymes actually present in the juice produced under normal conditions is not definitely known because the analyses which have been made on centrifuged fresh juice, on turbid cell-rich juice, and on bacterially contaminated scrapings of the intestinal mucosa have not yielded entirely satisfactory results.

The chief enzymes of the intestinal juice proper are erepsin (aminopeptidase and dipeptidase), enterokinase, lactase, maltase, sucrase, intestinal lipase, and amylase. Aminopeptidase acts on polypeptides, provided their amino groups are free. Dipeptidase splits dipeptides into amino acids, Enterokinase activates pancreatic trypsinogen. Lactase, maltase, and sucrase act upon the corresponding substrates to split them into monosaccharides.

One of the stimuli of prime importance to produce secretion of the intestinal juice proper is mechanical stimulation of the intestinal mucosa by the normal intestinal contents. The effect is local and operates through local reflexes involving the intrinsic nerve plexuses in the intestinal wall. Any regulatory effect exerted on secretion by extrinsic nerves is probably of an inhibitory nature as denervation results in increased flow (paralytic secretion). The latter effect may be due to vasodilatation. There is also evidence for a chemical control. Enterocrinin, a substance which has been isolated from the intestinal mucosa, is believed to be a specific secretory hormone which increases the quantity and denzyme content of the juice. Pancreatic juice, with trysin as the active agent, also seems to augment secretion. It is questionable whether the hormone secretion has a similar effect.

Movements of the Alimentary Canal

Mastication

The process of mastication, or chewing, occurs in the mouth and consists in breaking up food into small fragments and mixing it with saliva so as to form a smiliquid mass. The cheeks and the tongue serve to keep the food between the teeth. Contraction of the muscles of mastication which move the lower jaw on the upper bring the two rows of teeth together so that they can perform their function of tearing and crushing the food. The temporo-mandibular articulation articulation permits, not only up and down movements of the lower jaw, but also side-to-side movements. The muscles of mastication are skeletal muscles under voluntary control, but the act of chewing is usually carried out as a reflex, the stimulus being the presence of food in the mouth. A chewing center exists in the medulla oblongata to coordinate the actions of the masticatory muscles.

Mastication subserves a number of functions. The breaking up of food into small particles facilitates its being swallowed and makes it more accessible to the action of the digestive fluids. The stimulation of the taste buds and of the olfactory hair cells (by odors liberated) which occurs during mastication are factors affecting salivary and gastric secretion.

Deglutition

For purposes of description, the act of deglutition, or swallowing, is usually divided into three stages. The first stage, the passage of food from the mouth into the pharynx, is voluntary in nature. The muscles of the tongue roll the bolus, or mass of masticated food, to the back of the mouth cavity. The mylohyoid muscles then press the tongue against the hard palate and carry its base backward, thus projecting the bolus into the pharynx. In the latter action they are assisted by other muscles, such as the hyoglossi and styloglossi.

The second stage of deglutition, the passage of food from the pharynx into the esophagus, is involuntary. When the bolus of food comes in contact with the pharyngeal wall, a series of reflex contractions occur. The constrictor muscles of the pharynx contract to raise the intrapharyngeal pressure and at the same time the contraction of various other muscles serves to close off all the passages which communicate with the pharynx except the esophagus. The isthmus faucium is closed by the base of the tongue. The entrance to the nasopharynx is closed by the base of the tongue. The entrance to the nasopharynx is closed by the soft palate. The contraction of the thyrohyoid muscles raises the larynx under cover of the epiglottis and root of the tongue to close off the passageway to the lungs. A swallowing center exists in the medulla oblongata to coordinate the actions of all the muscles involved in this complex process of swallowing.

The third stage of deglutition, the passage of food from the esophagus to the stomach, is also involuntary. It is accomplished by peristaltic waves. Peristalsis is defined as a wave of muscular contraction above the bolus associated with relaxation below it. Recent observations, however, indicate that muscular relaxation does not occur below the bolus, but that the longitudinal musculature contracts to enlarge the lumen of the tube. Peristaltic contraction of the upper two-thirds of the esophagus, where the musculature is, for the most part, of the skeletal type, is dependent on extrinsic nerves. The effective innervation is para-sympathetic. In the lower one-third where the musculature is of the smooth variety, the initiation and transmission of the peristaltic wave are dependent on intrinsic or intramural nerve plexuses, but can be influenced by the stimulation of extrinsic nerves.

Movements of the Stomach

The stomach is divided into a fundus, a body, and a pyloric portion. The fundus and the cardiac half of the body constitute the cardiac pouch or cardiac one-half of this organ. The pyloric one-half includes the pyloric one-half of the body and the pyloric portion of the stomach. The latter is composed of the pyloric vestibule and the pyloric antrum.

The cardiac pouch is relatively free from peristaltic waves, but is the seat of tonic contraction. It exerts a steady pressure on food and forces it slowly into the active portion of the stomach. The pyloric one-half of the stomach exhibits peristaltic activity as well as cyclic changes in tone. Peristaltic waves commence near the middle of the body and pass rapidly toward the pyloric aperture, or pylorus. At the same time there are alternate phases of increased and decreased tone, and the gastric activity from one phase of decreased tone to the next is known as a gastric cycle. It is divided into gastric systole and gastric diastole.

The stomach, like the heart, is an automatic organ. The automaticity is dependent on an intramural nervous mechanism. The extrinsic innervation is regulatory. Vagal stimulation usually augments and splanchnic stimulation generally inhibits gastric tone and motility.

At either end of the stomach is a ring of smooth muscle known as a sphincter muscle or valve. The one about the cardiac orifice is called the cardiac sphincter, or cardia, and the one about the pyloric orifice is called the pyloric sphincter. The cardiac sphincter relaxes on the approach of an oncoming esophageal peristaltic wave. Otherwise it is tonically contracted except when the stomach is empty, at which time it is probably relaxed. The pyloric sphincter is normally relaxed or exhibiting low tone when the stomach is empty, and the pressures in the pyloric antrum and the first part of the duodenum (the duodenal bulb or cap) are equal. With food in the stomach, a strong peristaltic wave passing over the pyloric antrum raises the pressure of the intra-antral contents above that in the first part of the duodenum and some food then passes through the pyloric aperture until the peristaltic wave reaches the sphincter, which then closes. Emptying of the stomach would then cease until the chime was moved out of the duodenal bulb (first part of duodenum) and the intragastric pressure increased over that in the first part of the duodenum.

Various reflex and hormonal mechanisms of an inhibitory nature exist to retard empting of the stomach. Excessive filling of the duodenum or high acidity of its contents can serve as a stimulus for reflex (enterogastric reflex) inhibition of gastric motility and closure of the pyloric valve. The chief hormonal factor acting to regulate gastric motility is a substance called enterogastrone. It is produced by the duodenal mucosa and liberated when the mucosa is in contact with fat. It is transported by the blood to the stomach and inhibits the peristaltic activity of this organ. The purpose of this mechanism is to prevent a too rapid filling of the duodenum with fat which is digested more slowly than other foodstuffs.

Vomiting

The act of vomiting, or emesis, consists in the forcible evacuation of the gastric contents through the mouth. It is brought about chiefly by contraction of the diaphragm and the abdominal muscles which press upon the stomach and squeeze its contents out, the stomach itself being almost passive in the process. There also occurs relaxation of the cardiac sphincter and esòphagus, elevation of the larynx and soft palate, closure of the glottis, and inhibition of respiration. The actions of the muscles employed in vomiting are coordinated by the vomiting center in the medulla oblongata.

The vomiting center may be excited by impulses arising in the stomach (e.g., by peripheral emetics which irritate the gastric mucosa, such as tartar emetic, mustard, concentrated saline solution, etc.), in the other internal organs (e.g., uterus, intestines, heart, kidney, bladder, gall bladder, etc.), in the brain, or in almost any part of the body (e.g., the vestibular mechanism). The intravenous injection of such a drug as apomorphine (a central emetic) causes vomiting through its direct action on the center.

Movements of the Small Intestine

Three main types of muscular movements occur in the small intestine. They are peristalsis, rhythmic segmentation, and pendular movements. Antiperistalsis and contraction of the musculature in the intestinal mucosa also occur.

Peristaltic movements have already been described as occurring in the esophagus and the stomach. As in these organs, the same movement in the small bowel appears as a band of constriction which travels aborally and serves to propel the intestinal contests onward. The peristaltic movements are of two varieties: peristalsis and peristaltic rush. The first type of movement is a slow wave which moves food for a short distance along the intestine. The second type occurs at long intervals, has a greater rate of travel than the first, and moves food for a considerable distance. It often of is set off by the entrance of food into the stomach. Antiperistalsis may occur normally in the small intestine and block an oncoming peristaltic rush apparently to prevent a too rapid passage of the contents of the small intestine into the large intestine. Peristalsis is dependent on intrinsic nerve plexuses. The extrinsic innervation of the small intestine is regulatory. Vagal stimulation increases, and sympathetic stimulation inhibits peristaltic movements.

The rhythmic segmentation a length of bowel becomes divided by a series of regularly spaced, ring-like constrictions into a series of segments. The constrictions are due to contraction of the circular muscle. When the annular constrictions disappear, new ones appear in the middle of the former segments. The process is repeated in man at the rate of 10 to 30 times per minute. Segmentation is myogenic in origin. It serves to admix the intestinal contents with the digestive juices, facilitates absorption, and helps to squeeze blood and lymph out of the bowel.

Pendular movements are observed only at times in man. They consist of annual constrictions of the circular muscle which travel forward and backward over short lengths of intestine and cause loops of intestine to sway slightly in a pendular fashion. These movements are myogenic in origin and serve to mix food with the digestive juices which enter the small intestine.

The contraction of the smooth muscle, fibers in the villi causes them to exhibit several types of activity. They can shorten and elongate alternately; sway from side to side; or lash to and fro. The movements of the villi unquestionably aid in both digestion and absorption.

At the junction of the ileum with the colon is the ileocolic sphincter. It opens periodically to permit the passage of ileal contents into the large intestine. The valve retards the emptying of the small intestine and guards the ileum from the regurgitation of material laden with bacteria and products of putrefaction.

Movements of the Large Intestine

The main movements of the large bowel are mass or rush peristalsis and defecation. The most distinguishing movement of the colon is mass peristalsis. It occurs several times a day usually when food enters the stomach, and it serves to move the intestinal contents for a considerable distance. The type of peristalsis observed in the small intestine is rarely seen in the large intestine. Most of the musculature of the colon receives its motor innervation through the nervi erigentes. The sympathetic nerve supply is inhibitory except to the ileocolic sphincter to which it is motor.

Defecation is the ejection of feces from the rectum. The defecation reflex is initiated by the entrance of feces into the rectum as a result of mass peristalsis or overfilling of the distal colon. With the arrival of fecal material in the rectum, the intrarectal pressure rises. This acts as a stimulus for further peristaltic movements in the colon and the passage of more feces into the rectum. When the internal and external anal sphincters relax, feces pass out of the large bowel through the anus. The contraction of the diaphragm, the muscles of the abdominal wall, and the levator ani muscles increase the intra-abdominal pressure and assist in the act. The defecation reflex is governed by a medullary center located in the floor of the fourth ventricle and a subsidiary spinal center in the sacral cord.

Absorption of Food Substances from the Alimentary Canal

Most of the food substances are absorbed over the large area of the small intestine, the surface of which is further enlarged by the presence of the villi. Absorption through the stomach is very limited although alcohol is absorbed readily. In the large intestine absorption is mainly limited to water although some vitamins, produced by the bacteria, are absorbed, and in some herbivores e.g. the horse, the caecum absorbs the products of the breakdown of cellulose.

Absorption of fat

Views on the mechanism of fat absorption have tended to vaccilate during the last century. Early in physiological investigation it was found that after giving an animal a meal rich in fat, dilated white vessels (lymphatics) were seen in the mesentery of the small intestine. These were found to contain minute droplets of fat and led to the idea that much, if not all, of the fat in a meal could be absorbed directly without previous hydrolysis into glycerol and fatty acids. Bile salts were regarded as having an important role in emulsifying the fat into particles small enough to allow direct absorption. This early view was

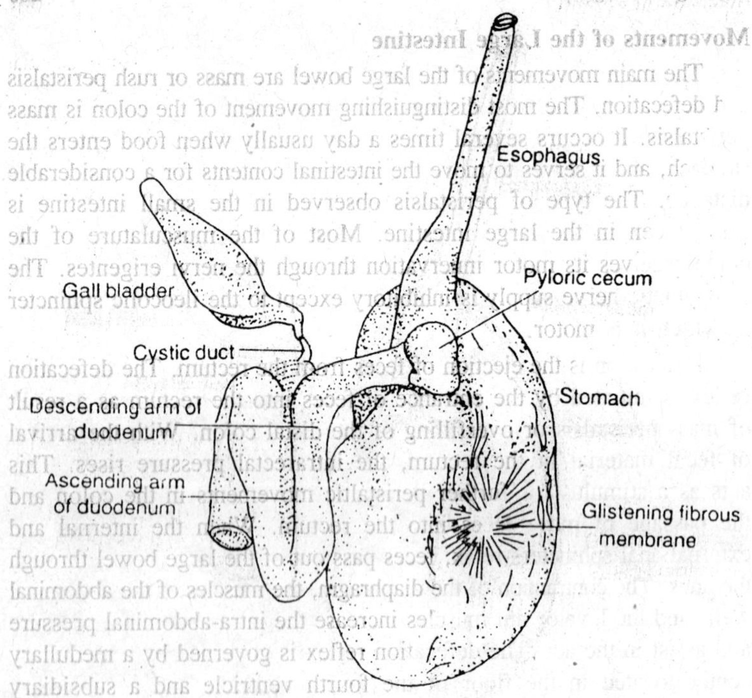

Fig. 11.14. Gizzardlike stomach and associated structures of a caiman, ventral view.

variously modified and it became accepted that although some of the dietary fat was absorbed directly in the form of minute globules, some fat was hydrolyzed into glycerol and fatty acids. The fat globules were regarded as being absorbed into the lacteals and lymphatic system, the glycerol and fatty acids being absorbed and transported to the liver by way of the hepatic portal vein. Bile salts in addition to their role in emulsifying fat also assisted the absorption of fatty acids by combining with them to render them water soluble so that they could cross the cell membranes of the intestinal cells. Fat globules are broken down into small particles, called chylomicrons, by the emulsifying action of the bile salts. After traversing the mucosa of the intestine the bile salts are freed from the fat and returned to the liver by way of the hepatic portal vein while the neutral fat is transported away from the intestine in the lymph. Some of the fat is hydrolyzed by the action of lipase into fatty acids and glycerol. The fatty acids, combined with bile salts to render them water soluble and the glycerol (o) are absorbed, freed from the bile salts and converted into phospholipids in the wall of the intestine and transported to the liver in the blood by way of the hepatic portal vein.

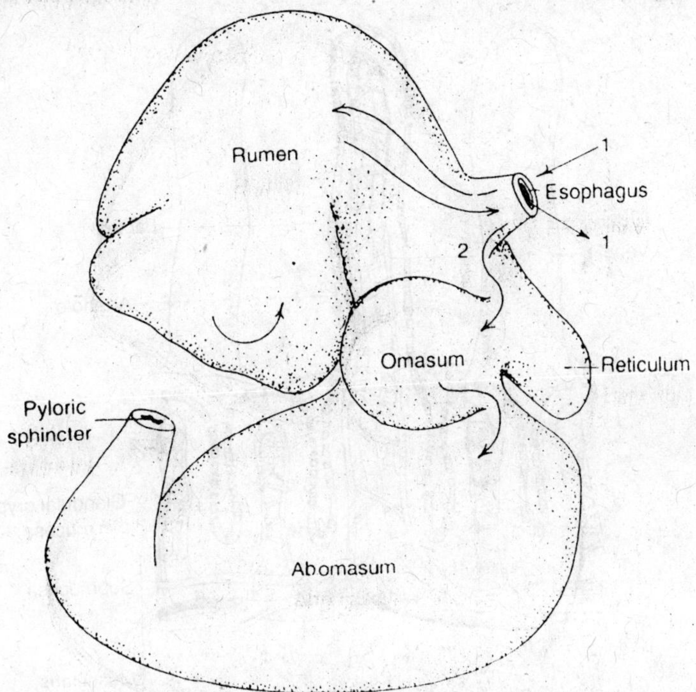

Fig. 11.15. Stomach of a ruminant (calf). Arrows indicate, 1–route of food when initially swallowed; 2–route after cud is swallowed. Cud goes directly to the omasum.

Recent work using fat specially prepared with glycerol 'labelled' with a radioactive isotope, and the fatty acids of a non-metabolizable nature (so that the fate of the fat can be readily followed) has cast some doubt on the above proposed mechanism. It seems now that all of the dietary fat is hydrolyzed to some degree, to glycerol, fatty acids and monoglycerides. The fat globules which appear in the lacteals seems not to be fat absorbed as such from the intestinal contents but new fat synthesized in the gut wall from the products of fat hydrolysis.

Absorption of glucose

During the absorption of glucose into the mucosa the glucose is probably phosphorylated. The absorption proceeds at a quicker rate than can be accounted for on the grounds of simple diffusion and substances which poison the energy-producing systems within the cells prevent the selective absorption of glucose. Glucose can be absorbed from weak solutions in the gut, another point of evidence that the absorption is not a simple diffusion along a diffusion gradient. The

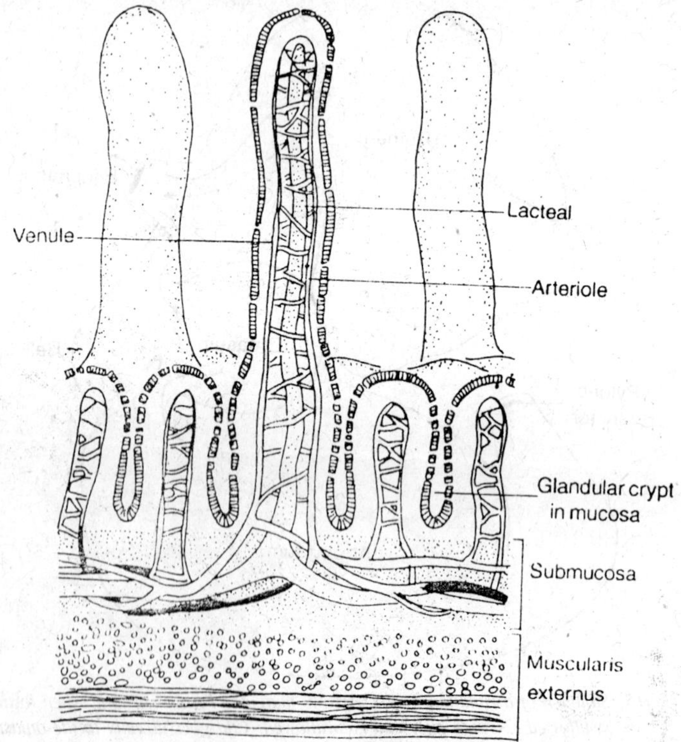

Fig. 11.16. Blood supply and lymph drainage of a villus, longitudinal view.

glucose passes from the mucosal cells into the capillaries draining into the hepatic portal vein which passes to the liver.

Absorption of proteins

As already described on proteins are not absorbed as such but in the form of their constituent amino acids. The absorption of amino acids, like that of glucose, involves active transport mechanisms. There are probably distinct active transport mechanisms for individual amino acids or for small groups of amino acids. These pass into the tributaries of the hepatic portal vein and thence to the liver.

Movement of Small Intestine

It is essential that the chime entering the duodenum from the stomach be thoroughly mixed with the digestive juices of the pancreas, duodenum and small intestine and with the bile. This mixing is achieved by the movements of the small intestine. Peristaltic movements push the contents of the small bowel onwards to the large bowel. In addition

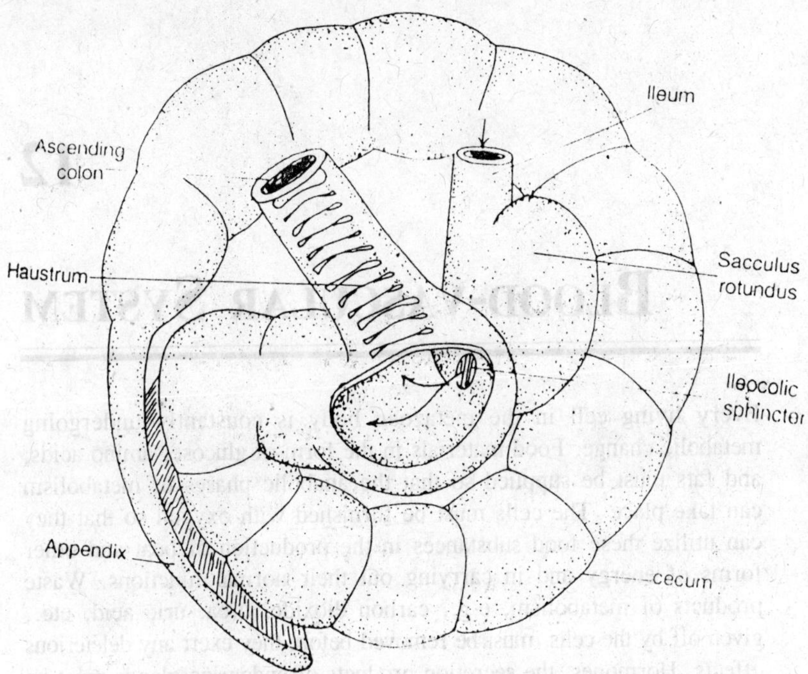

Fig. 11.17. Ileocolic junction, cecum, and vermiform appendix of rabbit. Arrow indicate available pathways.

there are segmentation movements of the small bowel brought about by the contraction of the circular muscles at intervals along the length of the bowel. These movements serve to mix the contents rather than propel it onwards. The contraction of the circular muscles at intervals give the appearance of strings of sausages to the segmenting bowel. After a time the muscles at the sites of contraction relax and the bowel then begins to segment at other sites along its length. The pattern of segmentation is altered in the active bowel about twenty times in each minute. In addition to these movements the villi of the mucosa are capable of an independent movement by virtue of slips of muscle which pass up into them from the muscularis mucosa. The movements of the villi help to ensure that the layer of fluid in contact with the intestinal epithelium is not a static one, and also helps to empty the lacterals of the villi into the lymphatic channels of the gut wall.

12

BLOOD-VASCULAR SYSTEM

Every living cell in the metazoan body is constantly undergoing metabolic change. Food materials in the form of glucose, amino acids, and fats must be supplied so that the anabolic phases of metabolism can take place. The cells must be furnished with oxygen so that they can utilize these food substances in the production of beat and other forms of energy and in carrying out their normal functions. Waste products of metabolism, e.g., carbon dioxide, urea, uric acid, etc., given off by the cells, must be removed before they exert any deleterious effects. Hormones, the secretion products of endocrine glands must be carried from one part of the body to other regions in order to be effective. Moreover, living cells must be kept moist in order to remain alive and to carry on their activities. This is necessary because every substance which enters or leaves the cells is in solution and can pass only through moist membranes or media. Cells that are exposed to air are usually dry and dead and are converted into non-living keratin, as in the corneal layer of the epidermis. It is constantly being worm off or shed and is replaced by keratinization of new cells which have proliferated from beneath.

In members of some of the lower phyla many of the above functions are carried out by individual cells which are in direct contact with the environment. However, in the evolution of higher metazoan animals, most of the cells making up their bodies have become farther and farther removed from the external environment, yet their metabolic requirements have undergone little or no change. Circulatory systems of various types have developed in the evolution of higher forms in response to demands created by the increasing complexity of their bodies. Fluids carried by the circulatory system transport substances to and from cells even in the most remote parts of the body. The fluids must circulate, for only by this means can supplied of food, oxygen, and water be replenished and harmful waste products removed.

The various cells of the body are not actually in direct contact with parts of the circulatory system and hence are not nourished directly, nor they directly supplied with oxygen. Likewise, the waste products of cellular metabolism do not enter the circulatory system directly from the cells. The fluid surrounding, or bathing, the cells is referred to as *tissue fluid* or *interstitial fluid*. It is this which serves as an intermediary in transporting nutritive materials, oxygen, and waste substances between the cells and the minute subdivisions of the circulatory system. Tissue fluid is intimately associated with various intercellular substances but its relationship to them varies in different parts of the body and also depends upon the type of intercellular substance in question. Tissue fluid is actually composed of water in which crystalloids of various kinds are in solution. It does not usually contain colloidal material. The large molecules of colloids fare generally confined to vessels of the circulatory system. Nevertheless, there normally seems to be considerable leakage of colloid into the tissue fluid probably because of pressure differences. Tissue fluid is thus much like sea water in its general composition.

The circulatory system in vertebrates is highly complex and actually made up of two systems of elaborately branched tubes which ramify throughout the entire body and thus are able to convey fluids indirectly to all living cells. They are the *blood-vascular* and *lymphatic systems*. The former is the better defined and the more obvious of the two.

Blood-vascular System

The blood-vascular system is composed of continuous tubes of various sizes, known as *blood vessels*, through which *blood*, a peculiar type of fluid connective *tissue*, is pumped to all parts of the body. This system consists of a muscular contractile organ, the heart, situated near the anterior end of the body and towards the ventral side. *Arteries* carry blood away from the heart. They subdivide into vessels of smaller caliber, the *arterioles*. These in turn branch into *capillaries*, which are extremely small vessels averaging from 0.004 to 0.012 mm in diameter and 1 mm or less in length. *Venules* collect blood from the capillaries. They combine to form *veins* which return the blood to the heart. Irrespective of the type of blood carried by the vessels, it should

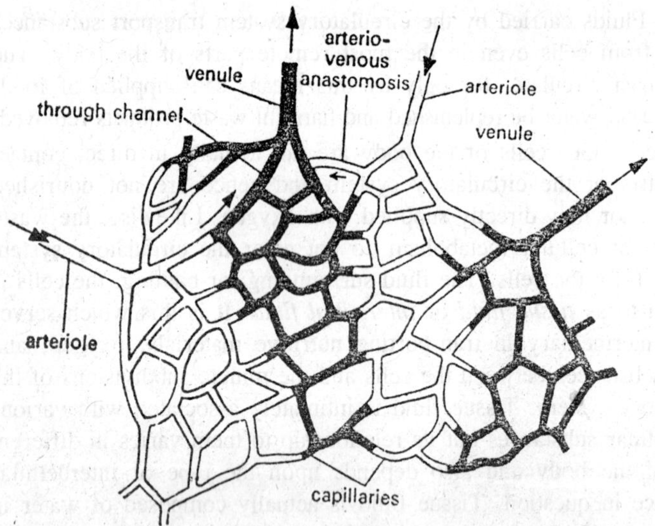

Fig. 12.1. Connections between arterioles and venules.

be clearly understood that *arteries carry blood away from the heart and veins carry it toward the heart*. The arteries and their branches make up the *arterial system*, and the veins with their tributaries form the venous system. The blood-vascular system is a closed system of blood vessels, having no direct connection with any other system of the body except the lymphatic system, with which it is closely associated.

The arterial system in higher vertebrates is separated into *pulmonary* and *systemic divisions*, the former carrying blood to the lungs, the latter carrying it to all other parts of the body. The venous system likewise has pulmonary and systemic divisions, but, in addition, two or three *portal system* may be present. A portal system is a complex of veins which begin in capillaries and end in capillaries before the blood which courses through them is returned to the heart. All vertebrates have a *hepatic portal system* in which blood collected from the digestive tract and spleen passes through capillaries (sinusoids) in the liver before rearing the heart. Adults of the lower vertebrate classes and embryos of higher forms have a *renal portal system* in addition, in which blood from the posterior part of the body passes through capillaries in the opisthonephros or mesonephros on its way to the heart. Another small but important portal system is found in association wit the blood vessels draining the hypophysis, or pituitary gland.

Lymphatic System

The lymphatic system if composed in part of delicate tubes, the *lymphatic vessels*. They have their origin in very fine *lymph capillaries* which are blind at their free ends and drain the tissue spaces. Lymphatic vessels may form connections in various parts of the body with relatively large spaces called *lymph sinuses*. Some also joint certain veins in the region of the heart, so that the connection between the blood-vascular and lymphatic systems is direct. The lymphatic system is a closed system, since the walls of even the lymphatic capillaries are composed of a delicate membrane made up of endothelial cells. They lymphatic system carries a fluid, *lymph*, the composition of which is, in general, rather similar to blood except that red corpuscles are lacking and the protein content is very much lower than that of blood plasma. In some animals, pulsating *lymph hearts* are present at points where lymphatic vessels enter certain veins. These aid in the propulsion of lymph into the blood-vascular system. In most forms, however, pulsations of adjacent blood vessels and movements of the viscera as well as of other parts of the body propel the lymph which, therefore, courses slowly through the vessels. Movement is in one direction since numerous valves are present which prevent back flow. In mammals, and to a lesser degree in birds, the lymphatic vessels are interrupted at certain places by the presence of *lymph nodes*, or *glands*. These are composed of a mesh work of reticular and collagenous connective-tissue fibers which support large masses of *lymphocytes*. Also present in lymph nodes are certain phagoeytic *reticuloendothelial cells* as well

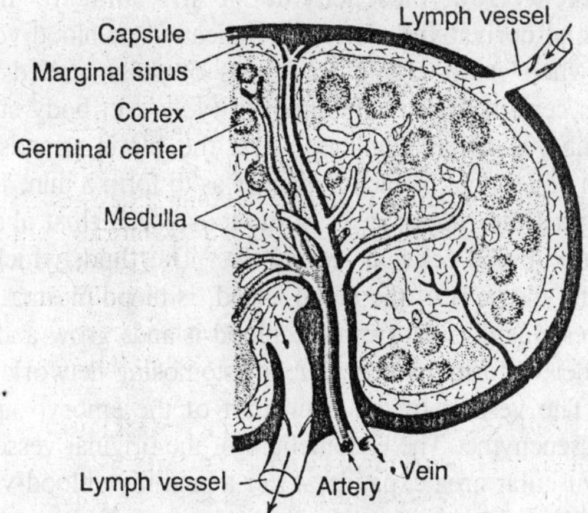

Fig. 12.2. Diagram of the structure of a lymph node.

as other known as *fixed macrophages*. These are derived from reticular cells. The lymphatic vessel breaks up into fine branches within a lymph node. The branches recombine on the other side of the node. Lymph filters slowly through the lymph nodes but in considerable quantity. Bacteria and other foreign substances are engulfed by the action of the fixed macrophages and reticuloendothelial cells. It is here, too, that lymphoeytes develop. These pass into the lymphatic stream and ultimately enter the blood-vascular system. Some of the macrophages also become free, wandering cells. The lymphatic system is of importance also in connection wit the regulation of both the quantity and quality of tissue fluid. It furnishes a means by which colloidal substances can get into the circulatory system since the endothelial walls of lymphatic vessels are somehow permeable to colloids, whereas those of blood capillaries are not. The lymph nodes are of further importance in the formation of specialized plasma cells, the function of which seems to be the production of antibodies. The actual function of lymphocytes is little understood. It is possible that they may serve to transport deoxyribonucleic acid (DNA) to cells in various parts of the body which are unable to synthesize this compound entirely by themselves. Deoxyribonucleic acid is a substance of great importance in determining the genetic constitution of cells.

Blood Vessels

Origin

The vessels of the blood-vascular system arise in embryonic development from mesenchyme *in situ* and are include among mesodermal derivatives. The first indication of blood-vessel formation is seen when small clumps or cords of cells, called *blood islands*, appear in certain parts of the embryo (yolk sac, body stalk, etc.). In a chick embryo, for example, at first the blood islands are solid but they soon hollow out in such a manner as to form a thin, flat endothelium enclosing a fluid-filled space. Some loose cells float about in the fluid and become blood cells, or corpuscles. The fluid, which is apparently secreted by the cells of the blood island, is blood plasma. The scattered, hollow vesicles derived from the blood islands grow and coalesce with one another to form an irregular, anastomosing network of small blood vessels. The vessels of the main part of the embryo appear as clefts in the mesenchyme. The endothelium of the original vessels proliferates, and the vascular area expands. After a primitive blood-vascular system is once established the formation of new vessels depends upon growth and branching of previously established channels. Every blood vessel, including the heart, has an endothelial lining.

Structure

At first there is no structural difference between arteries and veins, and the early blood vessels are histologically similar to the capillaries of later stages. From the mesenchyme surrounding the endothelium of the primitive vessels are derived the coating layers found in arteries and veins. The structures of the accessory coats differ in nature in these two types of vessels.

Arteries

Every artery has a wall composed of three coats: (1) an inner layer, the *tunica interna*, or *intima*, which is composed of the endothelium and an *internal elastic membrane*, the structure of which differs somewhat in arteries of different caliber; (2) a thick intermediate coat, the *tunica media*, consisting of smooth-muscle cells, arranged in a circular manner, together with an *external elastic membrane*, or elastic fiber network, which also courses circularly; and (3) an outer layer, the *tunica externa*, or *adventitia*, of varying thickness in vessels of different caliber, made up of loosely arranged connective tissue as well as longitudinally arranged collagenous and elastic fibers, Large, medium and small arteries differ to some degree in the detailed structure of their walls. In the tunica media of large arteries there is a preponderance of yellow elastic fibers in addition to smooth muscle cells, the latter some times being referred to as the *muscular media*. In general, the smaller the artery, the greater is the relative amount of smooth muscle and the less is the relative quantity of elastic connective tissue. The main distinction between large and small arteries, aside from actual size discrepancies, lies in the thickness of the tunica media.

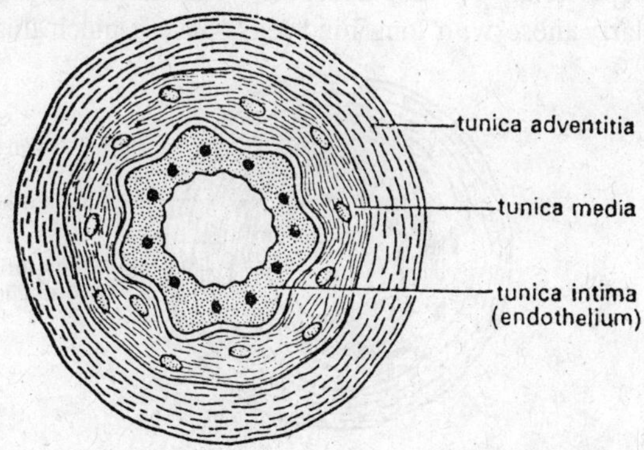

Fig. 12.3. T.S. of an artery.

The thick muscular and elastic walls of arteries are of great physiological significance. Blood, forced into the arteries by contraction of the heart, distends threes vessels. Since backflow is prevented by certain valves in the heart, tension of the elastic walls of the larger arteries tends to force the blood farther along the vessel. This accounts for the greater parts of the *movement* of blood as well as for the maintenance of diastolic blood pressure within the arterial system. The wall of the smaller, or *distributing arteries*, which are primarily muscular, are of most importance in regulating the *amount* of blood supplied to a given area. This is accomplished by varying the size of the lumen under different conditions, the smooth-muscle cells being capable of responding to nervous stimuli, whereas elastic tissue can only react passively. Arterioles, with their relatively thick muscular walls and small lumina, are largely responsible for maintaining the high degree of blood pressure within the arterial system. They are responsible, also, for the fact that blood which enters a capillary bed is under greatly reduced pressure. If this were not the case, the very thin and weak capillary walls would be damaged and incapable of functioning properly so far as diffusion of substances to and from tissue fluid is concerned.

Veins

The walls of veins are much thinner than those of arteries but consist, nevertheless, of three similar coats. The tunica media is, in some cases, difficult to distinguish, particularly in the larger veins. Elastic and muscular elements are poorly developed in veins, but connective tissue is more clearly defined than in arteries. The tunica externa, or adventitia, makes up the greater part of the wall of the vein. Superficial, or subcutaneous, veins of the legs of mammals, particularly those with long hind legs, have a much thicker muscular

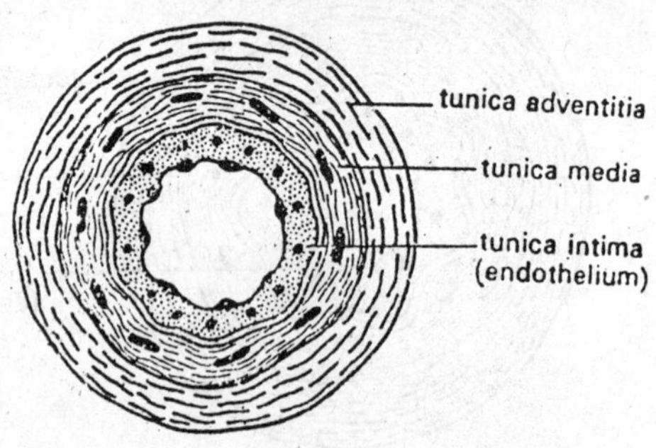

Fig. 12.4. T.S. of a vein.

portion of the tunica media than is usually the case. The walls of such veins, which lack the support of surrounding tissues, must be able to withstand the hydrostatic pressure produced by the relatively long column of blood being returned to the heart from the extremities. In long-legged birds and mammals, many of the medium-sized veins of the fore-limbs and hind limbs are provided with paired semilunar valves on their inner walls. They prevent backflow of blood in the veins and generally are located distal to the point where a tributary enters a larger vessel. The valves are formed by foldings of the tunica interna reinforced to connective tissue. In addition, they contain elastic fibers on the side toward the lumen of the vein. If the walls of superficial veins are inadequate to support the long column of blood, or if the semilunar valves break down, the veins become swollen and knotted and are referred to as *varicose veins*. Under certain conditions they are removed surgically. Valves have been demonstrated in the segmental veins of certain elasmobranchs and teleosts. It is believed that they may aid in increasing the return of venous blood to the heart. In general, the veins take the same course in the body as their corresponding arteries and usually are give similar names. A vein is always of greater diameter than its corresponding artery since its walls are thinner, are less rigid, and are more capable of being distended.

In both arteries and veins 1 mm or more in diameter, the cells making up their walls are supplied with nutrient blood vessels of their own. Only the cells in the walls of smaller vessels are able to utilize directly the nutrient materials and oxygen that are present in the blood coursing through the vessel proper. The nutrient blood vessels with which their walls are supplied are referred to as the *vasa vasorum* (vessels of the vessels). These arise from the same vessel or from a neighboring vessel some distance from the point where they are distributed. They are found in the tunica externa in both types of *vessels*. In general, the walls of veins have a more abundant supply of vasa vasorum than those of arteries. Furthermore, the vasa vasorum of veins may even penetrate to the tunica interna. The blood coursing through the veins is low in oxygen content and the cells in the walls can obtain very little oxygen by diffusion from the lumen.

Nerve fibers are abundant in blood vessels, particularly in arteries. They are of two types: medullated sensory fibers and nonmedullated, or sparsely medullated, autonomic motor fibers.

Capillaries

The endothelium of the capillary is its only component. Since it is at the capillary bed that exchange of gases, food materials, and wastes takes place between the blood and the extra vascular fluids, the thinness of the capillary wall is of obvious advantage. Diameter of the capillaries is relatively constant for each species and is related to the size of the red corpuscles which pass through them. In man the average diameter is about 0.008 mm. They are seldom more than 1 mm long. Lacking muscular and elastic coats, it has been a point of conjecture as to whether they are able to expand or contract of their own accord. In certain cases direct mechanical stimulation has been demonstrated to cause contraction of the endothelial cells, which have thus been thought to be contractile. It is possible however, that changes in diameter of capillaries may be passive rather than active and a reflection of alterations in internal and external pressures. In amphibians and some other poikilothermous vertebrates, peculiar cells having a number of long processes, called *Rouget cells*, surround walls of capillaries in certain regions. Contraction of the Rouget cells has in the past been thought to bring about contraction of capillaries, but this is now considered doubtful. Rouget cells may be nothing more than connective tissue elements which have become located on the capillary wall. Some cells similar to Rouget cells have been described for mammals. They are too few in number to have any significance in capillary contraction.

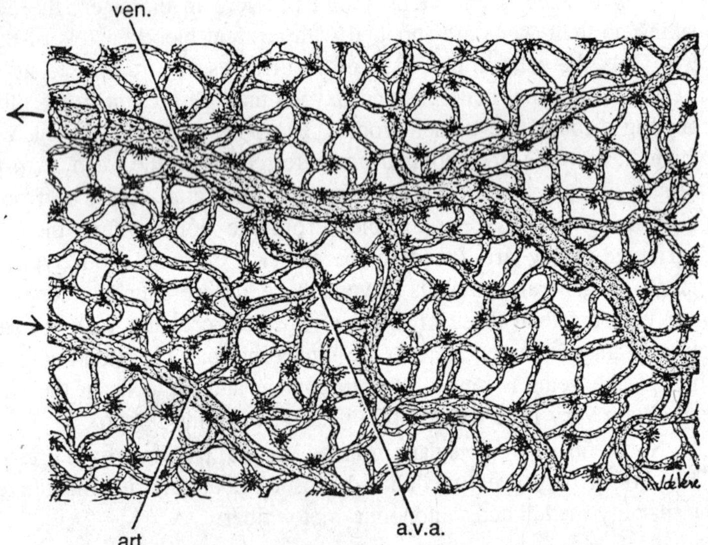

Fig. 12.5. Portion of a capillary bed from the web of a frog's foot, showing small arteriole (art.) and venule (ven.), a capillary network, and a direct arteriovenous anastomosis (a.v.a.).

Capillaries are so numerous that it is practically impossible to prick the skin in any part of the body without drawing blood. There are approximately 2,000 in a cubic millimeter of human muscle. It has been estimated that if all the capillaries in a man were stretched out in a single line they would measure nearly 150,000 miles.

The velocity of blood in arteries is high but when the blood reaches the arterioles, with their relatively thick muscular walls and small lumina, it encounters some resistance. This, coupled with the fact that the arterioles break up into great numbers of capillaries with a corresponding increase in the vascular area, brings about a considerable reduction in the velocity of the blood stream. Blood, therefore, courses more slowly through capillaries than through the larger vessels and its hydrostatic pressure is diminished. Naturally this is of advantage in the exchange of nutrient materials, gases, and wastes between blood and tissue fluid at the capillary bed.

Electron microscopic studies have brought some interesting facts to light concerning the detailed structure of the cells of which capillaries are composed. The capillary wall is only one cell thick. It usually takes no more than two cells to surround the capillary lumen at a given point. Adjacent cells are joined together by a cement substance. Extremely small vesicles, from 400 to 650 A in diameter, have been observed within the cytoplasm of these cells; some are clustered along the inner cell membrane and others against the outer cell membrane. The vesicles actually seem to be tiny invaginations of the cell membranes. Apparently they are capable of separating themselves from the cell membrane on one side of the cell and of moving to the opposite side where they release their contents. It is possible that tissue fluid is absorbed or replenished by means of these vesicles. The process is termed *pinocytosis*. Pinocytotic vesicles have also been observed in the cells making up the proximal convoluted tubules of kidneys. Variations in the hydrostatic pressures of blood inside the capillaries, and changes in the osmotic pressures of blood and tissue fluid under different conditions, determine whether tissue fluid passes in or out of the capillaries.

Since the molecules of colloids are apparently too large to pass through the capillary endothelium, it is possible that the vesicles concerned in pinocytosis are too small to handle colloidal materials and confine their activity to the transport of water and crystalloids in solution. Nevertheless, it seems that some colloid is able to leak from the capillaries into the tissue fluid. How this is accomplished is not clear.

Other vessels

It is generally held that contractile smooth-muscle cells are not associated with capillaries and, therefore, cannot be responsible for changes in their diameter. It has been claimed, on the other hand, that there may be certain "preferred channels" through capillary beds. These are of capillary dimensions but have smooth-muscle cells here and there along their walls. These *arteriovenous bridges* are said to be direct continuations of arterioles and to give rise to regular capillaries as side branches. According to this idea, blood normally passes through the preferred channels but at time, when requirements demand that more blood be distributed to a given area, the other side-channel capillaries are brought into use.

In some regions there may be direct connections between arteries and veins without the intervention of a capillary network. Such vessels, which are called *arteriovenous anastomoses*, are to be found in the toes of birds, ears of rabbits, and such areas in man as the terminal phalanges of the fingers and toes, nail bed, lips, tip of tongue, nose, and eyelids. At ordinary temperatures they are closed, but when critical temperatures are reached, dilation occurs. These vessels, according to one theory, serve in two capacities: (1) at high temperatures an increased volume of blood circulates at the surface, thereby permitting excess heat to be removed from the body by radiation; (2) at low temperatures increased circulation through these vessels prevents harmful effects which might be caused by cold. An arteriovenous anastomosis may be a straight or a tortuous branch of an arteriole connecting with a corresponding venule. The wall of the end connecting with the arteriole is similar to that of the arteriole, whereas that connecting with the venule is constructed like that of the venule. The midportion has a thicker and more muscular wall with an abundant supply of sympathetic nerve fibers which control the diameter of the lumen of the vessels, thus regulating the amount of blood that passes through the anastomosis. Arteriovenous anastomoses should not be confused with arteriovenous bridges. Small arteriovenous anastomoses, known as *glomi*, are present in the dermis of the skin of the fingers and toes, as well as some other areas. Some authorities are of the opinion that arteriovenous anastomoses serve as a mechanism which regulates blood pressure and circulation in various parts of the body. Their actual function is not completely understood.

Sinusoids make up another kind of connection between arteries and veins. They differ from capillaries in structure and arrangement and are to be found in such organs as the liver, adrenal glands, bone marrow, parathyroids, pancreas, and spleen. They consist of relatively large, irregular, anastomosing vessels which, unlike capillaries with their rather constant diameter, are not lined with a continuous endothelial layer. Phagocytes and nonphagocytic cells, irregularly disposed, are present in their walls. Sinusoids are considered by some to represent in primitive form of capillary.

In certain parts of the body, arteries or veins break up into capillary networks but the small vessels, which may or may not be tortuous, recombine to form larger vessels of the same type, i.e., arteries or veins. Such connections of capillary dimensions are referred to as *retia mirabilia*. Examples of these structures within the course of arteries are to be found in the red bodies and red glands in the swim bladders of certain fishes, and in the glomeruli of kidneys. The portal vein of the hypophysis, and the hepatic portal and renal portal veins which break up into sinusoids or capillaries in liver or kidney, as the case may be, are examples of venous retia mirabilia, since the capillaries in these structures recombine to form veins.

In the penis and clitoris a peculiar type of tissue, called *cavernous*, or *erectile*, *tissue* is to be found. It consists of large, irregular, vascular spaces lined with endothelium and interposed between arteries and veins. They are supplied directly by arteries. *Erection*, or distension, of the structure containing this type of tissue is due to a filling of the spaces with blood under high pressure and a partial closure or compression of the venous outflow. Erectile tissue also occure in the lamina propria of the mucous membrane covering the middle and ventral nasal conchae. Some authorities doubt whether the lamina propria of the nasal conchae should be classified as erectile tissue, which otherwise is confined to the penis and clitoris. This is because septa containing smooth-muscle fibers are lacking. At any rate, it consists of numerous thin-walled veins associated with smooth-muscle fibers which are arranged in both circular and longitudinal directions.

Composition of Blood

Blood consists of a fluid called plasma in which are suspended the various cellular elements viz. red cells, white cells and platelets. These two constituents of blood, plasma and cells, are easily seen if a tube of blood, which has been prevented from clotting is allowed to stand; the heavier cellular elements gradually settle to the bottom of the tube leaving the buff coloured plasma above. About 45% of blood is made up of cells, the remaining 55% by plasma.

Red Blood Corpuscles

The red corpuscles are among the most specialized structures of the body. In adult mammals they are bags of haemoglobin, without nuclei, devoted solely to the transport of oxygen and surviving in man only for 100-120 days; then they are destroyed by the liver and spleen and replaced by others formed in the red bone marrow. There are about 5 million red cells per milliliter of blood in a normal man (less in woman, more in children), and therefore some 35×10^{12} of them in the body.

The shape of the corpuscles is characteristic and is usually described as showing a biconave or dumb-bell outline. The effect of this shape is to give the corpuscle a 25 per cent larger surface area for its volume than should be presented by a sphere. It also allows rapid and equal diffusion from the surface to all inner parts. These are obviously efficient features for the purpose of oxygen transport. The surface is not one of minimum area, and it is difficult to see how the shape is produced; it must be maintained by some structure having considerable rigidity. This structure is presumably provided by the cell membrane. This is birefringent and contains phosphatide and protein components. The membrane can be pulled away as a distinct structure with micro-dissection needles. The consistency of the contents of the corpuscle is that of a rather soft gel; the water content (60 per cent) is lower than that of many tissues. Yet the mechanical structure of the whole is such that the corpuscle can be deformed in shape, for instance, during passage through a capillary, and immediately afterwards resume its biconcave appearance.

The red corpuscles vary little in size, the mean longest diameter being 7.2 μm and the thickness 2.2 μm as measured in fixed

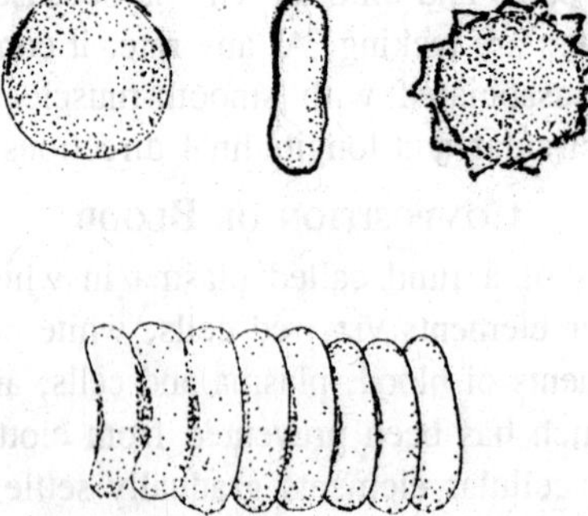

Fig. 12.6. Erythrocytes as seen under various conditions.

preparations. In the fresh state they are certainly over 8.0 μm in diameter and may show fluctuations with the composition of the fluid in which they are placed.

The surface membrane is permeable to water but certainly not to all solutes, so that the cell shape is markedly affected by the tonicity of the surroundings. The 'normal' shape is preserved in 0.9% sodium chloride; various signs of plasmolysis ('crenation') occur in stronger solutions and with weaker ones the corpuscles become inflated and may allow escapes of the haemoglobin (haemolysis) leaving only a red-cell 'ghost'. The contents approach those of other cells in inorganic composition, that is to say, they contain more potassium and less sodium and chloride than does the blood-plasma.

The red corpuscle of an adult mammal thus retains many characteristic features of cells but has lost its whole apparatus for synthesis, including nucleus, endoplasmic reticulum, ribosomes, Golgi bodies, and mitochondria, leaving a homogeneous 'cytoplasmic' material composed almost wholly of haemoglobin, which indeed makes up 95 percent of the dry weight of the corpuscles. The corpuscles of the foetus are, however, still nucleated.

Myeloid Tissue

The life of red cells, lacking a nucleus, is limited to about 3 months in man. Production of these, as of the other components of the blood, is basically a function of the connective tissues of the body, which are indeed allied both to the endothelial linings of the vessels and to the blood itself. In the earliest embryo the blood is formed from the connective tissue of various parts of the body, especially in the yolk sac, liver, spleen, lymph nodes, and bone marrow. The power becomes gradually more restricted, until in adult life all the haemopoiesis goes or in the *red marrow* of the ends of the long bones and especially in the flatter bones such as the bodies of the vertebrae, the ribs and sternum. The shafts of the long bones are filled with a fatty *yellow marrow*, which is not haemopoietic but may become so when there is a demand for more red cells.

The red marrow or myeloid tissue is a soft material having, like lymphoid tissue, a basic stroma of reticular tissue. This consists of a fabric of fine scleroprotein fibrils among which lie 'primitive reticular cells' which are presumed to give rise to all others in the tissue. In addition, myeloid tissue contains fat cells, may macrophages, and blood cells in all stages of formation. The blood-supply is peculiar in that,

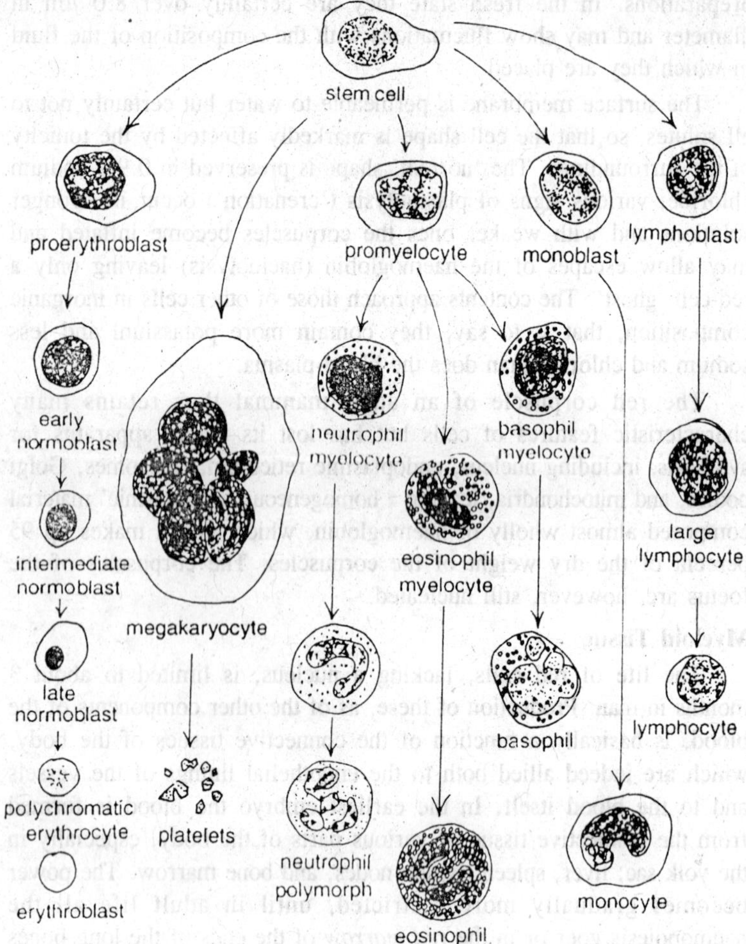

Fig. 12.7. Diagram of the processes of formation of red and white cells.

instead of capillaries, there are numerous large sinusoids, which have a very thin wall and allow free passage of cells.

Although much remains to be discovered about blood formation it is probable that, as the unitary theory holds, all types of blood cell can be produced from a single undifferentiated cells type of the mesenchyme. These stem cells have not been identified for certain but they are probably those that are some times called haemocytoblasts, which are rather like large lymphoeytes, with a deeply staining nucleus and somewhat basophil, non-granular cytoplasm. The red marrow has

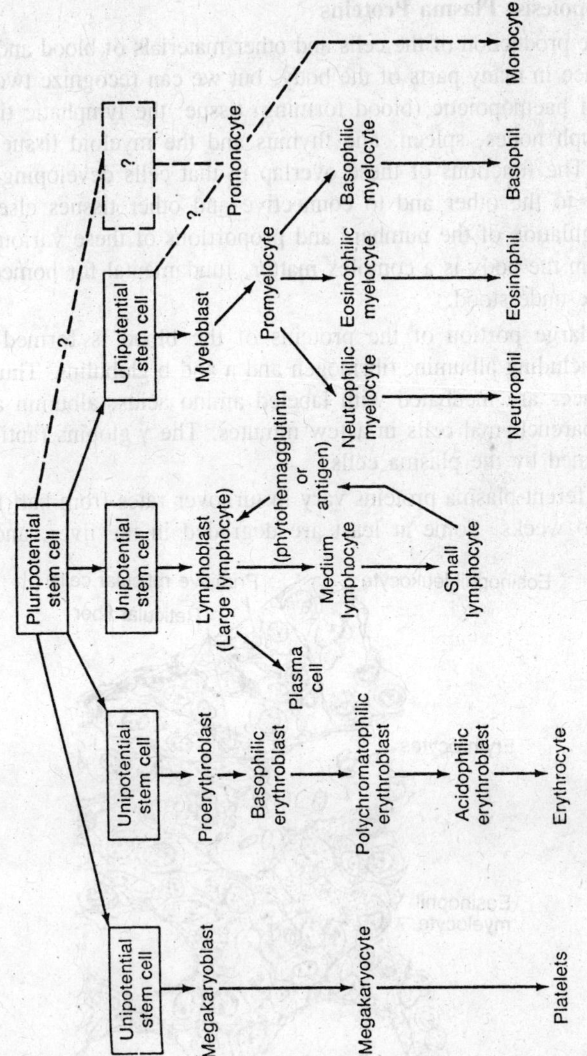

Fig. 12.8. Development of the various types of blood cells from pluripotential stem cells.

the double function of producing new cells, red and white, and removing the old ones. The mesenchymal stem cells therefore must be able to produce the precursors of the two main fines, those leading to production of new cells (known as myeloblasts) and those responsible for removing the old ones, the reticuloendothelial cells of haemopoietic tissue or macrophages of loose connective tissue. The myeloblasts in turn are the cells able to form the three main types: erythrocytes, granular leucocytes, and platelets.

Haemopoiesis: Plasma Proteins

The production of the cells and other materials of blood and lymph takes lace in many parts of the body, but we can recognize two major types of haemopoietic (blood forming) tissue: the lymphatic tissue of the lymph nodes, spleen, and thymus and the myeloid tissue of the bones. The functions of these overlap in that cells developing in one migrate to the other and to connective and other tissues elsewhere. The regulation of the numbers and proportions of these various types of cell in the body is a complex matter, fundamental for homeostasis, but little understood.

A large portion of the proteins of the blood is formed in the liver, including albumin, fibrinogen and a and b globulins. Thus when liver slices are incubated with labeled amino acids, albumin appears in the parenchymal cells in a few minutes. The γ globins (antibodies) are formed by the plasma cells.

Different plasma proteins vary in turnover rates from half-lives of hours to weeks. Some at least are degraded in the liver, and some

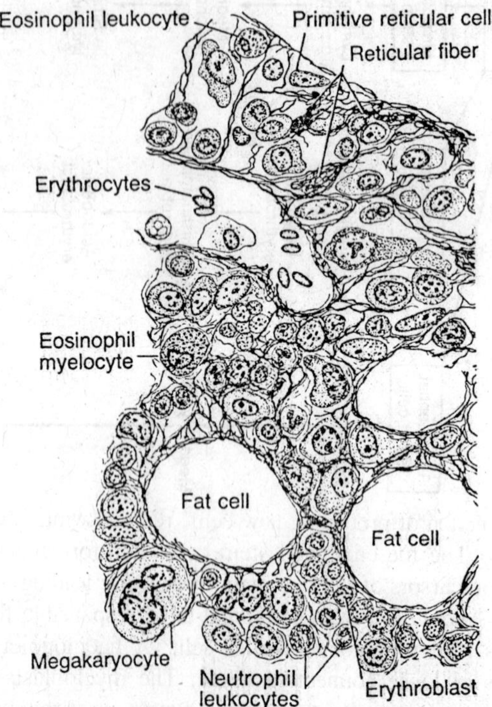

Fig. 12.9. A blood-forming tissue—bone marrow from a mammalian femur.

probably in the intestine. After injury or stress there is usually a fall in plasma proteins, followed by increased synthesis in the liver. The stimulus for this is not known by may be partly from the histamine and other substances produced by tissue injury. Adrenal corticosteroids, besides increasing the amount of protein degraded and excreted, also stimulate its production.

The net effect of injury is to increase the rate of plasma protein turnover, perhaps wit the effect of making amino acids more readily availably for repair.

Stem Cells of the Myeloid Tissue

Although the stem cell have not positively been seen, good evidence for their existence comes from lethally irradiated animals in these the number of neutrophils falls and, rather later, that of megakaryotes and red cells. Life can be saved by injecting small amounts of isogenic marrow from a normal individual. The cells form clones which appear as little nodules on the spleen. There is evidence that each nodule comes from a single cell, for when the injected material was itself sub-lethally irradiated then, in some of the colonies, all the cells could be seen to have the same chromosome abnormality.

Such splenic colonies are able to show differentiation, along all the three lines that derive from the myeloblasts. Probably the basic stem cells of the mesenchyme are similar. These cells are not yet fully undifferentiated since they retain the power of forming all the blood cells and perhaps other (such as fibroblasts) as well. We are beginning to understand how their multiplication is controlled and what causes the products to differentiate into the various cell lines as required. Control of this would be extremely valuable in the management of many diseases. The differentiation into the red-cell line is promoted by a hormone, erythropoietin. Presumably some such stimulus is responsible for initiating production of each of the lines deriving from the original stem cells. Probably the differentiation proceeds gradually, the first products being still capable of several further divisions to produce many cells of the given type.

Formation of Erythrocytes

The myeloblasts are 12-20μm in diameter, with a nearly round nucleus, not deeply basophilic. The cytoplasm is faintly basophilic and contains free ribosomes but little endoplasmic reticulum. The cytoplasm becomes more basophilic in the stages known as proerythroblasts and basophilic erythroblast, due to an increase in ribosomes. Then as haemoglobin begins to accumulate, the cytoplasm also takes up acidic dyes, hence these are called polychromatophilic erythroblasts. These may develop directly into erythrocytes, the nuclei degenerating (pyknosis). The cytoplasm then still contains some RNA, hence such cells show a basophilic network and are known as 'reticulocytes'. They occur in the blood when haemopoiesis is very active. More usually the polychromotophilic erythroblasts continue to divide and lose their cytoplasmic RNA. They are known as normoblasts and nucleus degenerates to leave a typical erythrocyte.

Control of Red-cell Production: Erythropoietin

There are probably separate control system for all the seven types of cell indeed for subtypes as well. However, there are conditions that influence many of them together, where they share a common stem cell. There are many of them together, where they share a common stem cell. There are various ways in which regulation could be achieved, and it is not certain which are used. It may be by single feed-back loops activated by increased demand and acting upon cell production. Alternatively, the situation may be more complex and demand may produce increased release from a storage compartment, whilst this might initiate maturation in a second compartment which in turn might activate a mitotic compartment. There is evidence that stem cells can divide only a limited number of times and there must therefore be some concatenation of compartments.

The control of the red cell production is mediated by a hormone, erythropoietin, produced by the kidney in response to anoxia. Production is increased following cell loss or decreased atmospheric oxygen and decreased by hypertransfusion of erythrocytes. The increase is faster than the decrease. Under normal conditions the feed-back is very precise, and no subtle waveform or other oscillations have been detected.

The evidence for the existence of the hormone is that plasma from anemic animals injected to normal ones produces increased red-cell formation, but not after nephrectomy. If one member of a parabiotic pair of rats breathes an atmosphere low in oxygen then the red marrow of both shows hyperplasia. The juxtaglomerular cells of the kidney are the main source of the hormone, but there may be others in the body. The hormone is a glycoprotein of molecular weight 40000-60000. It is knot known how its basic level is set or related to erythrocyte removal from the circulation.

The attainment of a certain low oxygen level is presumably sensed by the juxtaglomerular cells in such a manner as to promote the synthesis of more erythropoietin. The setting of this level of sensitivity by appropriate synthesis during development is the prime determinant of the rate of haemopoiesis. The level itself, of course, may be capable of later adjustment by change in the amount of some receptor that is synthesized or indeed simply by production or more (or less) juxta-glomerular cells. In some such manner it is ensured that at all stages of growth, activity, and environmental oxygen level there is an adequate supply of red cells.

The erythropoietin itself probably acts on the bone marrow by stimulating the formation of proerythroblasts from the myeloblast stem cells. Evidence of this is obtained by stopping haemopoiesis in mice by hypertransfusion (polycythaemia) and then giving a single dose of erythropoietin. This produces a crop of new proerythrocytes whose progress can be followed. After 36 hours most of them early normoblasts and after 72 hours they were reticulocytes ready for release. Single doses of actinomycin D, which prevents the formation of DNA-dependent RNA, if given 1 hour after the erythropoietin, prevented the process completely. Given at 16 hours or 24 hours it reduced the red-cell output, but less drastically. There is, therefore, certainly a marked influence of the hormone in inducing differentiation of stem cells. It may well be that this involves a derepression of the genes appropriate to the programme of differentiation into an erythroblast. If prolonged actinomycin inhibition is stopped red-cell production begins within 18 hours. The stem cells have therefore not been permanently damaged.

There is evidence that the key enzyme in haem synthesis is g-aminolevulinic acid synthetase (ALA). It appears in the spleen within 8 hours of a single dose of erythropoietin to polycythemic mice. This appearance is completely blocked by actinomycin given at the same time, but not if given 6 hours later. Therefore, this may in turn induce others involved, such as ALA dehydrase and iron protoporphyrin chelating enzymes.

The haemoglobin molecule is composed of four polypeptide chains, two known as a and two as β chans, together with four iron-containing haem groups. The molecular weight is 64500 and the a chains have 141 amino acids and the β chains 146. The molecule has long helical portions and the remainder are bent or kinked, giving an elaborate tertiary structure, approximately globular over all.

During haemopoiesis each polypeptide unit is assembled, like other proteins, by stepwise polymerization on a polysome starting from the amino end of the chain, the sequence being determined by the mRNA. Each amino acid is activated by ATP and transferred to a molecule of tRNA, three nucleotides of which are attached by hydrogen bonding to the complementary sequence on the mRNA. A new peptide bond is then formed enzymatically between the carboxyl group of the previous amino acid and the amino group of the incoming one, and the tRNA is removed. The complete polypeptide finally must be released and associated with its fellows and with haem, but it is not known how this is achieved.

In agreement with this model the nucleated erythroid cells of a rabbit, which synthesize Hb rapidly, have many ribosome, mostly in polysomes. As the nucleus is lost the rate of Hb synthesis decreases and the number of ribosomes falls.

Thus red blood corpuscles are produced continually in the red bone marrow, and after a life of a few months they are destroyed by the spleen. This continual replacement of the tissues of the body is another, clear example of the general rule that the living organism must be considered not as a particular set of matter carried from the cradle to the grave but as a way of so transforming energy that a particular plan of organization is preserved. It is calculated that, in man 10 million red corpuscles are formed and destroyed every second and this gives a clear idea of the great changes that are going on in spite of the constant appearance of the whole body.

Spleen

The spleen is a portion of the circulatory system having some characteristics of lymphoid tissue and others of bone marrow. It lies in the mesentery, close to the stomach, supplied with its own artery and vein. It assists in various ways in the adjustment of the blood. The spleen has a muscular contractile wall, but there is relatively less muscle in man than in many mammals. This capsule surrounds a pulpy substance in which the blood comes into contract with a network of macrophages that acts as a filter system. Around the branching arteries of the organ are little masses of lymphoid tissue ('white pulp'), which is a mass of connective tissue trabeculae and a mesh of reticular tissue. This is permeated by passages leading from the straight arteries called penicillary arteries to the venous sinusoids of the red pulp. Some of these channels have walls of reticuloendothelial cells. There has been much discussion as to the nature of the communication between arteries and veins in the spleen, and it is uncertain whether there is a system of open sinuses. Certainly its substance provides, like the lymphoid tissue, a means by which the blood is brought into close contact with the elements of the reticular tissue. Red cells in process

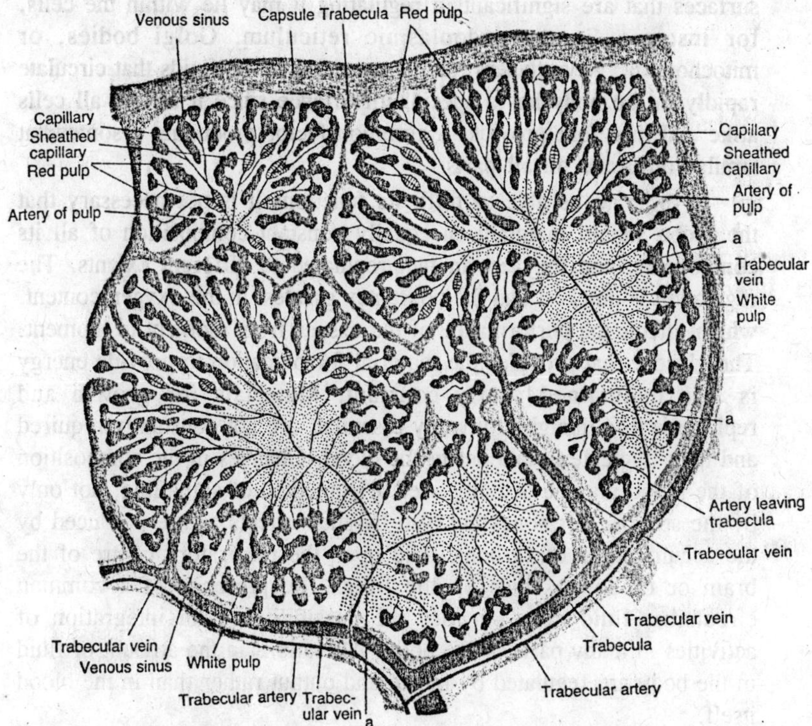

Fig. 12.10. Diagram of a part of the mammalian spleen.

of destruction can be seen at all times in the macrophages of the spleen, and bacteria or coloured matter injected into the blood-stream soon appear in the red pulp. The organ thus functions as a filter for the blood rather as the lymph nodes do for the lymphatic system.

The spleen also produces lymphocytes, monocytes, and plasma cells. In many mammals the spleen stores some of the available red corpuscles, and these can be expressed into the circulation when exercise or reduced oxygen supply make extra calls on the oxygen transport system. This action is performed by contraction of the smooth muscle in the walls of the spleen, under the control of the sympathetic nervous system. A considerable proportion (one-fifth in the dog) of all the red corpuscles of the body are stored in the spleen when the body is at rest.

Regulation of the Internal Environment

Our analysis has shown that the conception of a single internal environment in which the tissues lie is only of limited value. Exchange between the cells and their surrounding is exceedingly rapid and the surfaces that are significant in regulating it may lie within the cells, for instance, in the endoplasmic reticulum, Golgi bodies, or mitochondria. Nevertheless, the composition of the fluids that circulate rapidly in the body has a special importance since it affects all cells alike and therefore provides a common factor producing a somewhat similar influence on all parts.

In order to be able to fulfil its functions it is not necessary that the blood should maintain an equally constant composition of all its components, and they are in fact regulated to different extents. The regulation is quickest and the constancy greatest in the oxygen content, which affects the performance of the tissues from moment to moment. The glucose that provides the material for oxidation to provide energy is also maintained nearly constant. Materials for growth and replacement, although ultimately essential, are not constantly required and they fluctuate in amount in the blood. Changes in the composition of the blood may be actually used as a method of signalling, not only by the specific hormones but also by such effects as are produced by the accumulation of carbonic acid upon the respiratory centre of the brain or of amino acids on the liver. The presence of a common circulating fluid thus provides the possibility of an integration of activities in many parts of the body. Alterations in the amount of fluid in the body are regulated by intake and output rather than in the blood itself.

The oxygen-carrying power of the blood is one of its most important properties and is regulated both quickly by nervous action to anticipate rapid changes in demand and also by slower processes. The spleen provides a store of red corpuscles that can be discharged into the circulation during exercise. Moreover the speed of circulation of the blood can be increased and the direction of flow regulated by contraction of some of the arteries, so that blood is available where it is most needed. Slower adjustment of oxygen-carrying power is seen in the increase in number of red corpuscles produced by sojourn at high altitudes. No doubt there are still slower changes in the blood produced by the operation of the evolutionary processes of variation and natural selection. These have provided the characteristic features that give to mammalian blood its great capacity for transport of oxygen.

In its protective functions the blood is also regulated at various rates, but here the nervous system plays little or no part, since the regulation is slow. The capacity to form a clot is an inherited anticipation of the probability that vessels will at some time become damaged. Moreover the clot becomes invaded by fibroblasts and its fibrinogen molecules dissolved in a manner that provides for the laying down of collagen so that further stresses will be met.

The capacity to resist invasion by foreign substances and organisms is conferred ultimately by heredity but in each case the antibodies are produced in the blood as a response to the presence of a particular type of extraneous invader. The hereditary system confers on the blood the power to carry a specific memory of past 'experience'. This is only one example of a whole complex of changes that take place in the capacity of the organism to react to invasion and to stress. The entire white-cell system is continually changing, for example, in the number of macrophages or lymphocytes that are being produce according to the influences that have been falling upon the organism.

In many features therefore the internal environment provides a representation of the past and thus anticipates events that are likely to occur to the organism in the future. This anticipation is secured by selection among various possible reactions of those that conform to the conditions that have occurred in the environment. Thus, of the fibrin fibers laid down randomly in a clot, those that are not pulled upon are removed first. This has been ensured during evolution by the process of natural selection. Organisms are produced with a variety of different sorts of fibrin and those whose clots do not show this property are eliminated. We can see vaguely how by these selective mechanisms it is ensured that the internal environment can receive 'information' and thus adequately represent conditions outside. Far more study is needed to complete our understanding of the relationship and to make the formulation of it precise.

Protection and Defence of the Body

Turnover, Protection, and Defence

Organisms maintain their integrity by selecting certain environmental components which are then incorporated for varying periods of time. The instructions of the DNA ensure this continuity by organizing the formation of proteins of the numerous types needed. These include (a) those that prevent the entry of unwanted materials, for instance, the resistant keratins of the skin; (b) those that facilitate the entry of some molecules by providing the permeable surfaces of the lungs and gut and the appropriate enzymes inside and outside the cells; and (c) proteins that make possible the removal of unwanted material that has either entered in spite of the barriers or been produced within the body. This class will obviously include all the mechanisms usually called 'excretory', but we are concerned here with those that are considered as 'defensive' in that they deal with 'invasion' of bacteria, viruses, or other agents.

All of these living processes are interrelated, and it is really artificial to separate certain of them by making the military analogy with 'defence'. This is especially clear when we think that one of the chief agents is phagocytosis, by which the unwanted material is made part of the body itself, a process thus involving digestion and growth as well as defence.

In fact, since nearly all part of the body are involved in turnover and replacement, the question of what is part of the body and what is 'foreign' to it is less clear than it seems. The enzyme systems must be so regulated as to produce at all times an appropriate balance between incorporation and rejections of the materials that are presented. It usually is held that this involves mechanisms for identification of the body's own macromolecules, so that they are not rejected by the defence mechanisms designed to destroy 'invaders'. This is an attractive view, but it presents difficulties, for in fact there is continual turnover with breakdown and replacement. The question of the maintenance of its 'identity' is too subtle to be properly handled by picturesque metaphors taken from warfare.

Defensive Tissues

Nevertheless, we can recognize in a mammal a distinct set of tissues that are concerned with the removal of unwanted substances and particles from the tissues. They include the white cells of the blood (*leucocytes*) and the related macrophages of the *reticulo-endothelial* systems of the tissues. Defence is ensured either by engulfing particles (*phagocytosis*) or by producing *antibodies* that unite with an destroy invading agents such as viruses. There are distinct cells for these two types of function, but the responses overlap and have much in common. Both depend upon the presence of some system of receptors for recognition of what shall be destroyed. Moreover, the entire 'defence' system is assisted by a series of systemic humoral agents including some steroids of the adrenal cortex. Indeed the response of the body to a massive infection may involve a great part of its resources so that many organs show signs of 'stress'.

White Blood Cells

There are about 7000 white corpuscles per milliliter in normal human blood, divided as follows:

aranulocytes	lymphocytes	20-30 per cent
	monocytes	3-8 per cent
	neutrophils (polymorphs)	60-70 per cent
granulocytes	eosinophils	1-3 per cent
	basophils	0.5 per cent

Of these the monocytes and neutrophils become phagocytes, and the lymphocytes are concerned with antibody production.

The neutrophils are so called because they contain numerous granules that are neither acidic nor basic. Their more usual name is polymorphs (polymorphonuclues leucocytes) because their nuclei show several lobes. The cells are 10-12 μm in diameter in dry smears but in fresh blood are spheres of 7-9 μm. They are formed in the myeloid tissue of the bone marrow. Their function is to collect around foreign bodies that have penetrated the tissues and to destroy them with enzymes. These enzymes are produced by the granules of the polymorphs, which are lysosomes. These are the cells that collect around a splinter or in an infected wound. They are the first defensive cells to arrive, leaving the venous capillaries probably by passing between the endothelial cells. After they have discharged their enzymes the whole cell often dies, and the aggregate of living and dying polymorphs constitutes pus.

The polymorphs destroy bacteria by surrounding them and taking them into which digestive enzymes are then passed from the lysosomes. It is not known how the neutrophil 'recognizes' the bacterium as a foreign body, nor how it avoids ingesting particles of the body itself. The number of polymorphs in the blood is increased during a heavy infection. Some signal, therefore, must be sent from the infected site to the bone marrow, but its nature is not known.

Formation of Neutrophils (Polymorphs)

These cells differentiate in the marrow from myeloblasts, first into large promelocytes with a granules and then into smaller myelocytes with an indented nucleus and numerous granules. The nucleus then becomes increasingly indented, forming first a horse-shoe shape and then becoming constricted into five or more lobes.

Little is known about the control of polymorph production. Half the total in the body are stored in the marrow so that a sudden appearance of more in the blood (as after infection) does not necessarily indicate increased production. Such shifts to the blood may occur from adrenal medullary secretion or exercise. However, it is probable that formation is increased by demand; for the number in the blood is chronically raised in sevee infections or inflammations, where many are serving as phagocytes.

It is possible indeed that regulation of neutrophils is effected partly stimulation of the marrow by the 'endotoxins' that are produced by the membranes of bacteria. Injection of any material containing endotoxin increases the number of circulating neutrophils. Indeed, the difficulty of avoiding endotoxin contamination makes it hard to decide whether there are nay neutrophil-stimulating hormones. However, endotoxin stimulation cannot be the only method of regulation since animals kept free of bacteria still have polymorphs.

Eosinophils and Basophils

The eosinophils (1-3 per cent of the leucocytes), besides their acidophilic granules, large electron-dense granules. The cells are somehow concerned with immunological reactions and especially with allergy and with anaphylaxis. Hydrocortisone causes them to disappear from the blood and depresses allergic reactions. The eosinophils contain histamine, and either they produce it themselves or take it up from mast cells. The greatly increase during infection by parasitic worms.

Basophils are even less frequent in the blood (0.5 per cent). They resemble mast cells and like the eosinophils contain heparin and are concerned in allergic reactions.

Monocytes

These are the largest of the leucocytes, up to 20 μm in diameter when flattened in a blood smear. The nuclei are often kidney-shaped, and there is more abundant cytoplasm than in lymphocytes. There is a conspicuous Golgi body and some rough endoplasmic reticulum. Living monocytes have characteristic ruffled membranes, in continuous motion. The monocytes are thus motile but not themselves active in defence; they are immature cells, ready to develop into phagocytes when they enter the tissues. They may also be able to give rise to fibroblasts. The monocytes probably develop from monoblasts in the bone marrow as a line of cells distinct from the lymphoblasts. It is not known how the rate of formation is controlled. If a rat is injected with tritiated thymidine, any newly formed monocytes become labelled. If the animal is first irradiated with 750 rad no labeled cells appear, because no new ones are formed. But if the tibial bone marrow is protected from the radiation new cells are formed. Conversely, if labelled cells are infused into syngeneic recipients, only those of morrow (and to a lesser extent spleen) appear as monocytes.

Mononuclear cells appear in inflammatory exudates, where they differentiate into phagocytic macrophages. They appear later than the neutrophils (polymorophs), but probably partly in response to the same stimulating substances. There may be substances released by neutrophils that attract monocytes and others that are produced by the damaged tissues and act upon monocytes alone.

Macrophages

Some 3.6 × 10^6 monocytes are produced daily in a normal rat and then disappear from circulation with a half-life of 3 days. The cells leave the vascular system at random in a wide variety of tissues, passing between the endothelial cells intercellular space and basement membrane into the extravascular connective tissue. Here they become converted into the large phagocytic macrophages (histiocytes). They probably do not divide further and have been shown to remain in place for at least 50 days. These cells are important components of many organs, for instance, the Kupffer cells of the liver, sinusoidal macrophages of the spleen, microglia of the CNS, and 'dust cells' of the lung. Probably, but nor certainly, all these are derived from monocytes but some may be formed locally. Little is known of their permanence or of the control of their formation. Tritiated thymidine studies show that 1 per cent of Kupffer cells are labelled in 24 hours.

Monocytes kept in culture develop into large macrophages with much rough endoplasmic reticulum and Golgi apparatus and many lysosomes. No doubt this conversion is the source of the macrophages of the tissues. These are the cells that give the reticulo-endothelial tissues their power to take up bacteria and other foreign particles. They have very varied shapes. The nuclei are smaller and darker than those of fibroblasts and are indented. In haemopoietic tissues the macrophages line the snusoids as the 'reticulo-endothelial cells'. In loose connective tissue they are oval, but may be elongated or otherwise adjusted to their neighbours. The microglia cells of the CNS are highly branched macrophages, ready to round up and become phagocytes.

Macrophages protect the tissues by the processes of phagocytosis and pinocytosis, which essentially depend upon turning in some portion of the surface of the cell to enclose the foreign body in a vacuole. Here it is acted upon by enzymes, probably derived from the lysosomes of the cell. If it can be broken down the products are disposed of in various ways, mainly by liberation into the tissue fluids. It is not known how the macrophage 'recognizes' the material to be ingested. The first phase of phagocytosis is an attachment to the membrane of the cell, followed by formation of a vacuole. As has been remarked, 'the engulfment process *per se* is an astonishing event to behold under the microscope, the bacterium or other particle appearing to pass as if by magic through the cell membrane to reach a cytoplasmic site'. A wide range of materials can be ingested, including bacteria, virus particles, antigen-antibody complexes, and inert materials, such as trypan blue or Indian ink. Besides these 'foregin' invaders macrophages will also consume 'effete' red cells, fragments of broken tissue cells, and perhaps quite a wide range of the body's own components. In metamorphosis of the frog the tissues of the tail are consumed by macrophages.

Macrophages do not make specific immunoglobulins but probably pass to lymphocytes the information as to which antibodies are needed. Some macrophages are provided with characteristic 'dendritic processes' with which they embrace lymphocytes and then let them go. This may be the process by which information is transferred.

Lymphatic System

The lymph channels constitute an additional mechanism by which material is returned from the tissues to the blood-capillaries but they end blindly in contact with the cells or tissues spaces. These vessels therefore normally have no open ends, but their walls are permeable not only to water and crystalloids but also to colloids. Moreover, the finest branches of the lymph capillaries are modifiable structures, changing from moment to moment, so that gaps in their walls open and then close again. The lymph system is thus able to take up proteins and also any foreign particles or bacteria from the tissues; one of its main functions is to filter off and neutralize such intruders.

The lymph capillaries unite to form larger vessels, essentially like veins but with even thinner walls. At intervals along the lymph vessels are the lymph nodes, which are the main protective filters of the body. After passing through the nodes, the lymph is collected into still larger lymphatic vessels. The majority of these ultimately join a large sack, the *cisterna chili*, lying on the aorta below the diaphragm. From this sack the *thoracic duct* proceeds headwards to open into the left innominate vein. The lymph vessels of the intestinal villi are known as *lacteals* and serve to carry away the fat, which is otherwise unable to enter the blood-stream.

There is a continual slow movement of fluid through the lymphatics back to the venous system, and by this means material such as protein that cannot re-enter the capillaries returns to the blood. The lymph is therefore a fluid of variable composition, containing less protein than the plasma and having a salt composition similar to that of the blood. It contains few cells at the periphery but some lymphocytes in the main channel. The flow of the lymph depends to only a small extent on the pressure within the tissues produced by the heart. The propulsion comes mainly from the massaging actions produced by the muscles when they contract. This pushes the lymph in one direction because the lymph vessels are provided at frequent intervals with valves similar to those of the veins. The lymph flow is increased in conditions that lead to extra permeability of the blood capillaries such as increased muscle movement and the presence of certain poisons and of oedemas of varied origin.

The network of lymphatics thus consists of a set of channels collecting material from the tissues. It may be considered to have four main functions : (a) to return excess fluid (in some conditions); (b) to carry fat via the lacteals and chyle; (c) to circulate lymphocytes and produce antibodies; (d) to provide cell-mediated reactions against foreign bodies such as grafts. The system is highly complex including a network everywhere in the body; certain regions have special lymphatic functions, especially the thymus, the gut associated lymphatic tissues (GALT), the lymph nodes, and the spleen. Within this framework of channels and organs the lymphocytes proliferate, differentiate, migrate, and respond to antigenic stimuli.

Lymph Nodes

The lymph nodes and lymph nodules are scattered along lymph drainage pathways throughout the body in strategic situations at which infection is likely, such as the tonsils, the walls of the intestine and appendix, and the lymph nodes of the groin and axilla where lymph vessels from the limbs converge. They are places at which the lymph vessels break up into fine system of reticular and collagenous connective tissue, where it comes into close contact with macrophages and lymphocytes. The lymph nodes vary in size from minute invisible collections to masses that are large and easily felt. The afferent lymphatic vessels enter the nodes round its outer surface and the lymph then filters through an ill-defined system of spaces. The liquid collects at a hilum and is carried away by large efferent lymph vessels.

In between the trabeculae that carry the lymph sinues are solid masses of cells, the lymph nodules, each consisting of an outer mass of lymphocytes and a centre composed of lymph-producing tissue. This consists of a fine reticulum of supporting substance containing reticular cells, which, according to the unitary theory, are able to develop into either macrophages, lymphocytes, or erythrocytes. The lymph nodules, therefore, are composed of these active reticular cells and masses of lymphocytes and macrophages, presumed to be their descendants. The lymphocytes are of various diameters between 8 μm and 12 μm. They have a very large nucleus, containing deeply basophilic nucleoli, and relatively little cytoplasm. They are present in vast numbers in the lymph nodes and some pass into the lymph and bloodstreams. In the adult they are mostly formed by mitotic division of existing lymphocytes, germinal centres within the nodes becoming active for a while and then subsiding. In the young, and to some extent in the adult, they are formed by rounding up and division of reticular cells.

The lymph nodes are characteristic of mammals and birds; in lower vertebrates, blood-cell producing (haemopoietic) and leucopoietic tissues are usually found together. The presence of well-developed discrete lymph nodes is perhaps a sign of a greater need for defence mechanisms by the warm-blooded animals, but there is still much uncertainly as to how the lymphoid tissue functions in this capacity. The nodes are certainly efficient filters; if fluids heavily infected with staphylococci are injected into the lymphatics of a dog's leg the lymph in the thoracic duct remains sterile. This filtering function is performed mainly by the macrophages of the nodes. Granules of dye injected into the lymph are taken up by the macrophages.

Lymph nodes are agents for collecting small lymphocytes from the blood and transmitting them to lymph vessels. Antigen enters the node in phagocytes (macrophages) which filter foreign material from the lymph stream. These then inform the visiting lymph cells, which transform into immunoblasts (= large lymphocytes = plasmoblasts). These turn into plasma cells in the medulla of the gland and are discharged into the efferent lymph vessels to disseminate the immune response and produce a systemic immunity. It is a main function of a lymphatic system to collect from the tissues the large molecules that cannot enter blood-vessels, and these include antigens. These large molecules are then filtered out by the phagocytic cells of the regional lymph node. Obviously, particulate matter such as bacteria or viruses is more likely to be retained and produce a reaction than are soluble antigens, which may pass through the lymph node.

Blood Groups

It is appropriate to consider blood groups whilst we are concerned with red blood cells for it is because of the chemical properties of the surface of the red blood cell that the whole problem of blood groups arises. It is well known that if for any reason a patient has to be given a transfusion of whole blood then it is essential that he be given blood of the right group. There are many blood groups and sub-groups but the best known ones are groups A, B, AB, and O, and the rhesus groups. When blood samples from two persons with the same blood groups are mixed together, no change occurs in the blood, and the two bloods are said to be compatible. However, when two specimens of blood from different groups are mixed together there may be important and potentially dangerous changes in the red cells; the red cells may stick together to form clumps, and the damaged red cells may break down to release haemoglobin into the plasma. If this occurs during a transfusion of blood then the kidneys may be seriously injured by deposits of haemoglobin obstructing the renal tubules. If clumping of the red cells occurs when two specimens of blood are mixed together then the bloods are said to be incompatible. This incompatibility is the expression of an antigen-antibody reaction which is taking place in the blood.

The surface of the red blood cell has on it certain protein substances called antigens. An antigen is a chemical substance which causes the body to produce another chemical called the antibody, which neutralizes or diminishes the effect of the antigen. It is only possible then to define an antigen in terms of the antibody. If a chemical substance causes the body to make antibodies against it then such a chemical irritant is said to be antigenic. It is only possible to give a few examples of the many types of antigens. Many bacterial proteins are antigens and stimulate the body to produce antibodies which are very important in the defences of the body against bacteria. In some persons vegetable proteins such as pollen behave as antigens and cause the body to produce antibodies; these antibodies tend to be concentrated in the lining cells of the respiratory system (viz. nose, bronchi and bronchioles) and when these cells loaded with antibodies meet a dose of antigen, as when air containing pollen is breathed in by a susceptible individual, the antigen-antibody reaction which occurs on the surface of these cells irritates the cells in some way and so gives way to the symptoms of hay fever or of asthma. As described above, a transfusion in compatible blood may cause the body to produce antibodies which damages and destroys them. From the above examples it will be seen that the cells on which an antibody-antigen reaction is taking place are damaged or destroyed. A further interesting example of this reaction may be described which is encountered in skin grafting. When a large area of skin has been damaged, for example by a burn, the rate of healing of the skin from the healthy margin is so slow that thin slices of skin from another part of the body are laid by the surgeon in patches on the raw area; and these patches, by means of the growth of new skin cells from their edges, gradually form a new skin cover for the injured area. Now if the injuries are very extensive the patient may not have enough healthy skin from which to take patches to cover the wounds. If the skin from another person is used then the patches will remain healthy on the injured areas for a week or two but then they rapidly die. This is because the patient has made antibodies to these patches of 'foreign' skin, whose proteins have acted as antigens; the antigen-antibody reaction destroys the grafted patches of 'foreign' skin. Grafts of skin and other organs from one person to another usually fail unless the donor and the recipient are identical twins, in which case the proteins of the two individuals, donor and recipient, have identical structure.

Returning now to the red cells, there are two main kinds of antigen on human red cells called A and B. You may have antigen A on your cells and then you are said to have group A blood; if you have antigen B then you have group B blood. If your red cells have both antigens, then you have group AB blood and if you have neither antigen on your red cells then you have group O blood. In the case of blood groups the antibodies to these antigens are naturally occurring and are not the result of introduction into the body of antigens, although abnormal antibodies may develop in this way. These naturally occurring antibodies are dissolved in the plasma. Of course one cannot have antibodies in the plasma which are specific to antigens on the red cells; if one has group A blood then one cannot have antibody a in the plasma or ones own red cells would be destroyed, but one may have other antibodies in the plasma. Thus in group A blood there is antibody b in the plasma and in group B blood there is antibody a in the plasma. In group AB blood there is neither antibody present, and in group O blood both antibodies are present.

Let us now look at some possible transfusions and see which bloods are incompatible. Group A blood can obviously be given to other group A persons since the blood of the latter does not contain antibody a. However group A blood cannot be given to persons of group B or group O because both of the latter contain antibody a in the plasma which would cause clumping of the red cells of the donor.

If you study this table carefully you may be confused by the fact that you can mix group. A with group AB but not AB with group A. It is important to know which is the donor blood and which is the receiving blood. Thus blood of group A can be given to a recipient of group AB in spite of the fact that the donor blood contains antibody b; this is because the plasma containing the antibody b is rapidly diluted when given to the recipient group AB. For this reason persons of group AB have been called universal recipients, that is they can receive blood of any of the AB groups without ill effect. In practice however blood is always cross matched in the laboratory with a specimen of the receiver's blood before a transfusion is given; the presence of antibodies and antigens other than the A-B type make this essential. People of blood group O have been called universal donors, that is their blood can be given to persons of any of the A-B groups without ill effects; this is because their red cells have none of the A-B antigens, whilst their plasma containing antibodies a and b is rapidly diluted in the recipients own plasma. In England the most common blood groups are A (40%) and O (45%) whilst group B (10%) and group AB (5%) are less common.

Rhesus Groups

There is another antigen present on the red cells of some individuals called the rhesus antigen or rhesus factor. It is called rhesus factor because it was first discovered in the blood of rhesus monkeys.

Unlike the A-B antigens there is no naturally occurring antibody to the rhesus factor. Those persons containing the rhesus factor on their red cells are called rhesus positive, whilst those without the rhesus factor are called rhesus negative. Antibodies to the rhesus antigen only develop under certain unusual circumstances; firstly they may develop if an Rh-negative person is given a transfusion with Rh positive cells. The Rh positive cells stimulate the production of rhesus antibodies in the receiver's blood. The second way in which rhesus antibodies can develop is in the case of some pregnant women. The rhesus factor is inherited as a dominant gene and thus it may happen that if a rhesus positive man marries a rhesus negative woman, the rhesus negative woman may bear rhesus positive children. Now in the majority of cases of rhesus negative women bearing rhesus positive children there is no effect on the mother's blood. But in some cases it appears that small leaks of blood from the foetus, through the placenta, into the mother's blood cause the production by the mother of rhesus antibodies. Because of their molecular size these antibodies may gain access to the circulation of the Rh positive foetus through the placenta. When this happens the red cells of the foetus are progressively destroyed because of the antigen-antibody reaction on their surfaces. This effect becomes more important in successive pregnancies of an Rh negative woman bearing Rh positive children and after several Rh positive pregnancies there may be sufficient antibody produced by the mother to produce such a degree of an anaemia in the foetus that the foetus dies in utero. It becomes very important then to know the Rh group of blood before it is transfused to a woman, particularly if she is Rh negative. If by chance Rh positive blood is given to an Rh negative woman it will stimulate the production by her of rhesus antibodies which may damage the infant if the infant so happens to be Rh positive.

Plasma and the Mechanism of Clotting

When blood escapes from the body, as into a wound for example, it quickly changes from the fluid state into a thick jelly like material called a clot. This process of clotting, or coagulation as it is called, is an essential process, in that it prevents the whole of this very valuable fluid from draining from the body from the slightest cut.

We have now to study the way in which fluid blood is so quickly converted into a clot. The process of clotting depends upon the change in state of a protein constituent of plasma called fibrinogen. This protein fibrinogen is normally dissolved in the plasma but on shedding of blood there is a change in the state of the fibrinogen into long fibrous molecules called fibrin, which gradually contract to form a firm clot. As the clot shrinks a clear yellow fluid escapes from it called serum which consists of plasma minus the protein fibringogen, which has now been converted into fibrin. Within the meshes of the clot are trapped the various cellular elements of the blood. On the surface of the body this clot gradually dries to form a scab which forms a mechanical covering to a wound.

The process of clotting is a highly complex one and is dependent upon the presence of a great variety of factors, but the essential process, as described above, is the conversion of the protein fibrinogen from its corpuscular state into its fibrous state, called fibrin. When blood escapes from a blood vessel into a wound a variety of changes in the blood occur; first the blood is escaping from a series of smooth walled vessels into an area exposed to the air, with roughened, often dirty, surface sand escaping into the blood are tissue fluids containing substances derived from damaged cells in the area. Some elements of the blood break down upon the surfaces of the wound, particularly the elements called platelets. During this process there is the conversion of an inactive enzyme present in the blood called prothrombin into its active from, thrombin, and it is this thrombin which triggers off the change in the state of fibrinogen. The various substances which are liberated into shed blood from damaged cells and platelets are called thromboplastic factors and it is these factors which produce the change of the inactive enzyme prothrombin into thrombin. These changes are summarized in the diagram above. Prothrombin is converted into the active form thrombin under the influence of thromboplastic factors, and the thrombin then initiates the change in state of the protein fibrinogen.

Vitamin K

For the manufacture of prothrombin vitamin K is necessary, although it appears that vitamin K does not form part of the molecule of prothrombin. Anything interfering with the absorption of vitamin K interferes with the synthesis of prothrombin and may lead to some failure in the clotting mechanism. The presence of bile is necessary for the absorption of vitamin K from the bowel because it is fat soluble; thus, when the flow of bile is obstructed, for example by a stone in the bile duct, there is a failure to manufacture adequate amounts of prothrombin. The new born infant may also suffer from a deficiency of vitamin K which may lead to the appearance of spontaneous bleeding from various parts of the body. In part this deficiency of vitamin K in the new born is due to the fact that some of the vitamin K absorbed is derived from bacteria in the bowel and the young mammal does not have a full complement of bacteria in the bowel when it is born; it gradually acquires these. A further example of the importance of these bacteria in the bowel in the manufacture of vitamin K is seen in the treatment of patients with large doses of wide spectrum antibiotics, which tend to eradicate many of the organisms in the bowel, and so may lead to reduced prothrombin levels in the blood.

Dicoumarin

A substance called dicoumarin, which is present in spoiled clover, interferes with the manufacture of prothrombin in the liver. This then is a dangerous feeding stuff for cattle. Use of dicoumarin-like-substances is made in certain diseases, in which there is an increased tendency to form clots within the blood stream, such as coronary thrombosis, and here, regular does of these drugs reduces the clotting power of the blood by its effect on prothrombin synthesis, and so may halt the progress of the disease. Prothrombin is a very potent substance and 20 mg. in 100 ml. of blood is more than adequate to clot all the fibrinogen in the blood.

Prothrombin to Thrombin

It is not sufficient merely to have enough prothrombin in the blood but there must also be present a variety of factors which play a role in the conversion of prothrombin into thrombin.

Calcium Ions

Firstly there must be an adequate supply of free calcium ions. When blood is taken from donors it is collected into bottles containing a solution of sodium citrate; this precipitates out the calcium ions as insoluble salts and prevents the blood from clotting. If this blood is given in large amounts to a patient who is already bleeding then it is often necessary to give injections of calcium salts to enable the citrated blood to clot. Other salts will also prevent blood from clotting, including oxalates and fluorides, however both are poinsonous substances and are only used in laboratory tests.

Thromboplastins

Another group of substances necessary for the conversion of prothrombin to thrombin includes thromboplastic substances, sometimes called thromboplastins. They have been identified chemically as phospholipids and can be obtained in extracts from a great variety of tissues. In the clotting process, however, they are usually derived from injured cells and probably also from the breakdown of blood platelets in the area.

Globulins and Haemophila

A third type of substance necessary is included in the globulin fraction of the plasma proteins, and there are several types of globulins concerned in the conversion of prothrombin into thrombin which are given various names, such as accelerator globulin, anti-haemophilic globulin and Xmas factor. Haemophilia is inherited as a sex linked recessive trait, and the gene responsible is carried on the terminal portion of the X chromosome. The human female has two X chromosomes but the human male has only one X chromosome which pairs with a shorter chromosome called the Y chromosome. In the male the terminal portion of the X chromosome is unpaired since the Y chromosome is shorter than the X. It is in this unpaired region of the male X chromosome that the recessive gene responsible for haemophilia is carried.

If a recessive gene is carried on this unpaired portion of the X chromosome in the male, the character will be expressed in the phenotype. In the female however, since the X chromosome is paired along the whole of its length with another X chromosome, a recessive gene will always be paired with another gene; thus female will only have the disease haemophilia if she has a double dose of the haemophilia gene (i.e. the homozygous recessive state), one recessive gene on each X chromosome. If there is only one recessive gene present then the normal allele, or other gene partner, will be present on the other X chromosome, and this normal gene will govern the production of anti-haemophiliac globulin, and the disease will not manifest itself. However, although a female with a single dose of the gene responsible for haemophilia (i.e. heterozygous state) does not suffer from haemophilia she is still able to pass the disease on the male descendants. Females with haemophilia are very rare, because to produce such a female it would mean that a female carrier (or female haemophiliac) would have to marry a male haemophiliac; since the disease is relatively rare such a combination is highly unlikely. But the disease is not transmitted merely by inheritance in this way; there is a fairly high spontaneous mutation rate of the normal gene on the X chromosome, and thus it would be impossible to eradicate the disease by control of marriages. The mode of inheritance of haemophilia is illustrated.

Another type of haemophilia, sometimes called haemophilia B or Xmas disease is due to the absence of another factor from the plasma, necessary for the conversion of prothrombin into thrombin; this factor is called the Xmas factor. It is inherited in a similar way to classical haemophilia (or haemophilia A) and is due to a sex-linked recessive gene. Another rare type of haemophilia, called haemophilia C, is due to the deficiency of another plasma factor. Unlike haemophilia A and B it is inherited as a Mendelian dominant trait, and therefore affects males and females alike. In haemophilia there is a tendency to spontaneous bleeding or of bleeding on trivial injuries; thus minor injuries to joints or muscles may lead to the appearance of large collections of blood in these situations. Continuous bleeding may even occur after minor operations such as dental extraction. The treatment of these bleeding episodes consists in giving the patient plasma containing antihaemophilic globulin. Because normal freeze-dried plasma or normal stored whole blood loses its anti-haemophilic globulin it is necessary to use fresh-frozen plasma which retains its anti-haemophilic globulin.

Heparin

There are several substances present in normal plasma whose actions tend to prevent the development of thrombin within the blood stream and so the appearance of blood clots within the blood vessels. One of threes factors is called heparin. Heparin is produced from special mesenchymal cells present in the connective tissues called mast cells, and the liver is particularly rich in such cell. Heparin is a polysaccharide substance and is used in medicine as an anticoagulant and can be injected into the blood stream in cases where clots have appeared within the blood vessels and so helps to prevent extension of the clotting process.

Other factors have been mentioned which tend to prevent the clotting of blood within the vessels; one of these is the continuous smooth surface of the vessel walls to which the blood is continually subjected. When these surfaces become roughened by the degenerative processes of old age then clots may develop along the roughened surfaces; and when this happens in the coronary arteries supplying the heart muscle with blood the result can be disastrous.

The Heart

A great part of the success of the mammals depends on the efficiency of their circulatory apparatus, achieved by complete separation of the main or *systemic vessels*, as a greater circulation, from the pulmonary or lesser circulation, supplying the lungs. By this means well oxygenated blood is delivered to the capillaries at a far higher pressure than is achieved by a system such as that of a fish with a single circulation, in which the blood passes through two sets of capillaries under the influence of a single pressure source in the heart. The advantage of the separate circulations is probably especially marked in the case of the muscles, which are regions of high oxygen consumption. By means of this arrangement, quick and sustained movements become possible and the whole organization mammalian can reap the advantages of the high body temperature and active chemical processes. It is less important that the interchange between the blood and the internal organs should be rapid and in the liver a *portal system* still remains, in which blood passes successively through two sets of capillaries. Blood from the arteries to the gut is passed first through capillaries in its walls from which it is collected into the *hepatic portal vein*. This proceeds to the liver where the blood passes through the sinusoids around the cells and is then returned by the hepatic vein to the heart.

The separation of the two circulations has been achieved in the mammals by a method quite different from that adopted in the existing reptiles and birds. The original ventral arota of the fish-like plan of the circulation becomes completely divide into two. The Pulmonary artery arises as a single trunk from the right ventricle and supplies the lungs. The arotic arch is also single; arising from the left ventricle it curves over on the left side of the body.

With these modifications the general arrangement of the heart remains in mammals as in lower vertebrates. The organ is formed in the mesoderm below the gut by folding in such a way that it lies in a pericardium, whose inner wall (splanchnopleure) adhered closely to the heart surface, while the outer (somatopleure) makes a sac (the fibrous pericardium in the anatomical sense) separated from the inner layer by a small space containing coelomic fluid. The heart in its sac thus lies in a septum, the *mediastinum*, which runs down the centre of the thorax and separates the two lungs. The pericardium is attached to the diaphragm caudally. The surface of the heart and the inner surfaces of the pericardium are covered with smooth layers of mesothelium, allowing freedom of movement. The outer surface of the pericardium is loosely attached to the ventral thoracic wall and to the mediastinum.

The great veins (superior and inferior venae cavae) return their blood to the righ atrium, a thin-walled chamber, which leads into the right ventricle by an opening guarded by the flap-like right *atrio-ventricular valve*. This valve (the 'tricuspid' value of man) allows forward passage of blood but closes during contraction of the ventricle, being prevented from eversion into the auricle by the pull of a number of *papillary muscles*, attached to the valve by fibrous chordae tendiane. The opening from the right ventricle to the pulmonary artery is guarded by three semilunar valves, arranged to prevent reflux into the ventricle.

Blood returns from the lungs to the left atrium and from here passes through the left atrio-ventricular (*bicuspid* or *mitral*) *valve* to the left ventricle, the valve being checked by a series of papillary muscles even stronger than those on the right. The left ventricle is by far the most muscular of the four chambers and pumps the blood into the arota, whose entrance is guarded by semilunar valves.

It is essential to the proper functioning of this double circulation that there should be proper co-ordination of the working of the two sides of the heart. The heart musculature consists of a form of muscle in which the myofibrils are striated, but there is also much sarcoplasm and many mitochondria the nuclei lie at the centre of the fibers.

Frequent partitions, the *intercalated disces*, cross the fibers, which are thus divided into a series of cells. These discs are supported by desmosomes, but over part of their length they form tight junctions (*zonulae occludentes*). Another characteristic feature is that the fibers branch and anastomose, so hat the whole forms an elaborate net-like arrangement. This allows for much endomysium between the fibers, carrying an abundant blood-supply. The contraction of heart muscle fibers occurs rhythmically and continues without external stimulation from nerves. Characteristic of the action of heart muscle is its very long refractory period. Following one contraction no other will take place during the period of relaxation. This enables proper timing of the beat of the various parts. Contraction originates in one centre, the *sinu-auricular node*, a small patch of tissue in the wall of the right atrium, close to the point of entry of the great veins. This corresponds to the position of the sinus venosus of the fish heart, that is to say the posterior end of the region of the sub-intestinal vessel from which the heart has evolved.

In mammals the sinu-auriculr node acts as the *pacemaker* of the heart. It consists of a specially modified form of muscle fiber faintly striated, intermixed with nerve cells and fibers the vagal and sympathetic branches, both of which regulate the heart. The beat originates 'spontaneously' in the node, but its frequency is controlled by the nerves, the vagal fibers decreasing and the sympathetic increasing the rate. The contraction wave spreads waÿ spreads away from the sinu-auricular node at about 1 ms^{-1} by conduction along the muscle fibers that fan out from this region. The whole musculature of each chamber of the heart thus acts as if it were a single muscle cell. It is probable that electrical propagation between the individual cells occurs by means of the tight junctions. The atria thus contract, approximately together. Transmission to the ventricles does not take place in the same way, there being no muscular continuity, and the atria are separated form the ventricles by a ring of connective tissue, the annulus fibrosus. Control of the beat, therefore, is taken up another special region, the *atrio-ventricular node*, from which impulses are transmitted to the ventricles by a strand of tissue, the *atrio-ventricular bundle of His*. This is formed of modified muscle fibers, small near the node but larger distally and known as *Purkinje* tissue, which conduct impulses at 5 ms^{-1}. This bundle spreads out to join the musculature of both ventricles, whose beat is thus initiated nearly simultaneously at all points. After destruction of the bundle of His the ventricular muscle is able to initiate its own beats, but these are not coordinated with those of the atria.

Adaptation of the Circulatory System

At all times the circulation is being adapted to meet the need to supply an adequate quantity of blood at an adequate pressure to the various organs. These needs are for ever changing locally in that the different organs have varying rates of activity and the blood supply must be adapted to meet these changing needs. Adaptation to meet these needs can occur in one or both of two parts of the circulatory system, firstly the heart, secondly the peripheral blood vessels. We will first describe how the hearts' action is modified to deliver blood at the correct volume and pressure in spite of changing conditions in the peripheral blood vessels.

The heart. The output of blood by the heart may be modified in two ways; firstly by alterations in the rate of the heart beat and secondly by alteration in the amount of blood put out at each beat.

Regulation of Heart Rate

Although the origin of the heart beat is myogenic, modifications in heart rate occur through the influence of the nervous system and endocrine organs. The heart is under the influence of the two divisions of the autonomic nervous system, the parasympathetic nervous systems. The parasympathetic system exerts its effect on the heart by means of the vagus nerve. The sympathetic nervous system can influence the heart in two ways; firstly by means of sympathetic nerves to the heart itself and secondly by means of the hormone adrenaline produced in the adrenal medulla.

In the mammalian heart the vagus nerve spreads extensively in the fibers of the sino-auricular node, in the auricular musculature, auriculoventricular node and along the branches of the Bundle of His. The effect of vagal stimulation on the heart is to slow the heart rate and therefore reduce the amount of blood put out by the heart. This depressing action of the vagus nerve is brought into action by means of sensory endings situated in the arch of the aorta and in the carotid sinus. In many mammals these sense endings are in constant activity and by their connections within the brain stimulate the vagus to exert its restraining action on the heart. This restraining action of the vagus varies from one animal species to another and in man it tends to be highest in highly trained athletes who characteristically have a slow pulse rate. The sensory endings in the aorta and carotid sinus are increasingly stimulated by a rise in blood pressure, which produces a reflex slowing of the heart, thus tending to prevent an excessive rise in pressure. This reflex is described in Marey's law of the heart which states that the pulse rate varies inversely with the arterial blood pressure, and is the expression of vagal restraint upon the heart.

The heart is also supplied with branches of the sympathetic nervous system whose effect is augmented by the secretion of adrenaline from the adrenal medulla, which reaches the heart through the coronary arteries. The effect of stimulation by the sympathetic nervous system is to increase the heart rate and the force of each heart beat; this produces an increase in the output of blood and a rise in arterial blood pressure. The whole metabolic activity of the heart is raised, showing itself by an increased rate of utilization of glucose and lactic acid by the cardiac muscle. It will be seen that the effects of the sympathetic and parasympathetic nervous systems are in opposition and by variations in these two controls the heart's action can be adapted to meet a variety of circumstances.

Cardiac Muscle and Cardiac Output

A further way in which the output of blood from the heart can be altered is due to a property of heart muscle itself, when heart muscle is stretched it is capable of working harder. Thus, if there is an increase in the return of venous blood to the heart the output of blood from the heart can be increased because the cardiac muscle is stretched during diastole and therefore works harder. During exercise there is an increased flow of blood returning to the heart due to the effect of contractions of skeletal muscles around the veins, which pushes the blood onward to the heart. In these circumstances there is an increase in ilic diastolic volume of the heart which stretches the cardiac muscle and so the output of blood increases, thus meeting the increased needs of exercise.

Changes in the Arterioles

The second way in which the circulation can be adapted to meet varying needs is at the level of the arterioles. Like the heart these are under the control of the autonomic nervous system. The effect of stimulation of the sympathetic nervous system is to constrict the arterioles and so reduce the blood supply to the tissues whilst the effect of stimulation of the parasympathetic nervous system is to oppose this effect and to dilate the arterioles. The centre which controls these vasomotor nerves is in the medulla oblongata, with subsidiary centres in the spinal cord. The vasomotor centre itself is sensitive to the carbon dioxide content and pH of the blood passing through it;

when there is a rise in carbon dioxide or acidity then the sympathetic division of the vasomotor centre is stimulated, producing a constriction of the arterioles of the skin and gut with a consequent rise in blood pressure. Further the vasomotor centre is connected by nerves to special sense structures situated along the carotid arteries called the carotid bodies; these are also stimulated by a rise in the carbon-dioxide content of the blood, and reflexly stimulate a rise in blood pressure.

During physical activity the muscle tissues produce increased amounts of carbon dioxide. This increased output of carbon-dioxide, by means of the effects on the carotid bodies and vasomotor centre produces a generalized constriction of arterioles. But a local accumulation of carbon-dioxide has a direct effect upon the artrioles causing them to relax. Thus the blood is diverted into the dilated arterioles of the active muscles.

The arterioles are also affected by hormones, particularly by adrenaline. The effect of adrenaline in the circulation is to cause the constriction of the arterioles in the skin and gut, diverting the blood to more important regions.

Adaptation to Stress Situations

We can now consider some examples of the way in which these mechanisms are brought into action. First we will consider the 'stress' situation which occurs when an animal is confronted by a potential danger. In this situation there is an increased activity of the whole of the sympathetic nervous system and an outpouring of adrenaline from the adrenal medulla. By these means the heart rate is increased together with the cardiac output, and so the circulation is adapted to meet the need for increased activity. The adrenaline also stimulates the contraction of the arterioles of the skin and gut, diverting blood into the more important organs, brain, lungs and muscles.

Adaptation to Blood Loss

Secondly, we will consider adaptations of the circulation to blood loss incurred for example by an injury. When this occurs there is an immediate fall in blood pressure; this is a dangerous situation in which many vital organs e.g. brain and kidney are being deprived of blood. With the fall of blood pressure the heart is released from vagal restraint (Marey's law) and there is an increase in heart rate which, in itself, tends to promote a rise in blood pressure.

Further there is a reflex contraction of the arterioles in a variety of organs (mediated through the sympathetic nervous system) including

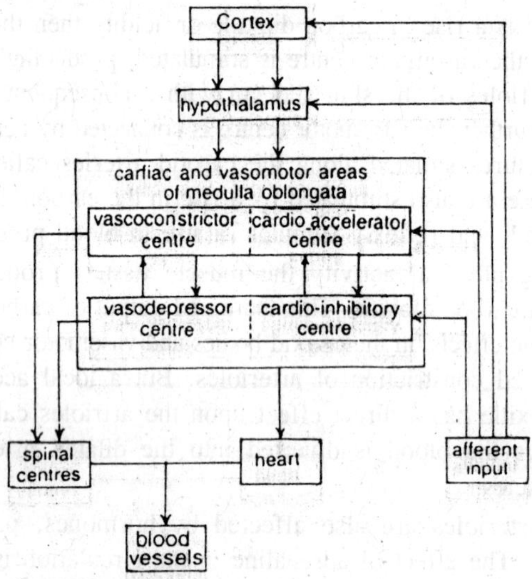

Fig. 12.11. Diagrammatic scheme for central control of the cardiovascular system in nervous control of the heart.

the skin and gut, so conserving blood for more vital functions. The effects of these are seen in man in cases of shock due to haemorrhage where the pulse is rapid and the skin pale, cold and clammy due to the constriction of the arterioles to the skin. Formerly, part of the treatment of shock due to haemorrhage consisted of warming the patient; but it will be understood that warming the skin may promote a reflex vasodilation of the arterioles and so oppose the blood conserving effect of vasoconstriction; this may result in further lowering of the blood pressure.

Regulation of the Blood-pressure

The work involved in the circulation is immense. A man's heart beats over 2600 million times in an ordinary life, pumping a total of 150000 tonne of blood from each ventricle. All this is done without any serious irregularity and with no rest longer than 0.75s.

The rate of the heartbeat varies greatly in different animals, being lower in larger than in smaller animals (roughly 25 per minute in the elephant, 70 per minute in man, 300-500 per minute in the mouse). It also varies with activity, with 'emotional' and other factors, and with 'training', being lower in athletes than in sedentary individuals. The means by which the 'basic' rate is determined are not known but the nervous control is reflex, there being receptor systems in the heart

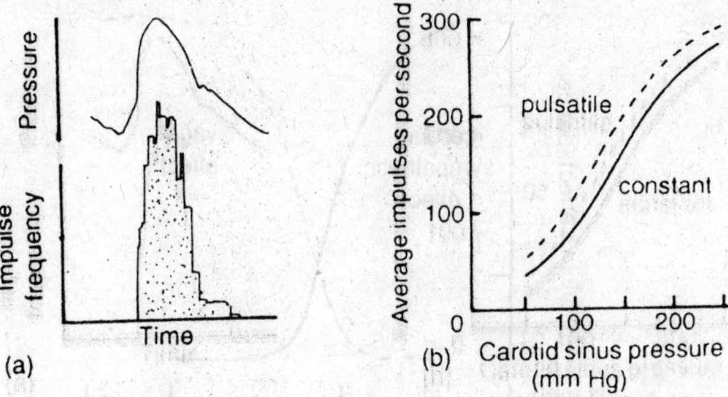

Fig. 12.12. Frequency of impulse firing in a carotid baroceptor and the pressure pulse at normal arterial pressure.

itself, in the main blood-vessels and in the lungs, whose discharge produces alternation of the rate of the heartbeat.

The vagus carries afferent fibers from receptors in the arch of the aorta. In the rabbit they run in a separate *depressor nerve*. They discharge when the pressure is raised. The effect of these impulses is then to lower the blood pressure by action through a centre in the medulla oblongata on the efferent sympathetic and vagal nerves, the latter reaching the sinu-auricular node through the cardiac plexus. If the depressor nerve is cut stimulation of its central end produces slowing of the heart; but stimulation of the peripheral and is without effect. Other receptors are those of the *carotid body* and *carotid sinus*, which

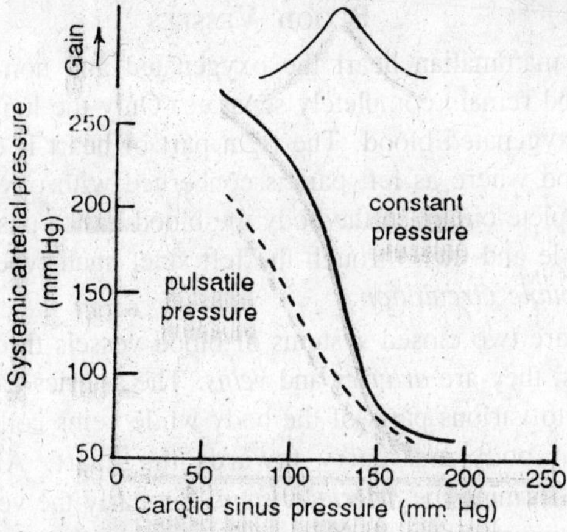

Fig. 12.13. Relationship of mean carotid sinus pressure and mean systemic arterial pressure when pressure in isolated sinus is either constant or pulsatile.

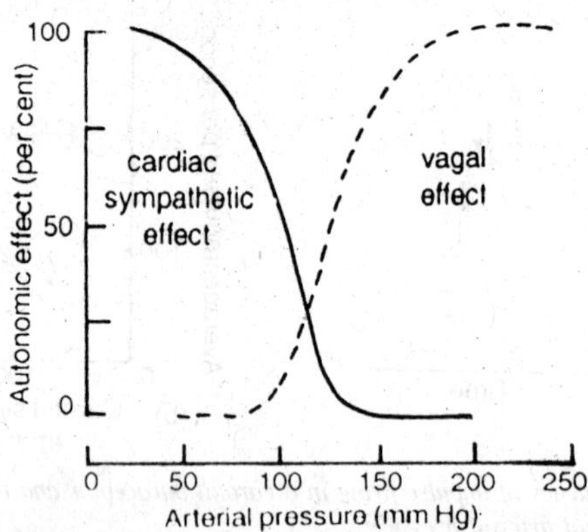

Fig. 12.14. Diagrammatic relationship of cardiac sympathetic and vagal autonomic effects at different arterial pressure levels.

lie close to the bifurcation of the carotid arteries and contain cells sensitive to changes of oxygen and carbondioxide content of the blood (chemoreceptors) as well as to pressure changes (baroreceptors). By the discharge of this complex set of receptors, as well as of others in the lungs, the heart-rate is adjusted to suit the requirements of the moment. The vagal efferent fibers slow the heart by inhibiting the atria (cholinergic). The sympathetic efferents accelerate it by direct action upon the pacemaker and ventricular muscle (adrenergic). When a man rises from the lying to the standing position there is an increase in systolic pressure.

Blood Vessels

In the mammalian heart the oxygenated and non-oxygenated or venous blood remain completely separate. Only the lungs are supplied with non-oxygenated blood. The right part of heart is concerned with venous blood where as left part is concerned with oxygenated blood. In one complete circuit in the body the blood has to pass once through the right side and then through the left side, such type of circulation is called *double circulation*.

There are two closed systems of blood vessels through which the blood flows, they are *arteries* and *veins*. The arteries carry pure blood from heart to various parts of the body while veins collect blood from all parts of body and carry towards the heart. All the arteries collectively forming the *arterial system*. Similarly the veins collectively form the *venous system*. Arteries supplying blood to particular organs ramify to form *arterioles* finally forming *arterial capillaries*. *Venous capillaries* unite together to form *venules* and finally to form veins.

Arterial System

The arterial system has two distinct vessels. (A) *Carotico-systemic aorta* (B) *Pulmonary aorta.*

Carotico-systemic aorta

Includes the aortic arch and its branches. The aortic arch arises from the left ventricle and turns towards the left. It loops round the left bronchus and reaches the dorsal side of heart, then it runs backward as dorsal aorta along the mid-dorsal line beneath the vertebral column in thorax and abdomen. The main arteries arise from aortic arch are as follows.

1. *Coronary arteries*. A pair of small arteries, *left* and *right coronary arteries*, are given from the carotico-systemic aorta. These arteries supplying blood to heart itself.
2. *Innominate artery*. From the arching place of aortic arch an innominate artery is given out. It soon gives off a *left common*

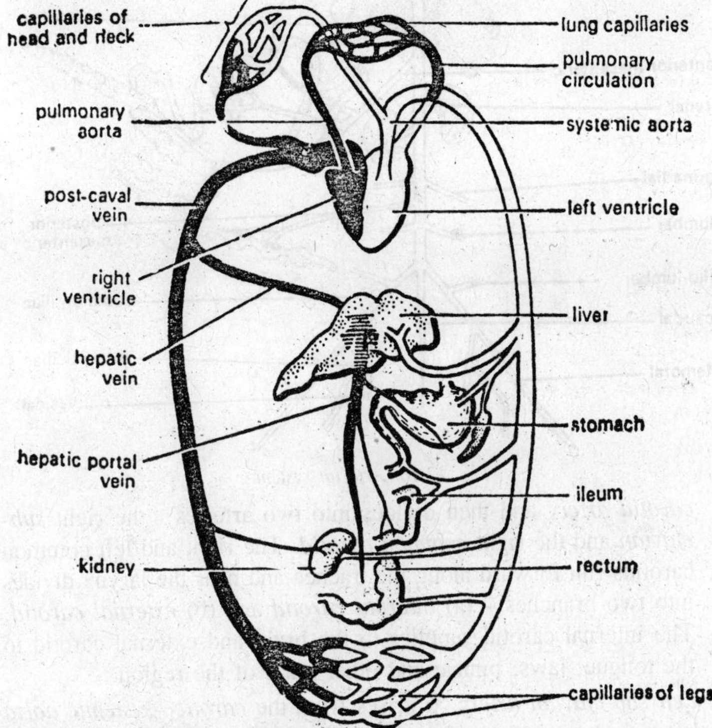

Fig. 12.15. Diagram showing double circulation of blood in a mammal.

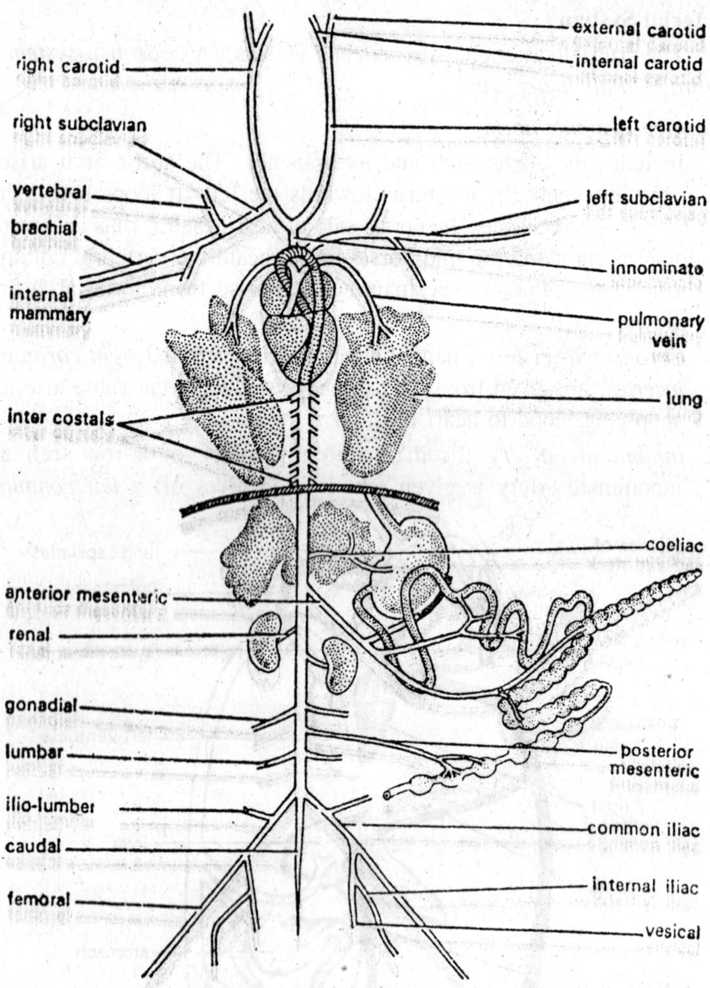

Fig. 12.16. Arterial system.

carotid artery and then divides into two arteries : the right *sub-clavian* and the right *common carotid*. The right and left common carotids run forward along the trachea and near the larynx divides into two branches – (a) *internal carotid* and (b) *external carotid*. The internal carotid supplies to the brain and external carotid to the tongue, jaws, pinnae and other parts of the region.

3. *Left sub-clavian artery*. It arises from the *carotico-systemic aorta* after it has given off innominat eartery and is just curving backwards. Both the sub-clavians i.e. right and left run into the

shoulder of their side and each is divided into following three branches.

(a) *Vertebral artery*. It enters the vertebraterial canal of the cervical vertebrae and supplies blood to the brain and spinal cord, neck and shoulder muscles.

(b) *Internal mammary artery*. It gives off small branches to the pericardium, diaphragm and extends over the ventral surface of abdomen as *superior eigastric artery*.

(c) *Brachial artery*. It enters the fore limbs of its side and supplies it with several branches.

4. *Intercostal arteries*. *Dorsal aorta* travels into the thorax piercing the diaphragm called *thoracic aorta*. From thoracic aorta a series of small paired arteries are given out called *intercostals arteries* supplying blood to the intercostals muscles between the ribs.
5. *Phrenic arteries*. These are a pair of fine arteries arising from the thoracic aorta behind intercostals arteries. They supply blood to diaphragm.
6. *Coeliac artery*. It is a median artery. It arises from the thoracic aorta about 12 mm, behind the diaphragm. It is bifurcated into:

   (a) *Hepatic artery*. Supplies blood to liver.

   (b) *Lieno-gastric artery*. Supplies blood to stomach by *gastric artery* and spleen by *splenic artery*.
7. *Anterior mesenteric artery*. Some distance after the celiac artery, the thoracic aorta gives off *anterior mesenteric artery* on the left side. Branches of this artery supply pure blood to duodenum, pancreas, ileum, ceacum and colon.
8. *Renal arteries*. Behind mesenteric artery a pair of renal arteries arises from the aorta. The right renal artery arises a little above the left and these supply blood to ovaries.
9. *Genital arteries*. They arise after the renal arteries. In male they are called *spermatic arteries* supply blood to testis and in female they are termed as *ovarian arteries* supply blood to ovaries.
10. *Posterior Mesenteric artery*. It is a small median artery that runs in mesentry. It supplies blood to hind part of colon and rectum.
11. *Lumbar arteries*. These are paired arteries supply blood to the dorsal body wall of the abdomen.
12. *Common iliac arteries*. These are a pair of arteries formed by the bifurcation of aorta. Each common iliac artery divides into:

    (a) *Internal iliac artery*. It runs backwards, and supplies blood to the dorsal wall of pelvic cavity.

(b) *External iliac artery.* It gives off following arteries.

(i) *Vesicular artery.* Supplies to the urinary bladder.

(ii) *Ulterine artery.* It supplies blood to the uterus in female and may arise from vesicular artery.

(iii) *Posterior epigastric artery.* Distributes blood to the ventral wall of the abdomen.

13. *Sacral artery or Caudal artery.* The dorsal aorta continues backwards into the tail and supplies blood there.
14. *Ilio-lumbar artery.* Near the origin of common iliacs, a small iliolumbar artery arises and supplies blood to the back muscels.

Pulmonary aorta

Gives rise pulmonary arteries to lungs

Venous System

The venous system of mammals is divided into four parts:

1. Pulmonary Venous System
2. Venae-cavae System
3. Portal System.
4. Coronary System.

Pulmonary venous system

It includes a pair of *pulmonary veins* which bring oxygenated blood in the left atrium of heart. Both the veins open into left auricle of heart by a single aperture.

Venae-cavae system

It consists of three caval veins and their branches. There are two anterior venae a right and left and a posterior vena cava.

Anterior venae-cavae

They receive six veins.

(i) *External jugular vein.* It is large vein running through the neck outside the trachea. It is formed by the union of *arterior* and *posterior facial veins* that bring blood from the face and ears respectively.

(ii) *Internal jugular vein.* It is a slender vein draining blood from brain. It runs down the neck along side the trachea and opens into the external jugular vein near its junction with the sub-clavian. It brings blood from brain and the muscles of the neck. It rabbit, the two external jugular veins are connected with each other by *transverse jugular anastomosis*.

(iii) *Sub-clavian vein.* Each brings blood from forelimb and shoulder and opens into jugular vein.

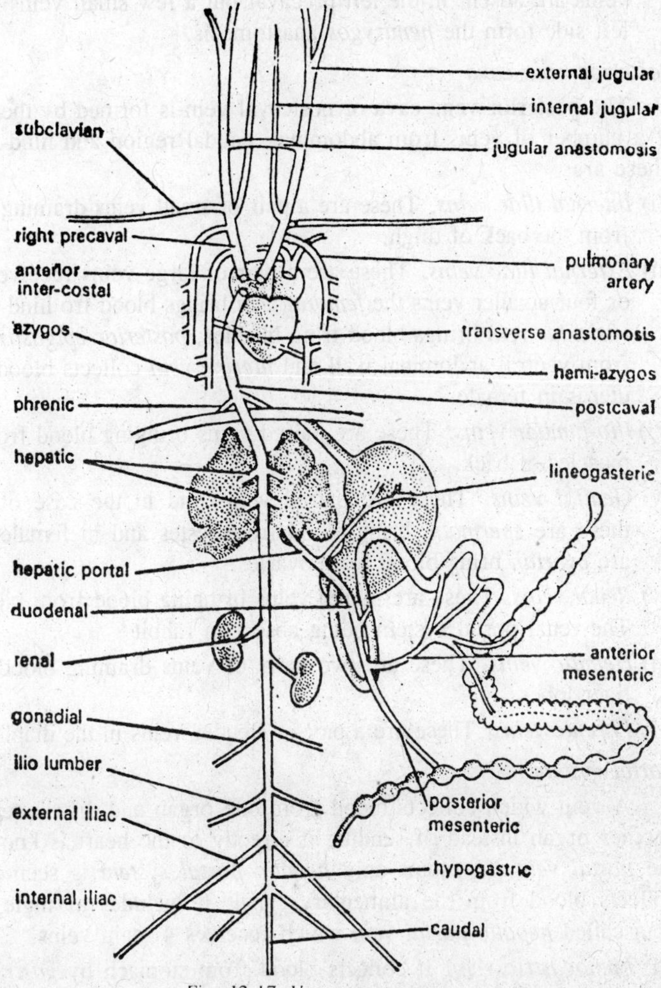

Fig. 12.17. Venous system.

(iv) *Intercosal vein.* It collects blood from first 4 or 5 intercostal muscles of its side and opens into vena cava behind the internal mammary vein.

(v) *Internal mammary vein.* It is a fine vein receiving blood from the ventral wall of the thorax.

(vi) *Azygos vein.* It is a median vein running forward along the mid-dorsal line of thoracic wall. It brings blood from posterior intercostals spaces and lumbar region. The azygos and intercostals veins are absent in the left precaval but a few small veins on the left side form the *hemizygos* anastomosis.

Posterior vena cava

The posterior vena cava or postcaval vein is formed by the union of a number of veins from abdominal, caudal region and hind-limbs. These are:

(i) *Internal iliac veins.* These are a pair of small veins draining blood from the back of thigh.

(ii) *External iliac veins.* These are a pair of large veins receive three or four smaller veins the *femoral vein* brings blood fro hind-limbs; *vesicular vein* brings blood from bladder; *posterior epigastric vein* from ventral abdominal wall and *uterine vein* collects blood from uterus in female.

(iii) *Ilio-lumbar veins.* These are paired veins bringing blood from the muscles of back.

(iv) *Genital veins.* These are paired veins and in the case of male these are *spermatic* bring blood from testes and in female these are *ovarian* bring blood from ovaries.

(v) *Renal veins.* These are paired veins bringing blood from kidneys. The renal portal system being absent in rabbit.

(vi) *Hepatic veins.* These are two pairs of veins draining blood from liver lobes.

(vii) *Phrenic veins.* These are a pair of slender veins in the diaphragm.

Portal system

A vein which collects blood from one organ and distributes it to another organ instead of sending it directly to the heart is known as the *portal vein.* In rabbit only *hepatic portal system* is seen which collects blood from the alimentary canal. It includes a single large vein called *hepatic portal vein* which receives 4 main veins.

(i) *Lienogastric vein.* It collects blood from stomach by *gastric* and spleen by *splenic veins.*

(ii) *Duodenal veins.* Receive blood from duodenum and pancreas.

(iii) *Anterior mesenteric vein.* It collects blood from the ileum, caecum, colon and anterior part of rectum by several small branches.

(iv) *Posterior mesenteric.* From lower part of rectum.

The hepatic portal vein divides into capillaries and supplies blood to various lobes of liver. Here, excessive sugar and other nutrients are retained and stored for the future use. Blood from liver is drained into postcaval by the hepatic veins.

13

RESPIRATORY SYSTEM

Every living cell in an organism consumes oxygen. A supply of oxygen is essential to that phase of metabolism known as *catabolism*. Oxidation of substances within the cells results in the liberation of heat and other forms of energy and in the production of carbon dioxide. This end product of metabolism, unless removed from the cell (or body), acts as a poison which is harmful to protoplasm. It has been estimated that the average human adult exhales about 20 liters of carbon dioxide per hour. In a poorly ventilated room occupied by several people, the percentage of carbon dioxide is not poisonous unless the gas is present in large quantities; air containing more than 15 parts in 10,000 is not fit for respiration under any conditions.

RESPIRATION

Aerobic Respiration

The term aerobic respiration is used to denote the exchange of oxygen and carbon dioxide between an organism and its environment. In this type of respiration oxygen is taken into the body and carbon dioxide is given off.

Anaerobic Respiration

In certain organisms *annerobic* respiration takes place. Carbon dioxide is given off, but no oxygen is taken in. The necessary oxygen is obtained, in such forms, by the incomplete breakdown of carbohydrates, and possibly of fats, in the body.

Role of Blood in Respiration

Although in members of certain lower phyla there is a direct exchange of gases between the cells and the environment, in vertebrates and some other forms the blood serves to bring oxygen from the environment to the cells and to carry carbon dioxide from the cells to the environment. It is the presence of the pigment hemoglobin that gives vertebrate blood its unusual capacity for carrying oxygen. Hemoglobin is present in the blood of all vertebrates and even in certain invertebrates. In vertebrate blood it is confined to the red blood corpuscles, or erythrocytes. It is interesting that the muscles of birds and mammals also contain a form of hemoglobin called *myoglobin*. There are minor differences in the hemoglobins of different forms and they vary somewhat in their properties. The importance of hemoglobin in connection with respiration lies in its ability to combine with oxygen through the medium of iron which it contains. Two atoms of oxygen, under favorable conditions, unite with each atom of iron in the hemoglobin complex. Thus $Hb_4 + 4O_2 = Hb_4O_8$. Hemoglobin which has combined with oxygen in this manner is spoken of as oxyhemoglobin. It is scarlet red in color and a very unstable compound, since it gives up its oxygen so readily. After having given off its oxygen it is known as *reduced hemoglobin*, or simply *hemoglobin*, and has a purplish-blue colour.

The red corpuscles with their contained hemoglobin also play a prominent role in transporting carbon dioxide. When carbon dioxide leaves the cells and tissues of the body it enters the bloodstream and diffuses into the red blood corpuscles. Here most of the carbon dioxide combines with water to form carbonic acid as follows: $CO_2 + H_2O = H_2CO_3$. An enzyme, *carbonic anhydrase*, present in the red cells, facilitates the reaction. The carbonic acid thus formed combines with potassium (formerly in combination with hemoglobin) to form potassium bicarbonate. Bicarbonate ions then diffuse out into the plasma in exchange for chloride ions which enter the cell. In the plasma the bicarbonate ions combine with sodium to form sodium bicarbonate. Thus, carbon dioxide is transported by the plasma as well as by the red corpuscles. At the respiratory surface oxygen enters the red blood corpuscles and the above processes are reversed. Carbon dioxide is then liberated as a free gas.

Phases of Respiration

In animals in which the blood is used in carrying oxygen and carbon dioxide, two phases are involved in respiration. They are referred to as external and internal respiration. The term external respiration is used to denote the exchange of gases between the blood and the environment. This usually takes places in the capillaries of the gills or lungs, but in some cases other structures such as the skin are utilized. In gills the capillaries are almost indirect contact with the water in which oxygen is dissolved. In lungs they are practically in contact with air that has been taken into the cavities of the lungs. Even though lungs are situated within the body, nevertheless the term external respiration is used. During external respiration the blood takes up oxygen and loses most of the carbon dioxide which it is carrying. Internal refers to the gaseous exchange between the blood and the tissues or cells of the body. Oxygen is taken from the blood, which in turn gains carbon dioxide given off by the tissues. The principles involved in internal and external respiration follow the "laws of gases" and are physical rather than biological in nature.

In vertebrates, respiratory organs are present which serve to facilitate external respiration. Certain requirements are demanded of respiratory organs in order that they may function properly: (1) a large, vascular surface area must be provided so that an ample capillary network may be exposed to the environment; (2) the membrane surfaces through which gaseous exchange occurs must be moist at all times and be thin enough to permit the passage of gases; (3) provision must be made for renewing the supply of the oxygen-containing medium (air or water) which comes in contact with the respiratory surface and for removing the carbon dioxide which has been given off from that surface; and (4) blood in the capillary network must circulate freely.

With few exceptions, the organs of respiration in vertebrates are formed in connection with the pharynx. In some forms, notably amphibians, the skin itself is an important respiratory organ. In a fish, the loach, *Misgurnus fossilis*, a peculiar method of respiration is utilized. The animal has the habit of swallowing air, passing the air bubble the length of the intestine, and voiding it at the anus, oxygen being absorbed en route by blood vessels in the extremely vascular intestine.

The function of the pharynx as part of the digestive tract is merely to serve as a passageway from mouth to esophagus and to initiates swallowing movements. However, the part it plays in respiration is of much greater importance. The internal gills of aquatic vertebrates and the lungs of air breathers are structures, derived from the pharynx, which are particularly adapted for respiration. In some cases the vascular epithelium of the pharyngeal wall itself is important in respiration. Other structures not associated with respiratory activity are derived from the pharynx. Among these are the swim bladders of certain fishes, tonsils, middle car and Eustachian tube, as well as such glands as the thyroid, thymus, parathyroids, and ultimobranchial bodies.

Nose

The nose is subdivided into two parts; the external nose or snout, which projects from the face; and the internal nose or nasal cavity, which is divided by a median septum into right and left nasal chambers or fossae. The nasal chambers communicate in front with the exterior through the anterior nares, or nostrils, and behind with the nasopharynx through the posterior nares, or choanae. Communications also exist with the sphenoidal sinuses, the frontal sinuses, the air cells of the labyrinths of the ethmoid, and with the orbit via the nasolacrimal canal (in man also with the maxillary sinuses).

The framework of the nose consists, in small part, of cartilage and, in large part, of bone. In the cat the roof of the nasal cavity is formed by the nasal bones and parts of the frontals; the floor by the maxillaries, premaxillaries, and the horizontal plates of the palatines; and the sides by the palatines, maxillaries, premaxillaries, lacrimals, and frontal bones. The nasal septum is formed by the perpendicular plate of the ethmoid and by a median cartilage which is continuous with it. The two nasal chambers in the cat are nearly filled by the ethmoturbinals or labyrinths of the ethmoid, the superior nasal conchae which project inward from the medial surfaces of the maxillary bones. Each nasal chamber is thus divided into three narrow spaces, the nasal meati. The ventral or inferior meatus lies ventrad of the inferior nasal concha. The superior or dorsal meatus lies ventrad of the superior nasal cocha. The middle meatus, which in the cat is nearly filled by the ethmoid, is merely the space between the superior and inferior meatuses.

The nasal chambers are lined by the nasal mucous membrane which is adherent to the underlying bone. This membrane is continuous with the mucous membranes of the paranasal sinuses, the nasolacrimal duct, and the nasopharynx. In the larger respiratory region of the nose, the epithelium of the mucosa is ciliated and pseudostratified columnar. In the smaller olfactory region, the epithelium is devoid of cilia and consists of supporting and olfactory cells. The olfactory cells are bipolar nerve cells. Their central processes, fibers of the fila olfactoria of the olfactory nerve, pass through the cubiform plate of the ethmoid bone and terminate in the olfactory bulb. Collectively the olfactory cells constitute the olfactory organ. The lamina propria of the olfactory mucosa contains the olfactory glands of Bowman whose secretion keeps the surface moist and furnishes a solvent for odoriferous particles. Odoriferous substances in solution stimulate the olfactory cells of the mucous membrane (Schneiderian membrane).

In addition to its primary function as an olfactory organ, the nose also has a respiratory function. The respiratory area serves as an air conditioner. In its passage through the nose, air is saturated with moisture by the nasal secretions; it is warmed by the highly vascular mucous membrane; it is freed of harmful particles by the mucous membrane; it is warmed by the highly vascular mucous secretions. The coat of mucus over the ciliated surface is moved by ciliary action into the pharynx whence it passes to the stomach where it is digested.

LARYNX

The larynx or organ of voice, is an irregularly tubular structure situated in the cranioventral part of the neck. It serves to connect the pharynx and the trachea. The framework of the larynx consists of

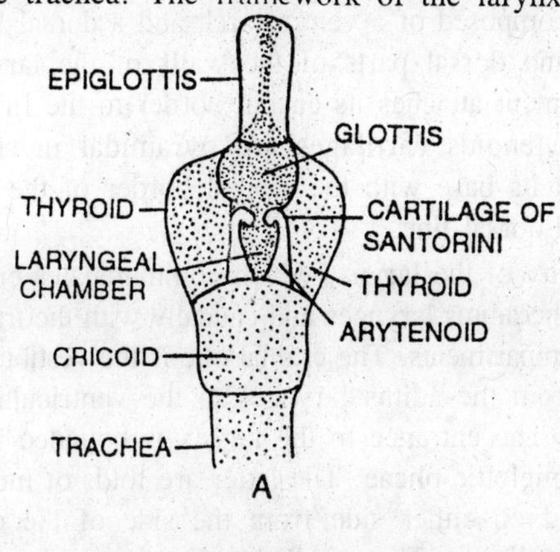

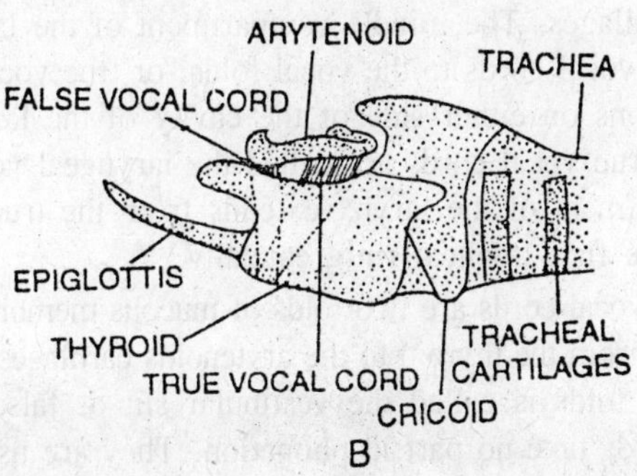

Fig. 13.1. A—larynx (dorsal); B—same (lateral).

several cartilages which are held together by ligaments and moved by muscles. The cavity of the larynx is lined by a mucous membrane. Sound is produced in the larynx by vibration of the true vocal cords which are set into vibration by expiratory currents from the lungs.

The principal cartilages of the larynx in the cat are the thyroid, the cricoid, the epiglottic, and the two arytenoids cartilages. The thyroid cartilage is the largest element in the laryngeal framework. It is shield-shaped and forms about two-thirds the circumference of a circle. It is so placed that it surrounds the other cartilages ventrally and laterally. The hypothyroid membrane connects its cranial border to the hyoid bone, and the cricothyroid ligament connects the middle of its caudal border to the cricoid cartilage. The dorsal border of the thyroid cartilage on either side projects craniad as the cranial thyroid cornu and caudad as the caudal thyroid cornu. The former articulates with the caudal hyoid comu and the latter with the cricoid cartilage. The leaf-shaped epiglottic cartilage is attached to the middle of the cranial border of the thyroid cartilage. The cricoid cartilage has the shape of signet ring and is composed of a ventral arch and a dorsal lamina. It forms the caudal and dorsal parts of the walls of the larynx. The crico-tracheal ligament attaches its caudal border to the first tracheal ring. The two arytenoids cartilages are pyramidal in shape, and each articulates at its base with the cranial border of the cricod cartilage near the mid-dorsal line.

The cavity of the larynx, which communicates cranially with the pharynx by the aditus larynges and caudally with the trachea, is divided into three compartments. The cranial one is the vestibule of the larynx. It extends from the aditus larynges to the ventricular folds or false vocal cords. The entrance to the larynx is bounded by the epiglottis and the aryepiglottic plicae. The latter are folds of mucous membrane which extend on either side from the side of the epiglottis to the arytenoids cartilages. The middle compartment of the larynx extends from the false vocal cords to the vocal folds, or true vocal cords. The lateral extensions on either side of the cavity of the larynx between the false and true vocal cords are called the laryngeal ventricles. The caudal compartment of the larynx extends from the true vocal cords cranially to the first tracheal rinfg caudally.

The false vocal cords are two folds of mucous membrane extending ventrodorsally from the thyroid to the arytenoids cartilages. The interval between these folds is called the vestibular slit or false glottis. The false vocal cords take no part in phonation. They are used in holding inspired air in the lungs so that the intra-abdominal pressure may be raised by contraction of the abdominal muscles to assist in the acts of micturition defection, and parturition.

The true vocal cords are two elastic ligaments covered by mucous membrane which extend ventrodorsally from the dorsal surface of thethyroid cartilage near the midline to the vocal processes of the arytenoids cartilages. The slit-like aperture between the true vocal cords is termed the glottis or true glottis.

The muscles of the larynx are subdivided into two groups: the extrinsic muscles which connect it with surrounding structures, and the intrinsic muscles which take origin and insert in the larynx. The extrinsic muscles of the larynx are described in the section on the muscular system. The intrinsic muscles of the larynx are the cricothyroid dues, the posterior cricoarytenoideus, the lateral cricoarytenoideus, the arytenoideus, and the thyroarytenoideus. They are innervated by the vagus nerve, chiefly by the inferior laryngeal branch.

The pentinent facts concerning the intrinsic muscles are as follows:

(a) *M. cricothyroideus*

Origin: ventral surface of the cricoid cartilage.

Insertion: caudal margin of the thyroid cartilage.

Action: tensor of the true vocal cords.

(b) *M. cricoarytenoideus posterior*

Origin: dorsal surface of the cricoid lamina.

Insertion: dorsal border of the caudal end of the arytenoids cartilage.

Action: widens the glottis.

(c) *M. cricoarytenoideus lateralis*

Origin: lateral part of the cranial border of the cricoid cartilage.

Insertion: caudolateral angle of the arytenoids cartilage.

Action: narrows the glottis.

(d) *M. arytenoideus*

Origin: arytenoids cartilage. Insertion: fellow of the opposite side.

Action: arrows the glottis.

(e) *M. thyroarytenoideus*

Origin: dorsal surface of the thyroid cartilage.

Insertion: caudolateral angle of the arytenoids cartilage.

Action: relaxes the true vocal cords.

Trachea and Lungs

The nasal passages with their vascular membranes for warming and cleaning the inspired air have already been described. The respiratory system proper may be said to begin where the air enters the *larynx*, whose inlet is guarded by the epiglottis.

The *trachea* is a tube strengthened by incomplete rings of cartilage which prevent is collapse during inspiration. At its lower end the trachea divides first into a pair and then into a number of secondary divisions or *bronchi*, leading in turn to smaller divisions, the *bronchioles*. From each bronchiole a system of irregular alveolar ducts leads to

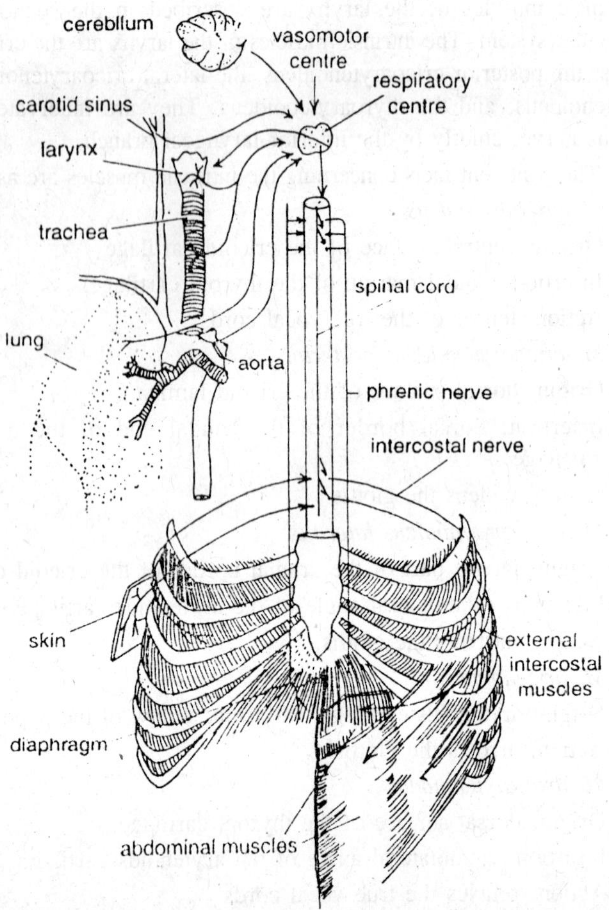

Fig. 13.2. Diagram of the respiratory muscles and the influences acting upon them.

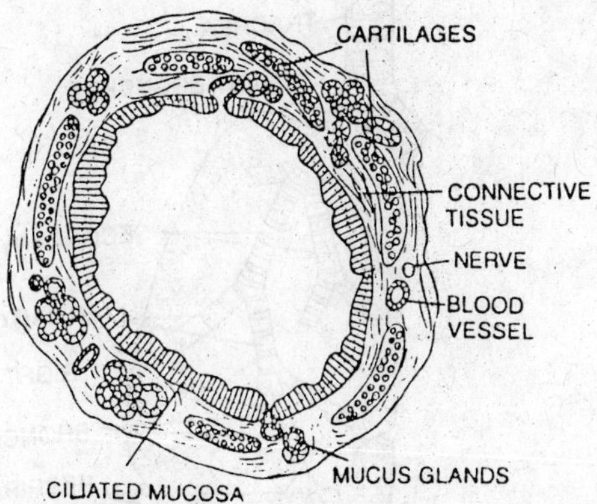

Fig. 13.3. T.S. through the trachea.

the terminal alveoli, chambers wider than tehbronchioles and each lobed to form a number of air sacs. The effect of this branching system of tubes is to allow air to penetrate to every portion of the lung, which is thus a spongy elastic tissue, honeycombed with passages. In the embryo the branching system of tubes is sharply defined and the terminal branches are lined by a definite epithelium. In the adult the finer branches are less regular and the structure of the tree is not apparent in sections.

The walls of the finest chambers are excessively thin, allowing a very close relationship between the air and the blood. Outside the capillary wall there is a further thin membranous layer of flattened pulmonary epithelial cells. Each sac is of the order of 100 m across and there are probably more than 700 million of them in the two lungs of man, providing an immense surface (at least 50 m^2) for gaseous interchange. The presence of this great number of tiny sacs and of the elastic tissue between them make the lungs compact organs, very different from the flimsy sacs of the frog, which collapse to almost nothing when punctured. The capillaries of the alveoli arise from branches of the pulmonary artery and discharge into the pulmonary veins. In addition the whole tree is served by bronchial arteries, bringing oxygenated blood from the thoracic aorta. The blood from the tree as far as the second-order bronchi is returned by bronchial veins to the right atrium.

A striking characteristic of the lung is its elasticity; the walls of the alveoli and bronchioles are abundantly supplied with elastic bibres, giving them the power to contract after they have been dilated, which is important because expiration is otherwise largely a passive process,

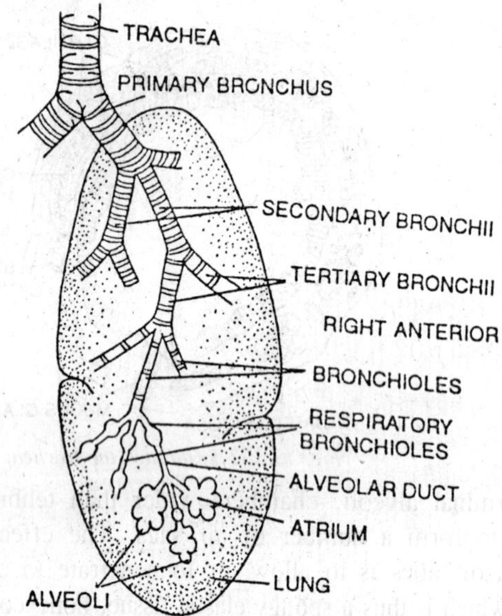

Fig. 13.4. Respiratory tree in a mammalian lung.

produced with only a slight muscular effort. The walls of the bronchioles also contain smooth muscles. These receive motor innervation from the vagus nerve and inhibitory sympathetic fibres. There are also many afferent fibres in the lungs.

The trachea and large bronchioles are lined by ciliated cells, which beat in such a way as to transfer substances upwards towards the pharyns, protecting the respiratory tree. Ciliated cells are highly specialized for their functions. Each cilium is a composite structure with an outer sheath and a central core of fibrils, which are attached to basal granules in the cell body.

When necessary the cleansing action of the cilia is reinforced by *coughing*. This action is initiated by the stimulus of particles in contact with receptors in the bronchi. Afferent impulses from these areas produce a sharp expiration, the glottis being at first closed and then suddenly opened to allow the particles to be swept out.

The lungs are covered by the mesothelium of the visceral pleura, the layer of splanchnic mesoderm covering the endodermal outgrowth that constitutes the embryonic lungs. The outer surface of the lung is thus shining and smooth and in life closely apposed to the parietal pleura, the layer of somatic mesoderm that lines the inside of the thorax and has a similar smooth structure. In life these two layers are in contact, with no space between, and this close contact is essential to the working of the lung. If air is present between the pleura, the lung does not expand and contract with the movements of the thorax.

DIAPHRAGM

A musculotendinous diaphragm, present only in mammals, separates the pleuropheritoneal cavity into two parts. That portion anterior to the diaphragm contains the lungs. Each lung is enclosed in a separate *pleural cavity* which is lined with a fibroelastic membrane called the *pleura*. The part of the pleura lining the wall of the pleural cavity is referred to as the *pariental pleura*. The lung itself is invested by a layer of *visceral pleura* reflected over its surface. The outermost layer of each pleura, or that facing the pleural cavity, consists of squamous mesothelial cells. Visceral and parietal pleurae are continuous with each other at the root of the lung, or hilus, where the bronchi and blood vessels enter. Pleurisy is a disease marked by inflammation of the pleura. Toward the median line the parietal pleurae of the two dies come close together so as to form a two-layered septum, the *mediastinum*, separating the two pleural cavities. The space between the two layers of the mediastinum is called the *mediastinal space*. It

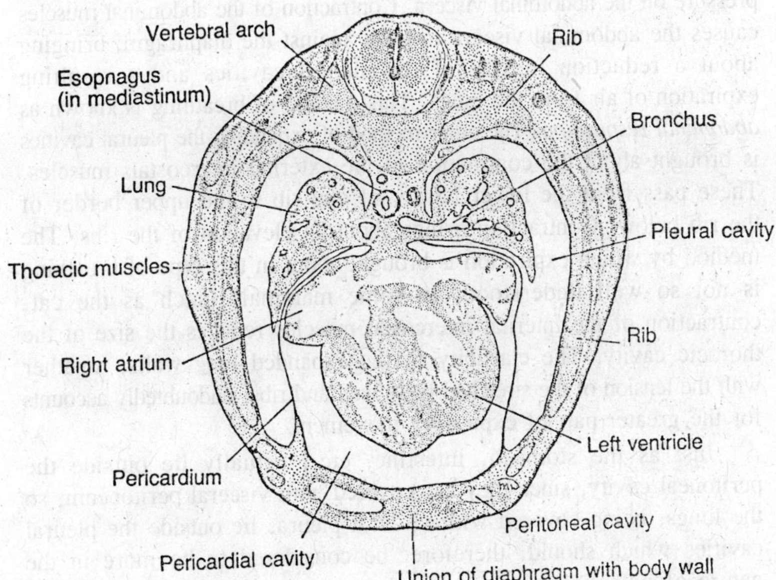

Fig. 13.5. Cross section of a human embryo to show the development of the lungs and of pleural and pericardial cavities and heart.

widens as it nears the diaphragm and contains the aorta, esophagus, and posterior vena cava. In the region of the heart the mediastinal membrane passes over the parietal layer of the pericardium. The pleural membrane is rich in capillaries and lymphatic vessels. A serous fluid, present in the pleural cavities, enhances the movement of the lungs within the cavities. Under normal conditions there are practially no pleural cavities in existence, since the spaces are completely filled by the lungs except for the small amount of fluid which serves as a lubricant.

The pleural cavities are airtight. Increasing their size results in an expansion of the lungs, and air from the outside is drawn inside, passively. Addition blood is concomitantly drawn into the blood vessels of the lung and those of the thorax proper. Reduction in the size of the pleural cavities forces air out of the lungs. The mechanism of respiration in mammals, therefore, consists of alternately increasing and decreasing the size of the pleural cavities. This is generally accomplished in two ways. Contraction of the muscles of the arched diaphragm flattens or lowers that structure. This increases the size of the pleural cavities, causing inspiration, and brings about an increased pressure on the abdominal viscera. Contraction of the abdominal muscles causes the abdominal viscera to push against the diaphragm, bringing about a reduction in size of the pleural cavities and a resulting expiration of air from the lungs. This method of breathing is known as *abdominal respiration*. The other method of expanding the pleural cavities is brought about the contraction of the external intercostals muscles. These pass from the lower border of one rib to the upper border of the rib below. Contraction brings about an elevation of the ribs. The method by which expiration is brought about in this type of breathing is not so well understood. In some mammals, such as the cat, contraction of the internal intercostal muscles reduces the size of the thoracic cavity. The elasticity of the expanded lung walls, together with the tension of the stretched body wall and ribs, undoubtedly accounts for the greater part of expiratory movement.

Just as the stomach, intestine, etc., actually lie outside the peritoneal cavity, since they are covered with visceral peritoneum, so the lungs, being covered with visceral pleura, lie outside the pleural cavities which should, therefore, be considered to be more in the nature of *potential* spaces rather than as actual spaces. If air enters the pleural cavity through a wound, or because of rupture of the lung wall in certain diseased conditions, the potential pleural cavity becomes a real cavity and the lung on the side will collapse since its tissues are under constant tension because of the elastic fibers present in them. Such a condition is known as a *pneumothorax*. As a therapeutic measure, collapse of the lung is induced artificially in certain pulmonary lesions (tuberculosis) in which the disease is confined to one lung. The theory behind such treatment is to permit the lung to rest for a time by making it functionless. A pneumothorax is usually induced by injecting air into the pleural cavity. The air is gradually absorbed, and the procedure must be repeated at intervals. The term hydrothorax refers to the presence of an abnormal amount of fluid in the pleural cavity. This occurs in certain diseased conditions and here, too, the potential pleural cavity becomes a real cavity. Although the two pleural cavities are completely separated by the mediastinum, that membrane is so pliant that a pneumothorax or hydrothorax on one side affects the other side to some degree.

In the condition known as asthma, characterized by difficulty in breathing, usually in response to allergic conditions of one kind or another, there is a contraction of the smooth muscles of the smaller bronchioles. This is commonly accompanied by a swelling of the mucous membrane lining these small tubes. Epinephrine (adrenalin) or some other drug with similar properties is often administered because of its effect in dilating the respiratory passages so that breathing can be more easily accomplished. The patient suffering from asthma usually has more difficulty in expelling air than in drawing it into the lungs.

Mechanism of Breathing

The lungs are passive elastic structures and the movement of air through them is caused by movements of the surrounding structures. The lungs are surrounded closely by a thin layer of epithelium which is reflected back on to the chest wall at the root of the lung, forming another continuous layer of epithelium covering the inner layer of the chest wall. These two layers of epithelium, one covering the lung and one lining the chest wall, are called the pleura. The space between the outer (or parietal) layer and the inner (or visceral) layer is only a one and contains a thin film of fluid. There is a negative pressure amounting to 4 mm. of mercury in the intrapleural space; this negative pressure gradually develops from birth and appears to be due to the unequal growth of the chest wall and the lung, the chest wall growing at a greater rate than the lungs. It is this negative intrapleural pressure which is responsible for keeping the elastic lungs expanded against the chest wall and if for any reason this negative pressure is obliterated

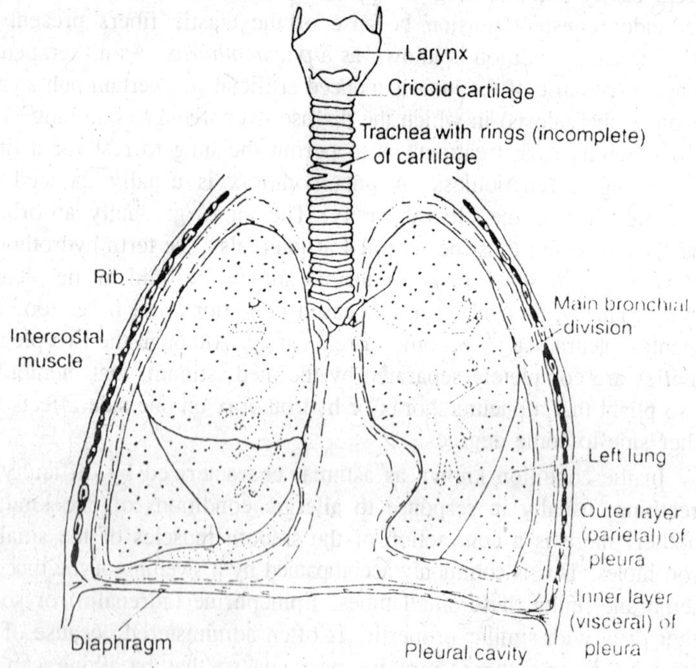

Fig. 13.6. Diagram of the human chest showing the respiratory organs.

e.g. by a wound penetrating the chest wall and pleura, so allowing the intrapleural pressure to rise to atmospheric pressure, then the lung collapses. Air may be introduced into the pleural space through the needle, producing collapse of a lung and this is used in medicine in order to rest a lung; this is called a pneumothorax.

The air in the alveoli is in direct communication with the atmosphere external to the body by way of the upper respiratory passages, trachea, bronchi and respiratory bronchioles, and is at atmospheric pressure. In order to draw air into the lungs the pressure of air within the alveoli must be reduced below that of atmospheric pressure so that air will flow into the lungs along a pressure gradient; since the pressure of a gas is inversely proportional to its volume (Boyle's law) the volume of the thorax must be increased in order to draw air into the lungs. Air is exhaled from the lungs by the reverse process, a reduction in volume of the thorax leads to a consequent rise in pressure of the alveolar air. There are two mechanisms for altering the volume of the thorax, by movements of the ribs and by movements of the diaphragm; these mechanisms are mainly concerned

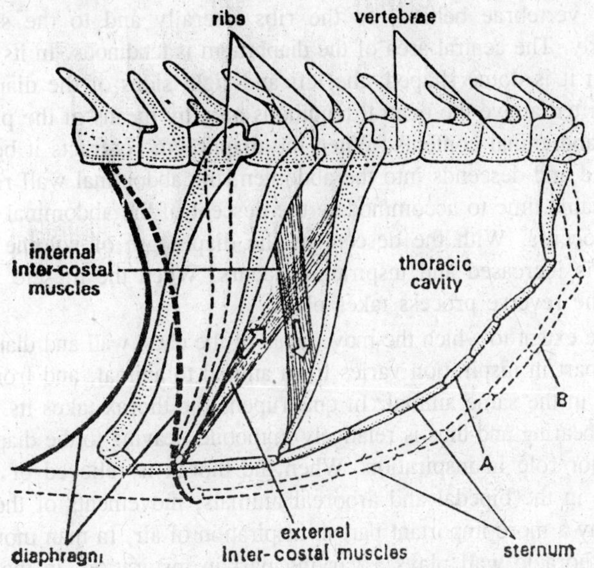

Fig. 13.7. Mechanism of breathing. A—Normal position. B—During inspiration.

in increasing the volume of the thorax, by movements of the ribs and by movements of the diaphragm; these mechanisms are mainly concerned in increasing the volume of the thorax. Reduction in the volume of the thorax at expiration is due mainly to passive elastic recoil of the chest wall.

The wall of the thorax is bounded laterally by the twelve pairs of ribs, ventrally by the sternum, and dorsally by the vertebral column. Each rib is jointed dorsally to the vertebral column and ventrally to the sternum. Since the ribs are curved they move outwards when they are raised, rather like the handle of a bucket does when it is picked up from a position resting on the side of the bucket. Thus when the ribs are raised by the effect of contraction of the intercostal muscles which pass from one rib to another, the diameter of the chest is increased, the volume of the thorax is increased and air passes into the lung so that its internal pressure comes into equilibrium with the atmosphere. When the muscles which move the ribs relax then the chest returns to the resting position because of the elastic recoil of the chest wall.

The diaphragm is a thin sheet of muscle which separates the contents of the thorax from those of the abdomen, but it is pierced by the blood vessels, nerves and lymphatics which pass from one cavity to the other. The margins of this sheet of muscle are attached to the lumbar vertebrae behind, to the ribs laterally and to the sternum anteriorly. The central area of the diaphragm is tendinous. In its resting position it is dome shaped, the left and right sides of the diaphragm being lifted above the central tendinous area by means of the pressure of the abdominal contents. When the diaphragm contracts it becomes flattened and descends into the abdomen, the abdominal wall relaxing at the same time to accommodate the descent of the abdominal organs which occurs. With the descent of the diaphragm of volume of the thorax is increased and inspiration occurs. When the muscle relaxes again the reverse process takes place.

The extent to which the movements of the chest wall and diaphragm play a part in inspiration varies from animal to animal, and from year to year in the same animal. In quadrupeds the thorax takes its part in weight bearing and thus is relatively immobile, leaving to the diaphragm the major role in inspiration. When the thorax is relieved of weight bearing in the bipedal and arboreal animals, movements of the chest wall play a more important part in inspiration of air. In man movement of the thoracic wall plays a varying part in inspiration; in the infant the ribs are nearly at right angles to the vertebral column, so that movement in any direction would tend to *reduce* the volume of the thorax, and diaphragmatic movements are more important: with the development of the downward slant of the ribs of the adult, movements of the chest become more important.

At the end of a normal passive expiration the lungs still contain large amounts of air, amounting in man to about 2.5 litres. During periods of greater activity the lungs can be further emptied of air by bringing into play certain accessory muscles; these include the abdominal muscles the contraction of which forces the abdominal viscera up into the relaxed diaphragm and further reduces the volume of the thorax. Other muscles situated around the thorax e.g. upper limb girdle muscles, can compress the thorax and reduce its volume during expiration.

Because the air passages are never emptied of air, the air in the lungs is not completely changed during an expiratory and inspiratory movement – it is merely 'freshened up'. In fact the air within the alveoli is of constant composition and its oxygen and carbon dioxide content does not fluctuate during the respiratory movements. Of course if the breath is held then the alveolar air gradually changes its composition, its oxygen content failing and its carbon dioxide content rising. Its constant composition is maintained by regulating the rate and depth of breathing movements as will be explained in the section on the control of breathing.

A man's lungs hold about 5000 ccs. of air and during an expiratory movement about 500 ccs. of air pass from the alveoli into the larger air tubes, which themselves hold about 150 ccs. of air. The volume of air inhaled and exhaled in a single breath (500 ccs.) is called the tidal air.

The composition of inspired and alveolar air is given in the following table:

	Composition of inspired and expired air in volumes per cent.		
	inspired	*expired*	*alveolar*
oxygen	20.71	14.6	13.2
carbon dioxide	0.04	3.8	5.0
water vapour	1.25	6.2	6.2
nitrogen	78.00	75.4	75.6

Respiratory Exchange

The movements of respiration producer in each cycle only a partial change of the air in the lungs. During expiration the lungs are not completely emptied, for the alveoli are maintained distended by the small negative pressure in the thorax. Much of the alveolar air, however, is swept into the bronchioles and mixed with fresh air at the next inspiration. Atmospheric air contains about 21 per cent of oxygen, 79 per cent of nitrogen, and 0.04 per cent of carbon dioxide. By special methods alveolar air can be collected and shown to contain about 14 per cent of oxygen and as much as 5.5 per cent of carbon dioxide, whereas expired air contains about 16 per cent of oxygen and 4 per cent of carbon dioxide, that is to say, it has a certain admixture of fresh air that has not reached the end of the respiratory tree.

The gases in the air sacs of the alveoli come into equilibrium with the blood by diffusion across the epithelium and capillary walls. On account of the thinness of these membranes the equilibrium is reached rapidly and the oxygen is taken up by the blood-plasma, and from this by the haemoglobin of the corpuscles, while the carbon dioxide is given up by the blood. The rates and extents to which these processes take place depend wholly on the partial pressures of the gases in the alveolar air and on the composition of the blood; it is not now thought that any secretion of oxygen across the lung surfaces is involved.

The solubility of oxygen in water is such that only about 0.3 ml or less is held in simple solution in 100 ml of blood. However, there are 15 g of haemoglobin is this volume, and this is able to hold 1.34 ml of oxygen per gram, that is to say, nearly 20 ml per 100 ml blood. The oxygen carried in simple solution is therefore only about 1 per cent of the whole and would not be adequate to maintain activity such as that of a mammal except by the very rapid of enormous quantities of liquid. The efficiency of the haemoglobin is therefore an essential feature of mammalian life, indeed for that of nearly all chordate animals.

The haemoglobin molecule is formed by the union of protein (globin) with an iron-containing *porphyrin* pigment or *haem* and it is this latter that gives it a power of loose union with oxygen to form oxhaemoglobin. Under the conditions in which the haemoglobin occurs in the blood it becomes 95 per cent saturated with oxygen at the partial pressure of that gas in the alveolar air (100 mm Hg). There is therefore little to be gained in man by adding further oxygen to the air breathed under ordinary conditions, though this may be valuable when there is a low oxygen tension, for instance, at high altitudes. The haemoglobin remains saturated to as much as 80 per cent if the oxygen tension is reduced to 50 mm Hg, but below this tension it begins to give up oxygen and this normally occurs in the capillaries. Surrounded here by an environment poor in oxygen the haemoglobin yields its oxygen store to the plasma, and through the capillary wall to the tissues. The shape of this *oxygen dissociation curve* of haemoglobin is obviously of great importance for the maintenance of a proper flow of oxygen to the tissues. The particular form of the curve depends on the composition of the blood; for instance, the presence of carbonic and lactic acids causes the oxygen to be give off more readily, shifting the curve to the right. Haemoglobin alone, in simple aqueous solution, retains 80 per cent of its oxygen down to pressures as low as 20 mm Hg, and this would not provide an adequate supply. In worms that use haemoglobin as a means of storing oxygen the dissociation curve is flat, considerable quantities of oxygen being retained down to quite low tensions.

Tissue Respiration

The process of taking oxygen into the lungs and its transport in the blood is only a preliminary to the essential feature of respiration, the using of oxygen by the tissues to provide energy by combustion. We can distinguish, therefore, between *external respiration* and the *internal respiration* of the tissues the latter takes place by a system of enzymatic exchanges, associated with the mitochondria, making the oxygen able to undertake chemical transformations that would not otherwise occur under these conditions. Ultimately much of the energy of muscular contraction, for instance, comes from the combustion of glycogen to carbon dioxide and water, a reaction that would not take place to any extent at 37°C in vitro.

A central position in this sequence of changes is taken by *cytochrome*, a haem-containing substance (or set of substances) resembling haemoglobin, and present in all animal cells. Cytochrome has a characteristic absorption spectrum which, like that of haemoglobin, changes when oxidation occurs. This requires the action of an enzyme, *cytochrome oxidase*, which activates the oxygen brought to the tissues and transfers it to the cytochrome. The *dehydrogenases* are enzymes able to combine the oxidized cytochrome with the reducing substances produced by the metabolism of the tissues. Thus reaction such as the breakdown of pyruvic to lactic acid are made to proceed by removal of the hydrogen, and in this way the whole system of equilibria is shifted so that, for instance, glycogen can break down through stages of the carbohydrate cycle through pyruvic to acetic acid and ultimately to carbon dioxide and water.

Various poisons act at particular points of these cycles. Thus cyanides inhibit respiration by preventing the action of cytochrome oxidase, leaving the cytochrome in the reduced state. Urethanes, on the other hand, by acting on the dehydrogenases, leave the cytochrome oxidized. Besides this cytochrome system there are other respiratory enzymes present, which are able to change the rate of the reactions in which oxygen is involved. Together these respiratory enzymes make possible the last stages of the flow of oxygen, which goes on throughout life from the air of the lungs to the blood and thence to the tissue fluids and into the cells, where it takes part in the reactions that are the life of the organism.

Control of Breathing Movements

The process of breathing is a complex one involving many series of muscles, which include the intercostals muscles, muscles surrounding the thorax, the diaphragm and abdominal musculature, and the muscles of the larynx; thus during inspiration the vocal cords open, the process of swallowing is inhibited, the intercostals muscles and diaphragm contract and the abdominal muscles relax, so that the increased volume of the abdomen can be accommodated. The integration of these various muscles depends upon a complex series of reflexes maintained by the action of the nervous system. Further this series of reflexes must be adapted to meet the changing needs of the body for oxygen.

Nervous Mechanisms in the Control of Breathing

The basic centre for the control of breathing movements is situated in the medulla oblongata and this centre consists of two parts, an inspiratory centre and an expiratory centre; when one centre is active the other one is inhibited. The cell of the respiratory centre are connected by nerves, both sensory and motor, to the breathing apparatus. Further the cells themselves are sensitive to changes in the chemical composition of the blood, particularly to the level of carbon dioxide and the hydrogen ion concentration.

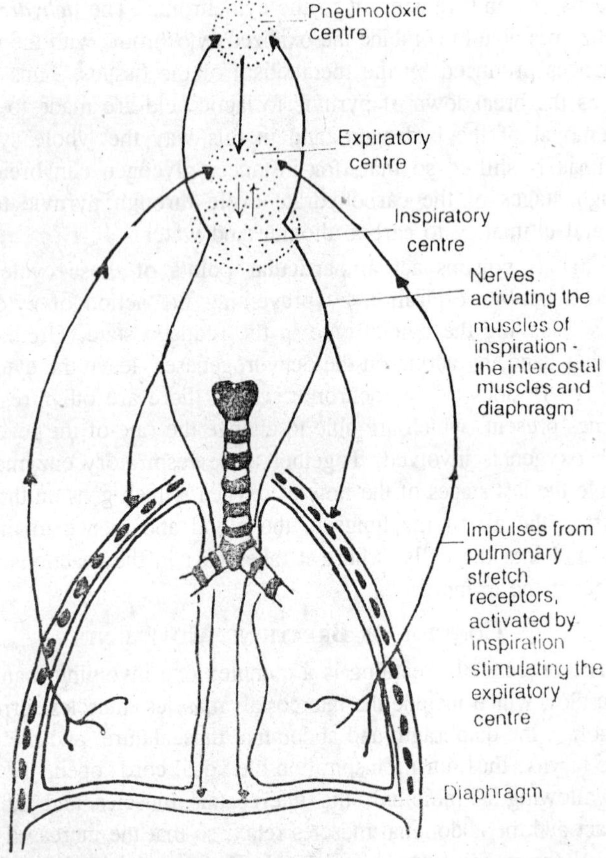

Fig. 13.8. Mechanisms maintaining the breathing rhythm. The arrows indicate the direction of the flow of nervous impulses.

Hering-Breuer reflex

There are sense endings in the walls of the respiratory bronchioles, air sacs and alveoli, which are stimulated by the change in tension in the walls of these air passages. At the height of inspiration these receptors fire off stimuli to the respiratory centre, via the vagus nerve, which inhibit any further inspiratory movement. Further there are deflation receptors which are stimulated by deflation of the lungs; these fire off nerve impulses which stimulate the succeeding inspiration.

Under normal conditions this periodic inhibition of inspiration by afferent impulses in the vagal nerves, from receptors stimulated by inflation of the lungs is largely responsible for the rhythm of breathing. But there is another inhibitory mechanism of inspiration in the form of an inhibitory nerve centre in the medulla connected to the inspiratory and expiratory divisions of the respiratory centre. During the phase of inspiration the inspiratory division of the respiratory centre discharges impulses to the inhibitory or pneumotaxic centre as it is called. The pneumotaxic centre is then activated and sends impulses to the expiratory centre. When this barrage of impulses becomes sufficiently intense the expiratory centre is activated and the inspirtory centre is reflexly inhibited.

These two mechanisms for maintaining rhythmic respiration, the Hering-Breuer reflex and the pneumotaxic centre are illustrated.

Chemical Control of Breathing

In addition to the nervous mechanism which maintains the rhythmic quality of breathing movements, the ventilation of the lungs must be capable of modification to suit the metabolic needs of the organism. This is achieved by chemical mechanisms. There are two ways in which the activity of the cells of the respiratory centre are modified to suit the metabolic needs of the mammal. Firstly the neurones of the respiratory centre themselves are sensitive to changes in the chemistry of the blood, and secondly there are special receptors in the carotid and aortic bodies which are similarly sensitive to changes in the chemistry of the blood; these chemoreceptors are connected to the respiratory centre of the medulla oblongata by way of the 9th and 10th cranial nerves.

The neurons of the respiratory centre are sensitive to changes in the oxygen and hydrogen ion concentration of the blood. When there is a rise in the carbon dioxide or hydrogen ion concentration of the blood the neurons of the respiratory centre are stimulated, and initiate an increased ventilation of the lungs, which corrects the change in blood chemistry by removing excess carbon dioxide from the lungs. This is

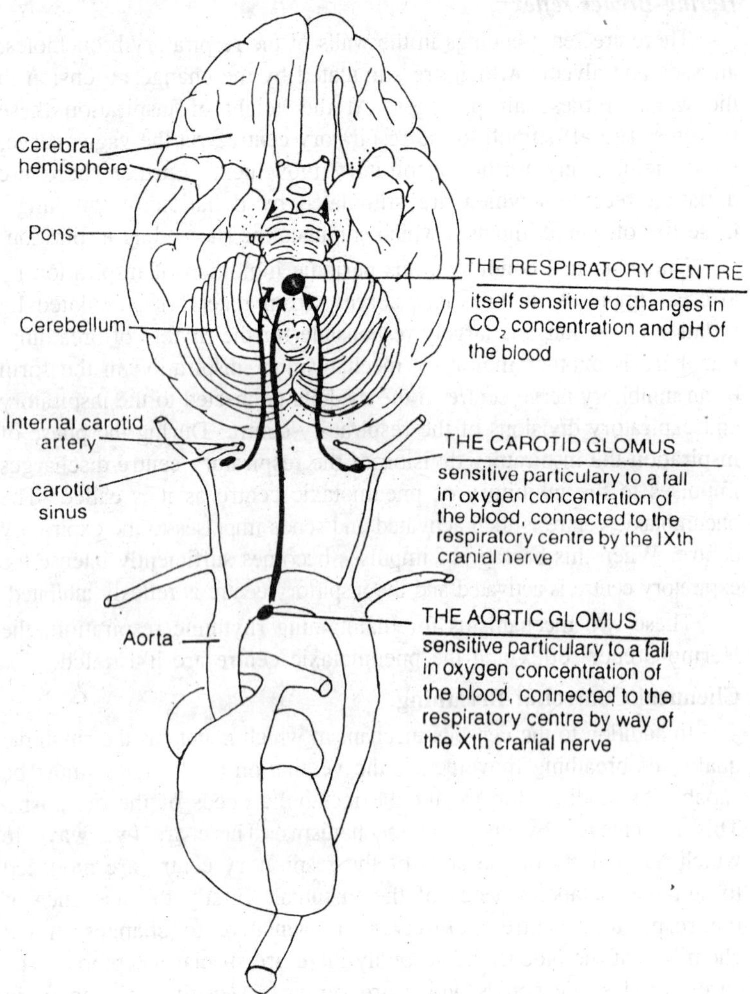

Fig. 13.9. The chemical control of breathing.

the most important cause for the increased ventilation of the lungs during exercise. If there is a fall of carbon dioxide content of the blood or a fall in the hydrogen ion concentration then ventilation of the lungs is depressed until sufficient carbon dioxide accumulates to rectify the change in blood chemistry. The sensitivity of the neurons of the respiratory centre is very great and very small changes in the carbon dioxide concentration of the blood will produce changes in the ventilation of the lungs.

If you take four successive rapid but very deep breaths forcing air out of the lungs at the end of each breath and then stop you will notice that there is a pause before normal spontaneous breathing movements return. You have over ventilated your lungs and reduced the amount of carbon dioxide in the alveoli and the blood and therefore the respiratory centre in the medulla is not getting adequate stimulation. The normal spontaneous breathing only returns when sufficient carbon dioxide has accumulated in the blood. (This experiment should only be tried by the healthy subject).

Since the concentration of carbon dioxide in the alveolar air does not fluctuate throughout the normal expiratory-inspiratory cycle only changes in the metabolic activity will produce changes in the activity of the respiratory centre, by way of changes in the carbon dioxidate and hydrogen ion concentration of the blood.

We have described receptors situated at the bifurcation of the carotid arteries and on the arch of the aorta, which are sensitive to changes in the pressure of the blood; these pressure receptors are important in the reflex control of the circulatory system. There is a further set of receptors which are chemoreceptors and sensitive to a decrease in oxygen pressure, a rise in carbon dioxide pressure and an increase in the acidity of the blood. These chemoreceptors are located in small epithelial bodies, the aortic and carotid glomi. The carotid glomi are situated on the external carotid artery, and the aorthic glomus is within the concavity of the aortic arch. The pressure receptors on the other hand are situated within the wall of the blood vessels, in the carotid sinus (the swelling at the base of the internal carotid artery) and the wall of the aortic arch and innominte artery.

The main role of the chemoreceptors in the carotid and aortic glomi is to stimulate breathing when there is a fall in oxygen pressure in the blood. Under conditions of oxygen lack most of the cells of the body, particularly those of the nervous system e.g. the respiratory centre, are depressed, under these conditions the chemoreceptors, by means of their nervous connections with the brain, act as an important drive to respiration.

Mountain Sicknew and Adaptation

When man climbs to a height of 3000 m. he enters a rarified atmosphere in which there is a much smaller amount of oxygen and he experiences a state which approximates to drunkenness. He can no longer calculate accurately, and he may experience a sense of elation or giddiness or sickness. It is this sort of sensation that made the early balloonists wish to go higher and higher once they got to high altitudes. If man ascends to 6000 m. he may do permanent damage to his nervous system through oxygen lack.

If the ascent is made slowly and the climbers have undergone a period of training at high altitudes it is possible to climb to altitudes higher than would be expected. On the Everest expedition of 1924 a height of 8500 m. was reached without any oxygen apparatus. The climbers are reported to have taken ten breaths for each pace forward. That training does lead to adaptation is shown by the fact that at 5000 m. one would expect, from the normal dissociation curve of oxyhaemoglobin, only 38 mm. pressure of oxygen in the alveoli; but experiments on trained climbers adapted to high altitudes showed an alveolar oxygen pressure of 52 mm. Adaptation to altitudes involves several factors including an increased number of red blood cells in the blood, increased ventilation of the lungs and an increased output of blood from the heart.

Flying in Aircraft

Although the composition by volume of air does not change much from ground level to 21,000 m. the actual amounts, and therefore the pressure, of the gases gets less as one moves away from the earth's surface. Variations in temperature and pressure cause mass movements of air which result in the weather. Above 10,000 m. the temperature is constant at -55°C; pressure changes also get less as one gets higher. Modern jet engines are more economical on fuel at higher altitudes and since pressure and temperature changes are less at these altitudes one can fly as it were above the weather. High flying may also be necessary for tactical reasons in military aircraft.

There are physiological difficulties to be overcome in flying at these altitudes. There is a lack of oxygen and a danger of decompression sickness. Because the partial pressure of nitrogen falls as one ascends there is a danger of nitrogen coming out of solution into the body fluids and accumulating as gas bubbles.

Royal Air Force regulations for the use of oxygen say that air crew flying at cabin altitudes above 3000 m. will always use oxygen. Oxygen should be used for passengers at 3700 m. Above 10 000 m. the efficiency of airmen using oxygen equipment falls and at 12 000 m. breathing 100% oxygen, man is in much the same state as regards oxygen supply as he is at 3000 m. breathing air. The maximum height of flight in unpressurized aircraft is therefore at 12 000 m. It is possible to pressurize the breathing equipment, or indeed the whole cabin, and this makes stratosphere flying above 12 000 m. possible.

It would be ideal physiologically to have the cabins of aircraft fun of air at atmospheric pressure (about 101325 Pa) but there are other considerations to be made e.g. weight of pressure equipment, and in military aircraft there is the danger of losing pressure rapidly from perforation of the cabin by missiles. Thus cabins are not so highly pressurized. V-Bombers are pressurized at 60 800 Pa and this gives conditions in the cabin similar to those at 2500 m. when the aircraft is actually flying at 14 000 m. Since one does not need oxygen at 2500 m. then there is no, danger of oxygen lack, or of decompression sickness, even when flying at 14 000 in. In civilian airlines high pressurization is also used e.g. 50 662 Pa giving an effective 2500 m. equivalent at 10 500 m.

Caisson Sickness

In engineering work under water, a diving bell or caisson is used. The water is pumped out of the bell by compressed air and then the divers are allowed to enter the bell and descend to the depth of the water where they are working. The pressure in the bell may be about 4 atmospheres and men can work safely at this pressure. The danger occurs when the men come to the surface again. At four atmospheres pressure the nitrogen in the air dissolves to a greater extent in the body fluids; but when the men return to conditions at atmospheric pressure the nitrogen comes out of solution in the body fluids. If the change from four atmospheres pressure to atmospheric pressure is sudden then the nitrogen comes out of solution in the form of gas bubbles which may obstruct the smaller blood vessels or cause damage to delicate tissues such as the brain. The condition produced is called 'the bends' and there may be severe abdominal pain, paralysis, sickness or even collapse and death. In order to avoid Caisson sickness the period of exposure to high pressures is restricted so that large amounts of nitrogen do not go into solution in the body fluids, and the change of pressure on returning to the surface is brought about gradually in special decompression chambers.

Diving Mammals and Respiration

Whales are very highly adapted to their existence in water, so much so that they are often mistaken for fish. They are of course mammals and not fish. Their streamlined shape, absence of hind limbs, conversion of fore-limbs into flippers, and the development of horizontal tail flukes, their loss of hair and development of blubber under the smooth skin are obvious adaptations. The adaptations which they have fro diving are very exciting.

Although many whales such as the whalebone whales (which filter off the plankton for food) do not dive to great depths, there are others such as the sperm whales (which possess teeth and hunt squids in deep water) which need to dive very deep for periods up to half an hour. The problem is that the whale is a mammal and must breathe using lungs and cannot use gills like a fish; it has a breathing apparatus for use in air but it lives its life diving deep in the sea. The large 21 m. fin whale can take down about 3350 litres of oxygen which is enough to last it about 16 minutes if it moved at about 5 knots. But these whales move faster than this underwater and have been known to stay down for half an hour. It seems therefore that they must use oxygen at a very low rate whilst they are submerged. The muscles work anaerobiclaly and are capable of building up a very large oxygen debt, which is paid off when the whale comes to the surface (for further discussion of oxygen debt). The nostril is single and easily pokes out of the water because of its position on top of the head; the whale then takes very deep breaths which ventilate the lungs rapidly. In expiration the breath which is very heavily laden with moisture is blown out and much of the moisture condenses in the atmosphere. This process is called spouting and it used to be thought that the whale was blowing out a jet of water which it used to be thought that the whale was blowing out a jet of water which it had drunk under water. The direction of the spouts and the number (for some whales have two nostrils) enable the whaler to recognize the whales.

We have seen that muscle tissue in whales can stand a very large oxygen debt. Whale meat has a very deep red colour because it contains large amounts of the respiratory pignient myoglobin in the muscle cells. Muscle hemoglobin or myoglobin has a greater affinity for oxygen than has blood haemoglobin and does not give up its oxygen until the partial pressure of oxygen in the surroundings is very low. This myoglobin is responsible for much of the oxygen which the whale takes down with it. In the fin whale of the 3350 litres of oxygen taken down 9% is in the lungs, 48% in the blood as oxyhaemoglobin, 42% in the myoglobin of the muscles and 7% dissolved in the tissue fluids.

Although muscle can live without oxygen for about half an hour because of its myoglobin and its ability to withstand oxygen debt, the brain cells cannot tolerate this deprivation of oxygen; the brain and spinal cord must have a good supply of oxygenated blood at all times. Diving mammals have networks of fine blood vessels around the brain and spinal cord called retia mirabilia; when the animal dives the circulation to the rest of the body, that is other than to the central nervous system, is reduced, and the retia mirabilia serve to take up the increased amount of blood which has been diverted from the rest of the body.

The rate of the heart beat of seals has been shown to fall from 180/minute to 35/ minute during a dive; in spite of the resulting fall in the output of blood by the heart, the arterial blood pressure is maintained by means of vasoconstriction in the large capillary beds in the seals' skin. It is obviously difficult to record the heart beat of a 21 m. whale during a dive and there is no reliable information on this point.

Thus we have seen that whales have special mechanisms to provide the brain with oxygen-rich blood whilst the rest of the body relies for survival on oxygen stored as oxy-myoglobin and upon the ability to develop considerable oxygen debts. This oxygen debt is paid off during the efficient ventilation which occurs during spouting. There are many problems still unsolved. For example, how does a whale hold its breath for periods up to half an hour? We have seen that in most mammals breathing is regulated by the amount of carbon dioxide in the blood. The respiratory centre of whales seems to be indifferent to the amount of carbon dioxide in the blood. And how does a whale avoid decompression sickness? The blood of a whale will absorb nitrogen just as easily as human blood; if a man stayed underwater breathing air and surfaced as rapidly as a whale he would get the 'bends'. The whale only takes down the air in its lungs, and the larger deep diving whales have proportionately smaller lungs. When the whale dives the pressure of water acting upon the abdomen forces the abdominal contents forwards so that they compress the lungs; this is possible because the diaphragm is very obliguely inclined, being attached to the dorsal body wall at a rather posterior position. Thus the air in the alveoli is compressed and forced into the dead space of the air passages. Now here little gaseous exchange occurs so that little or no nitrogen finds its way into the blood. Thus the whale is free from the dangers of decompression sickness.

14

EXCRETORY SYSTEM

Excretion is the process whereby the waste products of the body's metabolism and the substances which are in excess of requirements are removed from the body. The waste products of metabolism include carbon dioxide, produced during the oxidation of carbon containing compounds in tissue respiration, and mitrogenous waste products produced from protein metabolism. Urea is the commonest nitrogenous waste product in man. It is manufactured in the liver and then passes into the blood stream from which it is excreted by the kidneys. The lungs and the kidneys are the main excretory organs, the lungs excreting carbon dioxide, the kidneys excreting nitrogenous waste.

The definition of excretion includes the removal of substances which are present in the body in excess of requirements. The cells of the mammal are highly specialized and can only function properly in a restricted range of pH and osmotic pressure. The kidney, by excreting excess mineral salts and water is able to exercise control of the chemistry of the body fluids. Thus if we drink too many fluids the body fluids become in danger of dilution and the excess water must he excreted by the kidneys. Even an invaluable substance such as water is a danger when in excess, and this excess must be excreted.

The main excretory products may be grouped into four categories: (i) Carbon dioxide-derived from tissue respiration; (ii) Nitrogenous waste products-from protein metabolism; (iii) Water in excess of requirements; and (iv) Mineral salts in excess of requirements.

Excretion of Carbon Dioxide

This takes place mainly from the lungs. The carbon dioxide present in the capillary blood in the lungs diffuses into alveoli through the thin moist membrane of the aloveoli and is removed from the body in the process of breathing.

Some carbon dioxide leaves the body in the form of bicarbonate through the kidneys, and changes in the bicarbonate excretion by the kidneys plays an important role in the maintenance of the acid-base balance of the body.

Excretion by Kidney

The kidney is the organ which produces urine, containing all the important excretory products, except gaseous carbon dioxide. By alterations in the amount of water excreted and changes in the pattern of salt excretion the kidney plays a vital role in the maintenance of the constancy of the internal chemical environment of the body. The present chapter will be concerned with the kidney and its functions.

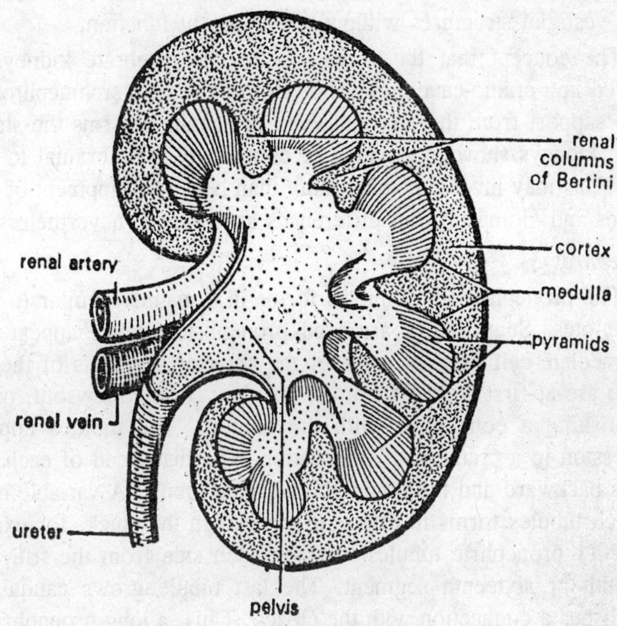

Fig. 14.1. Sagittal section of kidney.

Amniote Kidney

In reptiles, birds, and mammals, three types of kidneys are usually recognized: pronephros, mesonephros, and metanephros. These appear in succession during embryonic development, but only one, the metanephros, persists to become the functional adult kidney. Mesonephros and metanephros actually represent different levels of the opisthonephros of the anamniota, the metanephros being the equivalent of the posterior portion. In all forms, an anteriorly located pronephros is present during very early stages of development, but it soon degenerates and the more posterior mesonephros then develops. The duct of the pronephros, however, persists to become the duct of the mesonephros. This is actually the same as the archinephric duct, to which considerable reference has been made. Since most accounts refer to this duct as the *mesonephric*, or *Wolffian duct*, to avoid further confusion on the part of the student, we shall use the term Wolffian duct here in referring to the archinephric duct as it appears in amniotes.

The mesonephros persists for a time and then degenerates. In the meantime the metanephros has begun to develop from the region posterior to the mesonephros. Portions of the mesonephros may persist to contribute to the reproductive system in the male or to remain as mere vestigial structures without any apparent function.

The concept that the various kinds of vertebrate kidneys have evolved in a cranio-caudal sequence from an ancient archinephros gains some support from the fact that even in primitive forms the structure of the kidneys shows a gradient of complexity from cranial to caudal ends. This may involve nothing more than better development of kidney tubules and glomeruli, but differences are evident, nevertheless.

Pronephros

The pronephros in amniotes forms in a manner similar to that of anamniotes. Segmentally arranged pronephric tubules appear in the intermediate cell mass in some of the anterior segments of the body. These are at first solid structures, but they soon hollow out, one end establishing a connection with the coelom. The tubules appear in succession in a cranio-caudal direction. The distal end of each tubule bends backward and fuses with the adjacent tubule. A variable number of such tubules forms in different species. In the chick, for example, 10 or 11 pronephric tubules form on each side from the fifth to the fifteenth or sixteenth segment. The last tubule grows caudally and establishes a connection with the cloaca. Thus, a long pronephric duct is formed, the anterior end of which is connected to a series of tubules with coelomic connections. The tubules soon disappear; in fact the anterior tubules, may degenerate before the posterior ones even form. External glomeruli may or may not form. If they do, the sequence of their appearance may not even coincide with the time when the tubules are forming. In mammals, pronphric tubules appear only as the merest of vestiges. Hence, the pronephric duct can scarrely be said to be formed as the result of fusion of tubules. Nevertheless, it appears in the nephrotome region, first as a solid cord which grows back to the cloaca, hollowing out to become a typical pronephric duct. Thus the amniote pronephros is little more than a transient structure which undoubtedly has no value at all as an excretory organ. To the comparative anatomist it is of interest because its presence indicates that the amniote kidney, like that of the anamniotes, has evolved from the ancestral archinephric, or holonephric, type. When the duct of the pronephros becomes the duct of the mesonephros it is called the Wolffian duct.

Mesonephros

The mesonephros is sometimes called the *Wolffian* body. It extends over a greater number of segments than does the pronephros and develops after the pronephros has formed. Mesonephric tubules appear in the intermediate cell mass in the region where the pronephric duct is located, but posterior to the pronephric tubules. In some forms the anterior mesonephric tubules appear in the same segments as the posterior pronephric tubules, so that there is a slight degree of overlapping. The mesonephric tubules differentiate in the same manner

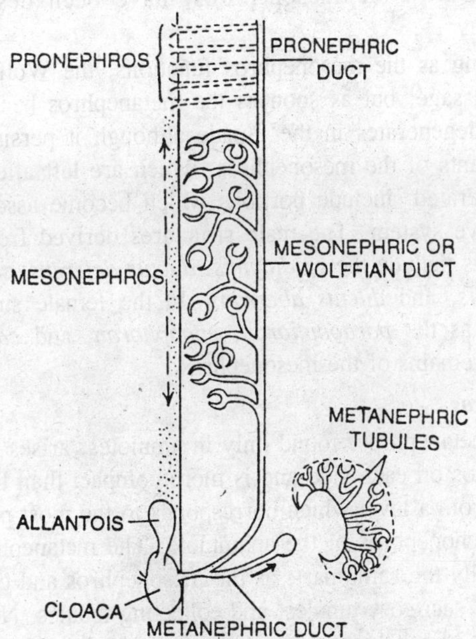

Fig. 14.2. Plan of amniote kidney in the embryo.

as the opisthonephric tubules described earlier. Some of the anterior tubules may form peritoneal connections, but this is unusual. Contrary to the condition in the pronephros, several tubules may form in a single segment. In the human embryo from 2 to 9 mesonephric tubules may develop in this manner, making a total of between 30 and 40 for each mesonephros. In other forms the number may be even greater. As a result, the mesonephros, at least in some forms, as in the pig, becomes rather voluminous, occupying an extensive portion of the body cavity.

In reptiles, birds, and mammals the mesonephros exists only temporarily and is followed in development by the metanephros, which becomes the functional adult kidney. In the embryonic stages of the mouse, rat, and guinea pig, the mesonephros is so poorly developed and appears so early that it can scarcely be regarded as functional at any time. Certain embryonic membranes in these animals may possibly help to rid the body of excretory products in the interval before the metanephros develops. On the other hand, in reptiles and in such mammals as echidnas and certainmarsupials, the mesonephros may persist for a time after birth. Well-developed peritoneal funnels, associated with the mesonephros, have been described for the monotremes.

As long as the mesonephros functions, the Wolffian duct is the urinary passage; but as soon as the metanephros becomes functional this duct degenerates in the female although it persists in the male. The remnants of the mesonephros, which are left after the main part has degenerated, include portions which become associated with the reproductive system. The male structures derived from mesonephric components include the *epididymis*, *ductus deferens*, *seminal vesicles*, *paradidymis*, and *ductus aberrans*. In the female such rudimentary structures as the *paraoophoron*, *epoophoron*, and *canal of Gartner* represent remains of the mesonephros.

Metanephros

The metanephros, found only in amniotes, arises posterior to the mesonephros on each side and is more compact than the latter organ. It comes from a level which corresponds to the most posterior portion of the opisthonephros of the anamniota. The metanephros is made up of essentially the same parts as the mesonephros and consists of renal corpuscles, secretory tubules, and collecting tubules. No nephrostomes are present. Each kidney has a twofold origin. A diverticulum, the *ureteric bud*, or *metanephric diverticulum*, from the Wolffian duct near

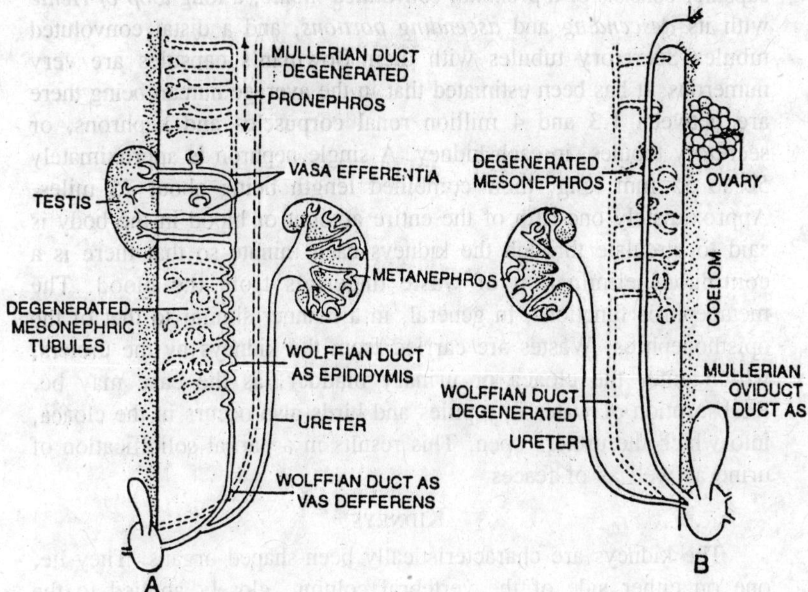

Fig. 14.3. A—Plan of kidney and genital duct in male; B—in female amniota.

its posterior end grows forward into the so-called *metanephrogenous tissue*, or *metanephric blastema*, which lies posterior to the mesonephros. The metanephric blastema is continuous with the nephrogenous tissue which gave rise to the mesenephric tubules. The diverticulum of the Wolffian duct, 'which is destined to form the ureter, branches and rebranches a varying number of times and ultimately forms large numbers of very fine *collecting tubules*. An expansion at the point where the ureter undergoes its primary divisions, at least in mammals, becomes the pelvis of the kidney. Condensations in the mesenchyme of the adjacent metanephric blastema soon give rise to follow *secretory tubules* which grow longer and become S-shaped. One end of each secretory tubule establishes a connection with a collecting tubule. The other end expands and is soon invaded by a glomerular tuft from a branch of the renal artery so that a typical renal corpuscle is formed. All blood going to the metanephros is arterial, since the renal portal system of veins does not exist in these forms. As development progresses, there is a great increase in the number of secretory tubules.

The differentiation of a secretory tubule of the metanephros in mammals and, to a lesser extent, in birds is somewhat more elaborate than in the case of the mesonephros. Each tubule, as it leaves Bowman's capsule, consists of a *proximal convoluted tubule*, a long *loop of Henle* with its *descending* and *ascending portions*, and a distal convoluted tubules Secretory tubules with their Bowman's capsules are very numerous. It has been estimated that in the average human being there are between 1.3 and 4 million renal corpuscles and nephrons, or secretory tubules, in each kidney. A single nephron is approximately 50 to 55 mm long, their combined length being about 75 miles. Approximately one-fifth of the entire amount of blood in the body is said to circulate through the kidneys each minute so that there is a continuous elimination of waste materials from the blood. The metanephros functions, in general, in a manner similar to that of the opisthonephros. Wastes are carried from the kidneys by the ureters, which enter the cloaca or urinary bladder, as the case may be. Reabsorption of water in reptiles and birds also occurs in the cloaca, into which the ureters open. This results in a partial solidification of urine as well as of feaces.

Kidneys

The kidneys are characteristically been-shaped organs. They lie, one on either side of the vertebral column, closely applied to the dorsal body wall of the abdomen and dorsad of its peritoneal lining;

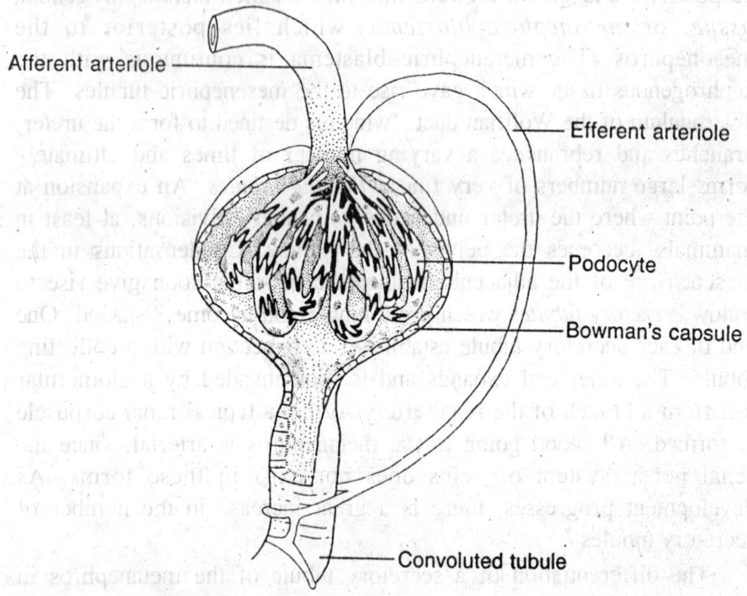

Fig. 14.4. A mammalian renal corpuscle.

that is to say, each kidney is retroperitoneal. The right kidney is placed further craniad than the left, its extremities reaching from the level of the transverse process of the second lumber vertebra to that of the transverse process of the fourth lumbar vertebra. The left kidney, on the other hand, lies between the levels of the transverse processes of the third and fifth lumbar vertebrae. Dorsad of the peritoneum, each kidney is cradled in a mass of fat, which is thickest at its borders and is prolonged through the hilum of the kidney into the renal sinus. The surface of the kidney itself is invested in a fibrous tissue capsule, the tunica fibrosa (renal fascia), which enters the renal sinus to blend with similar adventitious tissue of the renal pelvis.

Each kidney presents two surfaces, two borders, and cranial and caudal extremities.

Relations

The cranioventral surface of the right kidney is covered by the concavity of the caudate division of the right lateral lobe of the liver. The remainder of the ventral surface is in relation with the descending part of the duodenum and the head of pancreas. On the other hand, the ventral surface of the left kidney is closely approximated to the medial surface of the tail of the spleen and is generally traversed obliquely by the tail of the pancreas.

The lateral border of each kidney is convex and is directed toward the dorsolateral wall of the abdomen. The cranial portion of the lateral border of the left kidney is generally contiguous to the medial surface of the spleen.

The medial border of each kidney is centrally concave, becoming convex toward its cranial and caudal extremities. The concavity of the medial border presents a fissure, the hilum, through which are transmitted the renal artery, renal vein, nerves, and ureter. The medial border of the left kidney is in relation to the descending colon.

The cranial and caudal extremities of the kidney are convex. The cranial extremity of the right kidney fits into the concavity of the caudate division of the right lateral lobe of the liver. The cranial extremity of the left kidney comes to lie partially in relation to the medial surface of the tail of the spleen.

Internal Macroscopic Structure

A coronal section of the kidney shows the hilum to expand into a cavity, the renal sinus, in which is lodged the renal pelvis, the funnel-shaped enlargement of the cranial end of the ureter. In the renal sinus

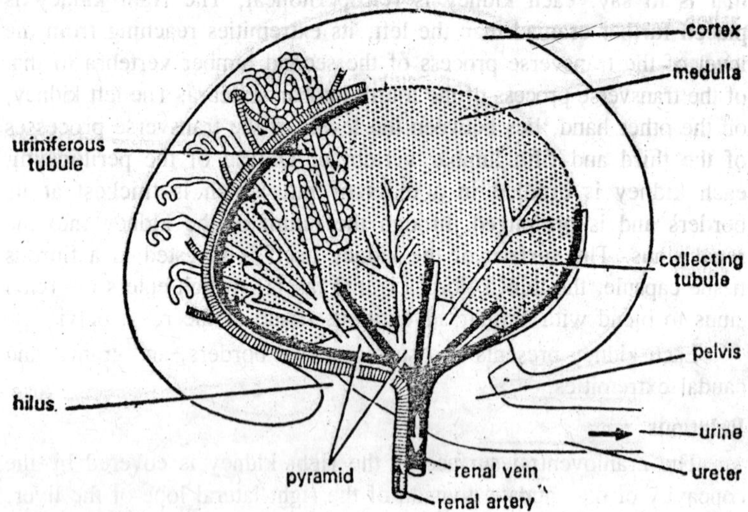

Fig. 14.5. L.S. of kidney.

are also contained the branches of the renal artery and tributaries of the renal vein. The sinus is lined by fibrous tissue, which represents a continuation of the tunica fibrosa. The fibrous tissue of the sinus ultimately blends into similar adventitious tissue of the renal pelvis. The substance of the kidney proper consists of an outer darker and rather granular portion, the renal cortex, which arches itself around a lighter, less granular, and more extensively developed portion, the renal medulla. The medulla converges itself into a pyramid, called the renal pyramid, the cone-like apex (renal papilla) of which is embraced by the renal pelvis.

Microscopic Anatomy of the Kidney

Histologically the kidney of the cat is a compound tubular gland. Essentially it is composed of numerous uriniferous tubules. Each uriniferous tubule consists of two types of tubules: the secretory portion or nephron, which functions in the secretion of urine; and an excretory portion or collecting tubules, which transport the urine to the papillary duct for discharge into the renal pelvis.

The nephron the structural and functional unit of the kidney, is comprised of the following parts in the order of their sequential occurrence: Bowman's capsule, proximal convoluted tubule, descenting limb of Henle, Henle's loop, ascending limb of Henle, distal convoluted tubule, and the junctional tubule.

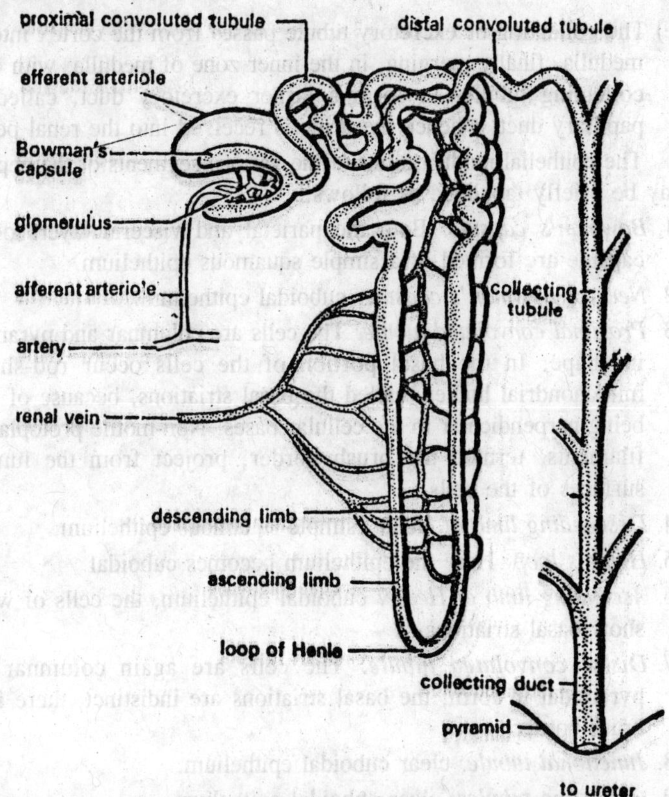

Fig. 14.6. Structure of an uriniferous tubule.

(a) The nephron begins in the cortex as a double-walled globular structure, called Bowman's capsule, the inner layer (visceral) of which intimately embraces a capillary tuft known as the glomerulus. Bowman's capsule and the glomerulus are collectively referred to as the Malpighian or renal corpuscle.

(b) The outer layer (parietal) of Bowman's capsule is continued into a neck whence arises, by cellular differentiation and enlargement, the proximal convoluted tubule.

(c) From the proximal convoluted tubule in the cortex, the descending limb of Henle passes to the medulla where it bends sharply to form Henle's loop from which recourses into the cortex the ascending limb of Henle.

(d) In the cortex, the ascending limb of Henle passes into the contortion of the distal convoluted tubule from which a short, junctional tubule originates to pass to the collecting tubule.

(e) The collecting or excretory tubule passes from the cortex into the medulla, finally merging, in the inner zone of medulla, with other collecting tubules to form a larger excretory duct, called the papillary duct, whence the urine is received into the renal pelvis.

The epithelial modifications of the various segments of the nephron may be briefly tabulated as follows:

1. *Bowman's capsule:* Both the parietal and visceral layers of the capsule are formed of a simple squamous epithelium.
2. *Neck of Bowman's capsule:* cuboidal epithelium.
3. *Proximal convoluted tubule:* The cells are columnar and pyramidal in shape. In the basal portion of the cells occur rod-shaped mitochondrial bodies, called the basal striations, because of their being perpendicular to the cellular bases. Non-motile protoplasmic filaments, termed the brush border, project from the luminal surfaces of the cells.
4. *Descending limb of Henle:* simple squamous epithelium.
5. *Henle's loop:* Here the epithelium becomes cuboidal.
6. *Ascending limb of Henle:* cuboidal epithelium, the cells of which show basal striations.
7. *Distal convoluted tubule:* The cells are again columnar and pyramidal in form; the basal striations are indistinct; there is no brush border.
8. *Junctional tubule:* clear cuboidal epithelium.
9. *Collecting tubules:* clear cuboidal epithelium.
10. *Papillary ducts:* simple columnar epithelium.

Blood Supply

The renal artery, after passing through the hilum, divides to form a number of so-called interlobar arteries. The latter radiate to the boundary zone between the cortex and medulla. Here the so-called interlobar arteries curve to form the arcuate arteries, which parallel in their courses the longitudinal axis of the kidney. From the arcuate arteries arise the interlobular arteries that extend radially through the cortex. In their course toward the renal capsule the interlobular arteries give off collaterally the afferent glomerular arterioles, each of which forms the glomerulus within the embrace of Bowman's capsule. From the glomerulus emanates the efferent glomerular arteriole that contributes to the formation a capillary network around the convoluted and collecting tubules of the cortex. The efferent glomerular arterioles from those glomeruli near the medulla of the kidney supply the medulla with straight arterioles called the arteriolae rectae.

Venous drainage in the cortex is accomplished by the interlobular veins, which in turn empty into the arcuate veins situated in corticomedullary area. The arcuate veins receive the venulae rectae, which effect venous drainage from the medulla. The arcuate veins subsequently open into the so-called interlobar veins, which converge toward the renal sinus to form the renal vein. The renal vein leaves the kidney at the hilum.

Innervation of the Kidney

The kidney receives its sympathetic innervation by way of the renal plexus, which is composed of postganglionic fibers from the celiac ganglion. The sympathetic innervation furnishes vasomotor fibers to the kidney. The parasympathetic innervation is represented in the vagus, which traverses the coeliac plexus in its route to the kidney. What role the vagus plays insofar as the innervation of the kidney is concerned cannot be indicated from the present controversial experimental data.

Ureters

The ureters are two musculomembranous tubes which convey urine to the bladder. Each ureter begins within the renal sinus as a funnel-shaped structure, the renal pelvis. The ureter then leaves the kidney at the hilum and passes caudad and slightly mediad over the ventral surface of the psoas minor muscle to open obliquely into the dorsocaudal angle of the bladder on its side.

Each ureter in its course lies dorsad of the parietal peritoneum (retroperitoneal) and is generally invested with some adipose tissue, as a result of which it manifests a whitish aspect throughout its course. In its passage to the bladder, the ureter is traversed obliquely by the internal spermatic artery and vein in the male and by the ovarian artery and vein in the female. Further caudad, the ureter crosses ventrad of the iliolumbalis artery and vein. Finally in the male, the ureter passes dorsad of the vas deferens and thence ventrad to open into the dorsocaudal angle of the bladder on its side. On the other hand, the ureter of the female, during the final parts of its course, turns slightly ventrad along the lateral surface the body of the uterus to attain the dorsocaudal angle of the bladder.

The ureters enter the bladder rather obliquely, forming on its internal surface two slit-like apertures around which the mucosa is elevated. The entrance of the ureters into the bladder in such manner results in the formation of a rather competent ureterovesicular valve, which thus precludes regurgitation of the urine into ureter, notwithstanding whatever conditions of pressure may obtain in the bladder.

Functionally the ureter itself, by virtue of its peristaltic contraction, serves to propel the urine through its lumen into the bladder. The volume of urine necessary to initiate peristalsis is very small.

Microscopically the ureter possesses three definite coats, which, from within outward, are the mucosa, muscular coat, and the adventitial or fibrous coat.

The epithelium of the mucosa is transitional. The subepithelial tissue, referred to generally as the tunica propria, consists of loose, fibroelastic tissue through which course many capillaries. The tunica muscularsis, or muscular coat, consists of inner longitudinal and outer circular layers of smooth muscle. In the caudal third of the ureter there is, in addition to the layers already mentioned, an outer longitudinal layer of smooth muscle. The adventitia consists of loose fibroelastic tissue, which is continuous with the surrounding adipose tissue.

Bladder

The bladder is a musculomembranous, pear-shaped sac lying in the abdomen, just craniad of the public tubercle; ventrad of the rectum in the male; ventrad of both the rectum and body of the uterus in the female. The caudal end of the bladder tapers into a neck or cervix, which passes into the pelvic cavity, lying dorsad of the symphysis pubis.

The bladder is clothed by peritoneum which extends craniad from the entrance of the ureters at its dorsocaudal angles to its apex, thence to its ventral and lateral surfaces whence it is continuous with the peritoneum on the ventral and dorsal body walls by means of the suspensory and lateral ligaments respectively.

The fixation of the bladder is accomplished by its neck and three ligaments. One ligament, the suspensory ligament, passes from the ventral surface of the bladder to the linea alba. The other two, called the lateral ligaments, extend from each side of the bladder to the dorsal body wall at a point just laterad of the rectum. In passing from the sides of the bladder to the dorsal body wall, the lateral ligaments form, between the bladder and rectum a peritoneal pouch, the rectovesical excavation, which communicates craniad with the greater peritoneal cavity.

Microscopic Anatomy

Microscopically the bladder is constituted, from within outward, of four coats: a mucosa, submucosa, muscularis and serosa. The mucosal epithelium is of the transitional type. The sub-epithelial fibroelastic tissue (tunica propria) blends into that of the submucosa. The muscularis is generally divided into three layers, which are not sharply definable in section; they are the inner longitudinal, middle circular (thickest), and the outer longitudinal. The musculature of the bladder wall, except for that in the trigonal area (the triangular area whose apices are the communications of the two ureters and urethra), is frequently referred to as the detrusor muscle. The circular musculature around the urethral orifice of the bladder is condensed to form a vesical sphincter (internal vesical sphincter) for the bladder. The serosa is comprised of loose, fibroelastic tissue covered externally by the mesothelium of the peritoneum.

Male Urethra

The urethra in the male extends caudad from the bladder to the external urethral orifice at the end of the glans penis. In the initial part of its course, it is embraced dorsally by the bilobed prostate gland; thence it continues caudad into the pelvic cavity, where it lies ventrad of the rectum and dorsad of the pubial and ischial symphyses, being cradled between the two pubiocaudales muscles as the latter ascend dorsad from origin. At the level of the ischial tuberosity, the urethra enters the bulbus urethrae of the corpus spongiosum through which it continues caudad to its external orifice at the extremity of the glans penis. The male urethra is, therefore, divisible into three portions: prostatic, membranous, and penile.

Prostatic portion

The prostatic portion of the male urethra is that part of the male urethra embraced dorsally by the bilobed prostate gland and lying dorsad of the public tubercle. On its dorsal internal surface may be seen the many minute apertures of the ducts of the prostate gland. Directly caudad of the orifices of the prostatic ducts, the prostatic urethra presents the two openings of the vasa deferentia. Between the latter openings there occurs a small elevation of the mucosa, termed the veru montanum.

Membranous portion

The membranous portion, about two and a half centimeters long, extends caudad from the prostatic urethra to the bulb of the corpus spongiosum. In its course through the pelvic cavity, it lies ventrad of the rectum and cradled between the internalsurfaces of the pubiocaudales muscles. The membranous portion of the urethra is completely surrounded by the compressor urethrae muscle, generally referred to physiologically as the external vesical sphincter.

On either side near the termination of the membranous urethra may be found the bulbourethral (Cowper's) glands.

Penile portion (cavernous urethra)

The penile portion is that part of the male urethra which is contained in the corpus spongiosum of the penis. It extends caudad, when the penis is relaxed, from the bulbus urethrae of the corpus spongiosum to the external urethral orifice at the end of the glans penis.

Microscopic Anatomy of Male Urethra

The male urethra possesses three coats: a mucosa, submucosa, and tunica muscularis which is confined to the prostatic and membranous portions.

The mucosal epithelium is transitional in the prostatic portion and pseudostratified columnar throughout the rest of the urethra except within the glans where the epithelium becomes stratified squamous. The subepithelial tissue (tunica propria) is of loose, fibroelastic tissue, which blends into similar tissue of submucosa wherein are also contained scattered bundles of smooth muscle. In the penile portion of the urethra the submucosa contains a plexus of venous vessels, part of the adjoining erectile tissue of the corpus spongiosum. The tunica muscularis is limited to the prostatic and membranous portions and is constituted of an inner longitudinal layer and an outer circular layer of smooth muscle. External to the outer circular layer of muscle of the membranous portion there obtains a layer of skeletal muscle, which represents the compressor urethrae muscle.

Female Urethra

The female urethra is represented in the elongated neck of the bladder. It extends caudad from the bladder, initially laying ventrad of the body of the uterus and spines of the pubes, and finally between the vagina and ischial symphysis. At the level of the cranial border of the ischium, the female urethra unites with the vagina in the formation of the urogenital sinus.

Microscopic Anatomy of the Female Urethra

The mucosal epithelium is generally stratified squamous. No distinct submucosa is discernible. The tunica muscularis consists of an inner longitudinal and outer circular layer of smooth muscle. The adventitial tissue is fibroelastic.

FORMATION OF URINE

Bowman's capsule, is the swollen thin walled blind end of the nephron; into it protrudes the glomerulus, which thus converts the

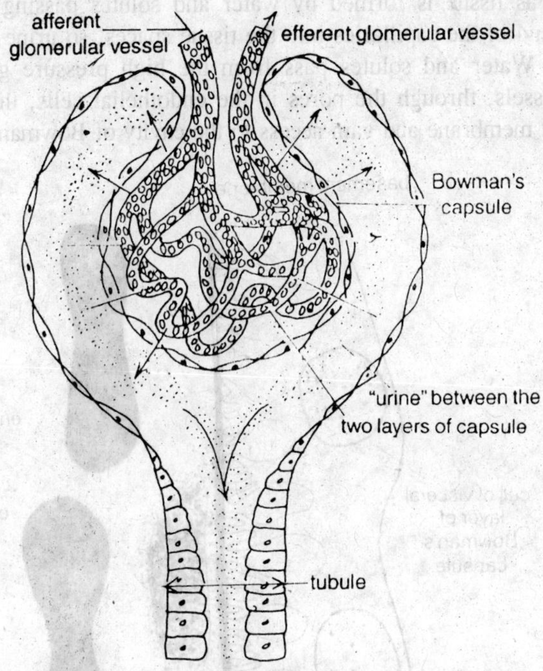

Fig. 14.7. The Malpighian body, showing that fluid and solutes pass from the glomerular blood vessels into the Bowman's capsule from which it drains into the tubule.

capsule into a hollow cup-shaped structure. The inner layer of the cup is closely applied to the glomerulus.

The layers which separate the plasma in the glomerular capillaries from the cavity of Bowman's capsule include the capillary endothelium, a basement membrane on which the endothelium rests, and the visceral layer of Bowman's capsule. Of these three layers only the basement membrane acts as a barrier to the free diffusion of substances. The cytoplasm of the endothelial cells is 'fenestrated' so that over certain areas of the endothelial cell the blood plasma is in direct contact with the basement membrane. The cells of the visceral layer of Bowman's capsule bear many branched projections, the various branches having somewhat swollen ends which about directly onto the basement membrane. Between these swollen terminals or foot processes, are gaps through which water and soluter can diffuse from the capillary lumen, through the basement membrane, between the cells of the visceral layer of the capsule, into the cavity of Bowman's capsule without having to traverse the cytoplasm of either the endothelial cells or the visceral layer of the capsule.

Just as tissue is formed by water and solutes passing from the arterial end of the capillaries into the tissue spaces, so urine is formed initially. Water and solutes pass from the high pressure glomerular blood vessels, through the pores in the endothelial cells, through the basement membrane and gain access to the cavity of Bowman's capsule

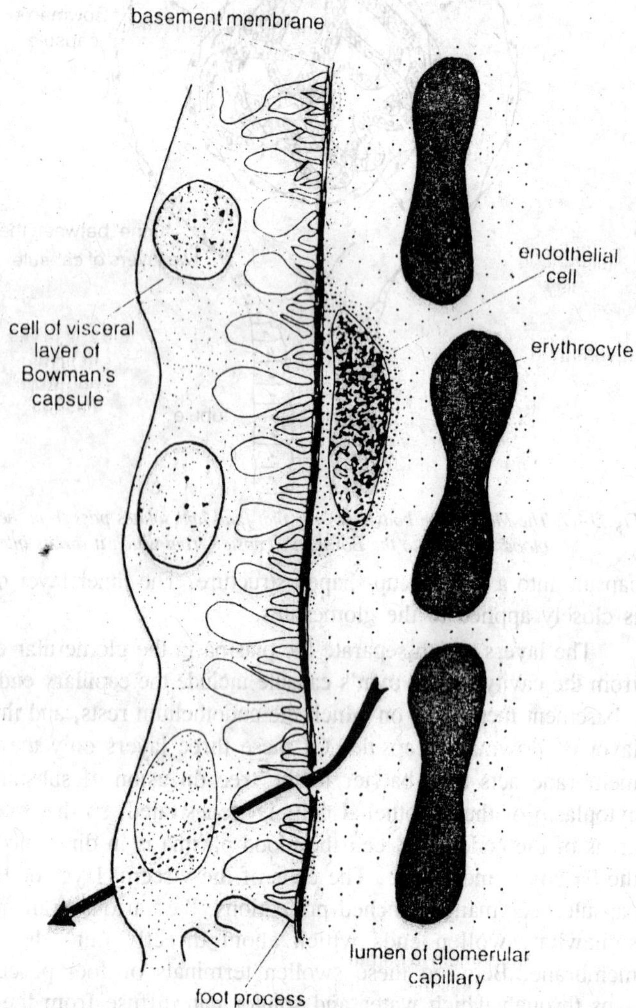

Fig. 14.8. Diagram of the structure of the layers separating blood in the glomerular capillaries from the cavity of Bowman's capsule. The arrow indicates the movement of water and solutes from the plasma in the glomerular capillary through fenestrations in the endothelial cells.

through the slit pores between the foot processes of the visceral layer cells. Because the blood in the glomerular capillaries is at high pressure throughout the length of the capillary and so much higher than the osmotic pressure of the plasma proteins there is no tendency for fluids and solutes to return to the glomerular blood vessel. This is unlike the condition at the venous end of the capillaries in other tissues where the hydrostatic pressure is lower and where there is a return of fluid into the capillary. As in the formation of tissue fluid, cellular elements of the blood and the substances of higher molecular weight i.e. proteins, are retained within the capillary network and only water and substances of lower molecular weight viz. mineral salts, glucose, amino acids, urea etc., pass out of the capillary. This processes is essentially one of ultrafiltration, the solutes of lower molecular weight being filtered, with water, through the basement membrane under the force of the blood pressure. If there is a fall in blood pressure below a critical level e.g. as in haemorrhage, then the filtration process stops and no urine is formed. It is possible to obtain samples of this glomerular filtrate by piercing the parietal layer of Bowman's capsule with a micro-pipette. Analysis of this fluid confirms that it is an ultrafiltrate of plasma containing water, salts, glucose, amino acids and excretory products, but none of the cellular elements of blood or the proteins of higher molecular weight.

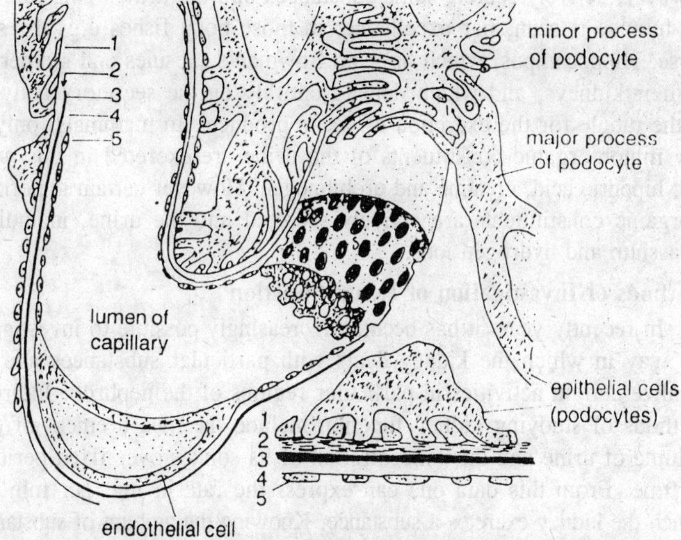

Fig. 14.9. Diagrammatic reconstruction of part of a capillary loop in the glomerulus.

Role of the Tubules in Urine Formation and in Homeostasis

A large volume of fluid containing solutes is filtered into the nephrons. The two million or so nephrons of both kidneys in man receive about 120 ml/min. of fluid filtered from all the glomeruli, i.e. over 7 litres an hour. If this fluid were excreted unchanged then the body would lose its entire water and salt content in less than half a day. In one day 600G of sodium, 35G potassium, 200G of glucose, 65G of amino acid and 180 litres of water are filtered into the nephrons. But in one day only 1 to 2 litres of urine containing approximately 6G sodium, 2G potassium, no glucose and only 2G potassium, no glucose and only 2G of amino acids are excreted. The prime role of the tubules is thus the conservation of water and solutes, and there are great changes in the volume and constitution of the fluid filtered off in the Bowman's capsules as it trickles down the nephrons. By varying the amount of water and solutes which are excreted the tubules play a very important role in the maintenance of the chemical composition of the internal environment. Thus the kidney is not only concerned with the excretion of nitrogenous waste products, e.g. urea and uric acid, but is vitally concerned in regulating the composition of the rest of the body.

The prime role of the tubules is thus the reabsorption of large volumes of fluids and large quantities of solute. The tubules can, however, actively secrete some substances into the urine. This role of the tubules is seen, par excellence, in those bony fishes e.g. the sea horse, Hippocampus, which have no Bowman's capsules and glomeruli in their kidneys, and which depend entirely on the secretory activity of the tubule for the excretion of waste products. In mammals, only a few minor organic constituents of the urine are excreted in this way e.g. hippuric acid, creatine and urobilinogen. However certain important inorganic constitutents are actively secreted into the urine, including potassium and hydrogen ions.

Methods of Investigation of Renal Function

In recently years it has become increasingly possible to investigate the way in which the kidney deals with particular substances and to localize certain activities in particular regions of the nephron. Indirect methods of studying kidney function include the measurement of the volume of urine and the concentration of its solutes over fixed periods of time. From this data one can express the rate in mg. per min. at which the kidney excretes a substance. Knowing the amount of substance which is delivered to the nephrons in unit time (calculated from a knowledge of the plasma concentration of the substance concerned and the rate of glomerular filtration) one can then assess the role of the nephrons in the 'renal handling' of this substance. Thus if the urine output was 1 ml.lmin. and contained 120 mg./ml. of a substance x, the plasma concentration of x was 1 mg./ml. and the rate of plasma filtration from all the glomeruli was 120 ml./min. then 120 mg./min. would be delivered to the nephrons and 120 mg./min. would be excreted in the urine. This would imply that the rubules neither reabsorbed nor secreted the substance x. Some substances are excreted in this way e.g. the polysaccharides inulin and mannitol. For these substances the following relationship holds:

plasma concentration (mg./ml.)	×	rate of plasma filtration (ml./min.)	=	amount excreted in urine (in mg./min.)

Knowing the plasma concentration and the rate of excretion in the urine one can calculate the rate of plasma filtration using these particular substances. Data from this type of experiment indicates that in man about 120 ml. of plasma is filtered off collectively by all the glomeruli of both kidneys in each minute. Provided that a substance is free to be filtered off in the glomeruli, that is it is not bound to plasma proteins or cellular elements, knowing the plasma concentration of the substance and the rate of plasma infiltration one can calculate the rate in mg./min. at which the substance is delivered to the nephrons. An analysis of urine will also provide information about the rate of excretion of the substance. If the amount excreted in the urine is less than the amount of substance delivered to the nephrons one can safely assume that the renal tubules are absorbing some of this substance. Conversely if more of the substance appears in the urine than was filtered off by the glomeruli then the tubules must have secreted some of the substance into the urine.

More direct methods of study are also available including the puncture of the renal tubules with micropipettes at various accessible sites. Analysis of the small amounts of fluids so obtained will reveal changes in the concentration of substances along the nephron. If one uses a double-barrelled micropipette one can introduce, under direct microscopic vision, a drop of coloured oil into a nephron, followed by a drop of a test solution, followed by a second droplet of coloured oil. This gives a column of fluid bounded in front and behind by coloured oil so that one can watch its progress along the nephron and note any changes in volume. Another direct method is to place a very fine polythene tube into one of the larger collecting ducts and allow the

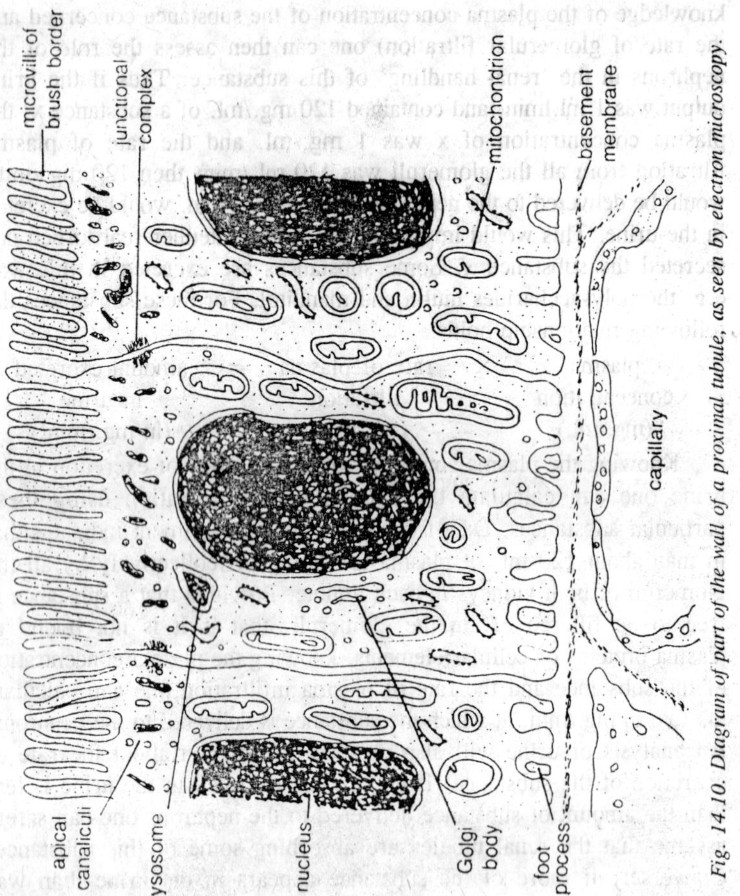

Fig. 14.10. Diagram of part of the wall of a proximal tubule, as seen by electron microscopy.

urine of drain through it. Then the tube is clamped so that urine cannot escape. The nephrons draining into that collecting duct become filled with a column of static fluid and each part of the tubule can more completely alter the fluid with which it is in contact. Then the clamp is released and small specimens of urine are collected in quick succession. The first sample obtained is fluid derived from the collecting ducts, the second sample is from the distal convoluted tubules, and so on. Analysis of these samples will reflect the activities of different parts of the tubules.

Reabsorption of Solutes

Much of the solute filtered off in the glomerular filtrate is reabsorbed in the proximal convoluted tubules. This reabsorption of solute is accompanied by the passive reabsorption of water so that of the 120 ml./min. of fluid delivered to the proximal tubules about 100 ml./min. is reabsorbed. There is no glucose or amino acid and a greatly reduced potassium content in the fluid which is passed on from the proximal convoluted tubules into the more distal parts of the system. The reabsorption of these solutes is against a concentration gradient since the solute concentration of the glomerular filtrate soon falls below that of the blood plasma as the solutes are reabsorbed. This reabsorption against a diffusion gradient is an active process involving the supply of energy by the cells of the proximal tubule. If the energy liberating processes of the tubule cells is hampered by cooling of the kidney or by metabolic poisons such as cyanide then the proximal tubules lose the ability to reabsorb solute and the urine soon resembles scarcely modified glomerular filtrate. The high absorptive capacity of the cells of the proximal tubule is reflected in their structure. The luminal surface area of the cell is vastly increased by the presence of many finger-like processes, the brush border, and the basal plasma membrane of the cell is infolded. The cytoplasm of the cell contains large numbers of mitochondria, the sites of energy release by oxidative metabolism, and there is a rich endoplasmic reticulum.

Glucose

In normal conditions all the glucose which is present in the glomerular filtrate is reabsorbed in the proximal tubules. There is, however a maximal capacity of 230 mg. per minute for glucose reabsorption. If more than this quantity of glucose/min. is delivered to the proximal tubules then the amount in excess of 230 mg./min. is excreted in the urine. Amounts in excess of this maximal tubular capacity for glucose reabsorption are reached when the glucose concentration of the plasma is abnormally high, a state of affairs which occurs in diabetes mellitus.

Amino acids

In addition to glucose almost all of the filtered amino acids are reabsorbed in the proximal tubules and appreciable quantities of amino acids do not appear in the urine, unless there is damage to the proximal tubule cells or there is an abnormally high concentration of amino acid in the blood, when the filtered load of amino acids exceeds the reabsorptive capacities of the tubule cells.

Sodium

Much of the filtered load of sodium is also reabsorbed by active processes in the proximal tubule. Negatively charged anions such as chloride move with the positively charged sodium ions and electrochemical neutrality is maintained. Water passively accompanies the movement of sodium and chloride ions to maintain osmotic equilibrium. Since both water and solute are absorbed by the proximal tubule cells the fluid remaining in the tubular lumen stays isotonic with plasma, although its volume is considerably reduced.

Bicarbonate

Bicarbonate ions are also conserved by the proximal tubule, although in a rather indirect fashion. The tubule cells are capable of producing amounts of carbonic acid from the carbon dioxide generated in their metabolism because they are rich in the enzyme carbonic anhydrase. This carbonic acid dissociates into hydrogen ions and bicarbonate ions. The hydrogen ions reach the tubular fluid they react with bicarbonate ions which have been filtered off from plasma in the glomerular filtrate to form water and carbon dioxide. Thus although the bicarbonate ions of the glomerular filtrate are destroyed by the secretion of hydrogen ions by the tubule cells, for each bicarbonate ions destroyed one new bicarbonate ion is added to the plasma by the tubule cell. If the enzyme carbonic anhydrase is inactivated by given an animal an inhibitor such as acetazoleamide then the tubule cells can no longer secrete hydrogen ions into the urine at the normal rate, bicarbonate cannot be conserved and up to as much as half of the filtered load of bicarbonate appears in the urine, making the urine alkaline.

Regulation of Body Water

The kidneys, by altering the amount of water excreted, play a vital role in controlling body water, in the face of large variations in water intake. When mammals are deprived of water the urine produced is scanty and concentrated and when excess water is taken in the urine becomes copious and dilute. Of the 120 ml./min. filtered into the nephrons in man, about 100 ml. is always reabsorbed in the proximal tubules. This is a passive reabsorption of water which moves along an osmotic gradient as solutes are actively reabsorbed. The remaining 20 ml. or so of water is passed on to the more distal parts of the nephron and it is this fraction of the filtered water which the kidney modifies according to the state of body hydration. If the body is dehydrated then most of this 20 ml. is reabsorbed and only about 0.5 ml./min. may appear as urine. If the body is over hydrated then almost all of this 20 ml./min. can be excreted in the urine. There are two questions which need answering about this facultative reabsorption of water, as it is called. First, how does the kidney absorb this water and second how is the kidney's activity regulated in relation to the state of body hydration?

Mechanism of Concentration of Urine

The ability of mammals to conserved body water by producing a highly concentrated urine varies greatly between species. Desert living mammals, such as the kangaroo rat, can produce a urine which is ten times more concentrated than the urine of an animal such as the beaver. If one looks at the kidneys of these animals one sees that the kidney of the beaver has a shallow medulla and all the nephrons have short loops of Henle' whereas the kidney of desert living mamm has a deep medulla and many nephrons with long loops of Henle which reach to the tips of the elongated pyramids. Kidneys of amphibian have no loops of Henle in the nephrons and these animals are unable to produce a concentrated urine.

The basis of the renal mechanism for the concentration of urine is a 'sodium pump' located in the ascending limb of the loop of Henle' which extrudes sodium ions from the tubular fluid into the tissue fluid between the tubules, so producing a hypertonic tissue fluid in the region of the medulla. The collecting ducts draining all nephrons have to pass through this medullary region of hypertonic tissue fluid in order to empty the urine into the pelvis of the kidney, and water diffuses out of the urine in the collecting ducts along the osmotic gradient. In order that the ascending limb of Henle can perform this function of producing a hypertonic interstitial fluid it must be itself impermeable to water otherwise water would diffuse out with the sodium ions and maintain osmotic equilibrium. And in order that the urine concentrating ability of the kidney can be varied to meet the requirements of the body's state of hydration, the permeability of the distal convoluted tubule and collecting ducts to water must also be capable of regulation. If the body is dehydrated then the distal tubules and collecting ducts remain permeable to water so that urine diffuses out from the lumen of the tubule into the hypertonic interstitial fluid of the medulla, water is thereby conserved and a scanty concentrated urine is produced. If the body is overhydrated then these distal parts of the tubules become impermeable to water so that water cannot move along the osmotic gradient and a copious dilute urine is produced. The arrangement of the loop of Henle as two closely aligned tubes in one of which is located a sodium extrusion mechanism imparts a special character to the mechanism of urine concentration. As the glomerular filtrate trickles down the descending limb of the loop, sodium ions diffuse into the limb from the surrounding hypertonic interstitial fluid and adjacent ascending limb. Thus the isotonic fluid delivered to the loop by the proximal convoluted tubule becomes increasingly hypertonic as it gains sodium ions. The ascending limb of the loop thus receives a fluid rich in sodium ions and by the expenditure of a small amount of energy the cells are able to extrude sodium ions into the interstitial fluid, so maintaining the hypertonicity of the interstitial fluid in the region of the medulla. If the tubular fluid of the descending limb of Henle did not equilibrate with the interstitial fluid then a fluid isotonic with plasma would be delivered required in order to extrude sodium ions from the isotonic tubular fluid into the hypertonic interstitial fluid, because of the net tendency of sodium ions to diffuse into the ascending limb along a diffusion gradient. The final concentration of sodium ions in the interstitial fluid would be much smaller for an equivalent energy expenditure of the cells of the ascending limb.

This system is called a countercurrent multiplication system. 'Countercurrent' describes the organization of flow in which the incoming fluid (in the descending limb) equilibrates with the outgoing fluid (in the ascending limb). Because of this countercurrent exchange the effect of work done by the sodium pump in producing a hypertonic interstitial fluid is multiplied. The longer the tubes in which Countercurrent exchange occurs the greater will be the effect, because more time is given for diffusion to reach equilibrium; thus the long loops of the nephrons of desert rodents enable a much more hypertonic interstitial fluid to be developed in the medulla of these kidneys than in, for example, the kidney of the beaver which possesses a shallow medulla and nephrons all of which have short loops of Henle. The concentration of sodium ions in the tubular fluid is greatest at the apex of the countercurrent exchange system, and a gradient of increasing tonicity is built up, increasing from the cortico-medullary junction to the tips of the pyramids.

The more hypertonic the interstitial fluid of the medulla, the greater will be the diffusion gradient of water from the distal convoluted tubules and collecting ducts to the interstitial fluid, and the more concentrated the ultimate urine will be.

The ascending limb of Henle delivers a hypotonic fluid to the distal convoluted tubules. This is because sodium ions have been extruded from the tubular fluid and water has remained in the tubular fluid. Thus if the walls of the distal convoluted tubules are permeable to water then water will diffuse out of the tubules into the isotonic interstitial fluid of the cortex. This renders the fluid in the distal convoluted tubules isotonic with blood plasma. This isotonic fluid which is now delivered to the collecting ducts can now lose increasing amounts of water to the hypertonic interstitial fluid of the medulla, provided that the tubules remain permeable to water. Thus in passing down the length of the nephron various changes in tonicity occur. The isotonic fluid delivered to the proximal tubules remains isotonic because both water and solute are reabsorbed. In the descending limb of Henle the fluid becomes hypertonic because water diffuses out into the hypertonic interstitial fluid of the medulla and sodium ions move in along their diffusion gradient. The hypertonic fluid delivered to the ascending limb from the descending limb gradually becomes hypotonic as sodium ions, but not water, are extruded. In the distal convoluted tubules this hypotonic fluid becomes isotonic as it loses water, and in the collecting ducts the fluid becomes increasingly hypertonic to plasma as water is lost to the hypertonic medullary interstitial fluid.

The concentration of urine then depends upon the ability of the kidney to develop and maintain a hypertonic interstitial fluid in the medullary region. The flow of blood through this region however threatens the maintenance of this mechanism because of the net tendency of sodium ions to be lost by diffusion from the medullary interstitial fluid into the blood plasma. This threat is reduced to a minimum by two mechanisms. First, there is air absolutely low rate of blood flow through the medulla and the cells of the medullary region rely heavily on anaerobic metabolism for their supplies of energy. Second, the arrangement of the blood vessels themselves in longitudinally aligned vessels, the vasa rectac and venae rectae, form a counter-current exchange system. This means that as blood which has collected sodium ions leaves the medulla it circulates past blood entering the kidney, and the concentration of sodium ions in the blood entering the medulla increases by exchange from blood leaving the kidney. This quantity of sodium derived from the outflowing blood effectively reduces the otherwise steep diffusion gradient for sodium ions between the hypertonit medullary interstitial fluid and the blood plasma. This countercurrent exchange of sodium ions means that for a given rate of blood flow the amount of sodium lost from the medulla is reduced.

Regulation of The Permeability of The Collecting Ducts by Anti-diuretic Hormone

The mechanism for water reabsorption by the kidney is the development and maintenance of a hypertonic interstitial fluid in the medullary region into which water diffuses from the collecting ducts along an osmotic gradient. However, the kidney must be able to alter the amount of water reabsorbed depending upon the state of body hydration. This is achieved by altering the permeability of the distal convoluted tubules and collecting ducts to water. The permeability of the distal parts of the nephron to water is determined by the presence of a hormone, called anti-diuretic hormone, liberated from the posterior nervous lobe of the pituitary gland. The name of the hormone describes its action on the kidney. The term diuresis describes the condition in which large volumes of dilute urine are passed. In anti-diuresis, water is reabsorbed from the urine and the urine passed is reduced in volume and is concentrated. The hormone increases the permeability of the distal parts of the nephron to water, permitting water to diffuse out of the nephron along the osmotic gradient, and the urine becomes concentrated. In the absence of the hormone these parts of the nephron become impermeable to water and no concentration of the urine occurs. There is a condition called diabetes insipidus in which the posterior pituitary gland does not liberate the hormone, or insufficient amounts of it, and so conservation of body water by concentration of the urine is markedly deficient. These patients pass large quantities of dilute urine, and need to drink large amounts of water to replace the losses of water in the urine.

The water content of the body is reflected in the osmotic pressure of the circulating blood. When the body is deprived of water, losses of water from the skin, respiratory tract and in the faeces continue. Further production of urine still continues because of the need to excrete nitrogenous waste products, although the urine passed is maximally concentrated. As water is lost from the body the osmotic pressure of the blood rises. Conversely when large volumes of water are drunk there is a dilution of the plasma and extracellular fluids and there is a fall in osmotic pressure of the blood. The osmotic pressure of the blood is being constantly monitored by nerve cells in the floor of the third ventricle of the brain, in the hypothalamus. These nerve cells, called osmoreceptors have long axons which travel downwards to the posterior nervous lobe of the pituitary gland where they terminate around blood vessels. These neurons secrete the antidiuretic hormone which passes down the axons, in the form of protein bound granules, towards the posterior lobe of the pituitary. A rise in osmotic pressure of the blood, indicating a deficit of body water, promotes a release of ADH from the neurons and this gains access to the circulating blood which transports it to the kidney where the hormone promotes an increased permeability of the distal part of the nephron to water. The urine becomes more concentrated and body water is conserved. Conversely if the osmotic pressure of the blood falls, indicating an excess of body water then the secretion of antidiuretic hormone ceases, the distal parts of the nephron become impermeable to water, and the excess water is excreted in the urine. The osmoreceptors of either hypotonic or hypertonic solutions injected into the hypothalamie are of the brain produce marked changes in the secretion of antidiuretic hormone and thus the renal handling of water, in spite of the fact that these minute local injections do not significantly alter total body water content or its osmotic pressure.

The osmotic pressure of body fluids obviously cannot be the only factor which determines the secretion of ADH from the posterior pituitary gland. If this were so then drinking a salt solution isosmotic with body fluids would produce progressive expansion of the volume of body fluids. This does not happen. The reason for this is that the secretion of ADH is also influenced by changes in the volume of fluid in the body and in particular the volume of blood in the vessels. This type of reflex response to changes in the volume of fluid within the blood vessels requires the presence of pressure-sensing elements (baroreceptors) in the vascular system. There are indeed baroreceptors in various parts of the vascular system – in the great veins, aorta carotid sinus and within the heart itself. Many of these baroreceptors have been found in influence the secretion of ADH. When there is a loss of body fluids e.g. because of haemorrhage, the fall in blood pressure stimulates the secretion of ADH by altering the discharge of impulses from the baroreceptors. The ADH which is released causes an increased retention of water by the kidney, thus helping to compensate for the loss of blood. The fall in blood pressure also stimulates the chemoreceptors of the carotid body; this stimulation is caused by anoxia of certain sensitive cells i.e. a deficiency of oxygen due to reduced blood flow. Sense cells within the carotid body increase their rate of 'firing' of nerve impulses to the central nervous system, which results in an increased rate of secretion of ADH.

The various factors which influence the secretion of ADH from the posterior pituitary gland.

Aldosterone and Sodium and Postassium Regulation

In addition to controlling body water the kidney regulates the plasma concentration of the important electrolytes sodium and potassium. The osmotic pressure of the plasma is determined mainly by the sodium ion (with the associated anions e.g. Cl). If the body loses sodium ions, then water is also lost because of the overriding necessity to maintain a normal osmotic pressure of body fluids. Conversely if sodium is retained by the body then water is also retained. Of the filtered load of sodium in the kidney, the bulk (90%) is reabsorbed in the proximal tubules. This active reabsorption of sodium is also responsible for absorption of water which occurs in this region, water diffusing passively with the sodium ions to maintain osmotic equilibrium.

Only 10% of the filtered load of sodium is passed on to the more distal parts of the nephron. In the loop of Henle sodium accumulates in the descending limb as the tubular fluid equilibrates with the hypertonic interstitial fluid. However in the ascending limb the accumulated sodium is extruded from the nephron by the 'sodium pump' mechanism in the tubule cells. Of the filtered load only a small fraction (up to 10%) eventually reaches the distal tubule. In the distal convoluted tubule some of the sodium ions are reabsorbed in exchange for hydrogen ions and potassium ions which are being secreted into the tubule in this region. All of the filtered load of potassium is reabsorbed in the proximal convoluted tubules and the potassium which appears in the urine is derived by active secretion of the ion into the fluid of the distal convoluted tubule in exchange for sodium ions. It is this small fraction of the filtered load of sodium, which is absorbed in exchange for secreted potassium ions, that the body regulates to achieve sodium balance. The exchange process which occurs in the distal tubule is stimulated by the presence in the blood of the hormone aldosterone secreted by the adrenal cortex. An excess of the hormone stimulates this exchange process so that there is a loss of potassium from the body and a retention of sodium. A deficiency in the supply of the hormone as in Addison's disease causes a net loss of sodium from the body and a retention of potassium. In Addison's disease there is thus a continued loss of sodium ions in the urine and with them water is also lost. This leads to a progressive shrinking of the volume of circulating plasma and extracellular fluid and eventually leads to a circulatory collapse. When a normal animal is deprived of sodium in the diet there is an increased outpouring of the hormone aldosterone from the adrenal cortex, which results in a conservation of sodium by the exchange mechanism of the distal convoluted tubule mechanisms regulating the secretion of aldosterone.

In order that the adrenal cortex can regulate the sodium balance of the body by way of the ettect of the hormone aldosterone on the kidney, it must be provided with information about the state of the body's sodium economy. One possibility is that the adrenal cortex 'senses' the concentration of sodium in the blood. This possibility was examined by means of an ingenious technique which made the adrenal gland readily accessible for experimental study. An adrenal gland of a sheep was removed from the abdomen and transplanted to the neck of the sheep was removed from the abdomen and transplanted to the neck of the sheep within a skin flap, the adrenal artery and vein being joined to blood vessels in the neck. With the adrenal gland in this position it was possible, in the conscious animal, to study the effect of changes in the sodium content of blood on the secretion of aldosterone. A needle could be inserted into the adrenal artery to perfuse the gland with blood of known sodium content and another needle in the adrenal vein could provide specimens of blood for analysis of aldosterone content. Using this technique it proved possible to show that the production of aldosterone by the adrenal gland does not increase when the sodium content of the blood entering the gland was lowered. Further in sheep which were already responding to a deficiency of body sodium by increased secretion of aldosterone the perfusion of the transplanted adrenal with sodium-enriched blood did not reduce the amount of aldosterone which has being secreted.

It seems then that the adrenal gland does not directly 'sense' the sodium content of the blood. Some other part of the body must perform this function and then relay the information to the adrenal gland. This information appears to be in the form of one or more hormones. This was shown by depriving a sheep of sodium until its own adrenal glands responded by increased secretion of aldosterone. A small amount of blood remove from this sheep and injected into the adrenal gland in a neck pouch of a sheep in normal sodium balance caused this animal to produce increased amounts of aldosterone.

We still do not have complete story of the way in which the adrenal cortex, receives information about sodium balance. However, studies which have been made since the last edition of this book was published indicate that the kidney itself produces one of the hormones which regulate the secretion of aldostrone. The kidney, like the adrenal cortex, does not appear to directly 'sense' the concentration in the blood. Rather does it 'sense' the effect of sodium upon blood volume. We have seen that sodium and water tend to be handled together by the body. When sodium is lost from the body water is also lost and when sodium is gained water is also gained; these reciprocal fluxes of sodium and water being determined by the need to maintain a normal osrnotic pressure of body fluids. Thus changes in body sodium content are reflected in water content - i.e. the volume of body fluids. It is the change in blood volume (and blood pressure) which the body senses. The sense organ in the kidney i.e. the baroreceptor - is in the afferent arteriole of the glomerulus, in the so-called juxta-glomerular apparatus.

As the afferent arteriole approaches he glomerulus the smooth muscle cells in the wall of the arteriole become replaced by large epithelial type cells containing secretory granules. This tissue forms a cuff in the afferent arteriole which butts against the wall of Bowman's capsule. There is a localized region of the distal convoluted tubule of the nephron containing enlarged cells which is associated with the polar cuff. This region of the distal convoluted tubule is called the macula densa. Together with the polar cuff the macula densa form the juxta-glomerular apparatus.

We do not yet know the function of the macula densa but the polar cuff appears to produce a hormone in response to changes in blood pressure in the afferent arteriole of the glomerulus. Illustrates the way in which we imagine the juxta-glomerular apparatus regulates the secretion of aldosterone by the adrenal cortex.

1. The renal afferent arteriole forms the 'volume receptor', the signal to the receptor being the degree of stretch of the arteriole.
2. When there is a decrease in stretch of the afferent arteriole caused by a decrease in blood volume (= sodium deficiency) the juxta-glomerular apparatus releases the hormone rennin into the blood stream.
3. Renin reacts with a member of the plasma proteins to form another active substance called angiotensin.
4. Angiotensin stimulates the adrenal cortex to produce more aldosterone.
5. Aidosterone is transported in the blood stream to the kidney where it stimulates the reabsorption of sodium in the distal convoluted tubules.
6. The retention of sodium (and water) causes an expansion of blood volume, a rise in blood pressure, an increased renal blood flow ... and a decline in the production and release of renin.

Returning to our original example - the loss of blood - it can be seen the fall in blood pressure which results stimulates the reabsorption of both salt and water by the kidney. The secretion of ADH (producing retention of water) is brought about by the effect of a fall of blood pressure on baroreceptors in the neck and chest. The secretion of aldosterone (causing retention of sodium by the kidney) is brought about by the effect of a fall in blood pressure on baroreceptors in the kidney itself.

Acid-base Balance, and the Kidney

There is an ever present tendency for the body fluids to become increasingly acid in reaction. In one day, in man, metabolic processes liberate 13 000 m. moles of carbon dioxide, which if converted into carbonic acid would form an amount of acid equivalent to thirteen litres of N hydrochloric acid. Various mechanisms exist which regulate the reaction of the body fluids, and among these the kidneys have an important role to play.

Acidity, Alkalinity and Neutranity

A solution is defined as acid in reaction when there is a net excess of hydrogen ions (H^+) over hydroxyl ions (OH^-). A solution is alkaline when there is an excess of hydroxyl over hydrogen ions, and neutral when there are equal concentrations of the two ions.

Expression of Degrees of Alkalinity and Acidity

At a particular temperature the product of the concentration of hydrogen ions [H^+], and the concentration of hydroxyl ions [OH^-], in a particular solution is constant. This means that one can express either acidity or alkalinity in terms of the concentration of one of these ions alone, and the concentration of hydrogen ions is used for this purpose. In any but the most acid solutions the concentration of free H^+ is so small that one would have to use inconvenient numbers to express them. So instead one uses the logarithm of the reciprocal of concentration (i.e. dilution). This expression, the log of the reciprocal of hydrogen ions concentration, is called the pH value.

$$pH = \log 1/[H]^+$$

In the pH scale, the neutral point is 7.0. Acid solutions have a pH value less than 7, the smaller the pH value the more acid the solution. Alkaline solutions have a pH value above 7, the higher the pH value the more alkaline the solution. The pH of body fluids is kept fairly constant around 7.4, although the range of survival covers 0.4 of a pH unit on either side of this point. The pH scale is a logarithmic one, a change of one pH unit implying a ten-fold change in concentration, so that this range of tolerance of tissues from pH 7.0-7.8 is a fairly wide one.

Acid and Alkalis

An acid is a substance which tends to give up a proton i.e. the hydrogen atom after losing its single electron. The more readily the acid gives off a proton the stronger it is. One can characterize this in the following way.

$$HB \rightleftharpoons H^+ + B^-$$

Acid proton base

A base is a substance which tends to accept a proton, the greater the avidity for protons the stronger the base. In the above formulation of the reversible dissociation of an acid it can be seen that a base is left after an acid has given off its proton. In this reversible reaction the stronger the tendency for the reaction to proceed to the right i.e. the stronger the acid, then the weaker is the base, and vice versa. The base left after dissociation of the acid is called the conjugate base of the acid in question and the two, acid and base, form a conjugate pair. Thus hydrochloric acid is a strong acid because of its marked tendency to dissociate into free hydrogen ions and the weak conjugate base (chloride ions).

$$HCl \rightleftharpoons H^+ + Cl^-$$

hydrochloric acid proton chloride ion.

A variety of acids are present in body fluids including.

1. *Carbonic acid.*

$$H_2CO_3 \rightleftharpoons H^+ + HCO_3^-$$

carbonic acid bicarbonate ion (base).

2. *Primary phosphate ion.*

$$H_2PO_4 \rightleftharpoons H^+ + HPO_4^{--}$$

secondary phosphate ion (base).

3. *Ammonium ion.*

$$NH_4^+ \rightleftharpoons H^+ + NH_3$$

ammonia (base).

The reaction of the body fluids (pH 7.4) is alkaline and this is determined by the presence of various bases. The chloride ion (a base) is the conjugate base of such a strong acid (hydrochloric acid) that is, it has very little tendency to accept protons. But in addition to chloride ions there are smaller amounts of three stronger bases which contribute to maintaining the alkaline reaction of the body fluids. These bases include bicarbonate ions, secondary phosphate ions and proteins (which exist in plasma as negatively charged anions capable of accepting protons). These three bases are the conjugate bases of very weak acids which are important because changes in pH around the neutral point can alter equilibrium point of dissociation of the weak acid into the proton and the stronger base.

$$\text{Weak acid} \rightleftharpoons \text{proton} + \text{strong base}$$

$$HB \rightleftharpoons H^+ + B^-$$

The equilibrium is shifted to the right if the pH of the solution rises (i.e. the concentration of free hydrogen ions falls) so that more hydrogen ions are formed. The equilibrium is shifted to the left if the pH of the solution falls (i.e. the concentration of free hydrogen ions increases). Because of this tendency to release or take up hydrogen ions depending on the pH of the solution these acids and bases tend to stabilize the pH of a solution containing them. For this reason they are called buffers.

pH of Buffer Solutions

One can calculate the pH of a solution containing a buffer by means of the modified law of mass action called the Henderson-Hasselbach equation.

$$pH = pK + \log \frac{\text{concentration of buffer base}}{\text{concentration of buffer acid}}$$

The pK value is a constant, under certain specified conditions, for a particular buffer system.

Three important buffer systems exist in blood to stabilize the pH.

1. *Carbonic acid-bicarbonate buffer system.*

$$pH +6.2 +\log \frac{[HCO_3^-]}{[H_2CO_3]}$$

$$= 6.2 + \log \frac{20}{1}$$

In this buffer system if there is a rise in the concentration of free H^+ ions then these react with the buffer base to form carbonic acid. E.g.

HCl	+	HCO_3^-	$\rightleftharpoons$	H_2CO_3	+	Cl
hydrochloric acid	+	bicarbonate ion	$\rightleftharpoons$	carbonic acid	+	chloride ion

In the above example electrochemical neutrality is maintained in buffering by the loss of one negative charge on the bicarbonate ion which is replaced by a negative charge on the chloride ion. The strong acid HCl is replaced by the much weaker carbonic acid, H_2CO_3 (i.e. there is much less tendency for this acid to dissociate to give rise to free hydrogen ions).

The concentration of bicarbonate ions, in the process of buffering, falls since the pH of the solution is determined by the ratio of HCO_3^- / H_2CO_3.

However, the pH falls much less than if free hydrochloric acid remained in the solution and was unbuffered. If an alkali is added to this system e.g. sodium hydroxide, then this would also be buffered:

$$NaOH + H_2CO_3 \rightleftharpoons NaHCO_3 + H_2O$$
$$\Updownarrow$$
$$Na + HCO_3^-$$

In this reaction the strong base sodium hydroxide is replaced by the much weaker base HCO_3^-. As the numerator of the equation

$$pH = 6.2 + \log\frac{[H_2CO_3^-]}{[H_2CO_3]}$$

increases, the pH of the solution rises but much less than if the sodium hydroxide had not been buffered.

2. *The phosphate buffer system.*

$$pH = 6.8 + \log\frac{[B_2HPO_4]}{[BH_2PO_4]} \text{ where B = Na or K}$$

Buffering of acids by this system occurs in the following way

$$B_2BPO_4 + H^+ \rightleftharpoons BH_2PO_4 + B^+$$

Electrochemical neutralityis maintained by the freeing on one cation (either a sodium or potassium ion). Buffering occurs because free hydrogen ions are replaced by the weak acid BH_2PO_4 which has little tendency to give up a proton.

3. *Proteins*. One cannot give a single pK value for proteins because of the large number of ionisable groups on the protein. At the pH of plasma, proteins exists as colloidal anions which can give up protons becoming more negatively charged or take up protons becoming less negatively charged.

$$\text{Buffering of alkali: } HProt^{(n-1)-} + OH^- \xrightarrow{\text{Alkali}} Prot^{n-} + H_2O$$

$$\text{Buffering of acid: } Prot^{n-} + H^+ \rightarrow HProt^{(n-1)-}.$$

The proteins form a very important system, both intra and extraecuularly. In the blood the protein haemoglobin is a potent buffer which enables most of the carbon dioxide generated in metabolism to be transported from the tissues to the lungs without change in the pH of the blood.

Sequence of Events on the Introduction of Hydrogen ions into the Blood

When hydrogen ions are introduced into the circulation, either artificially or from acids generated in metabolism, they are first diluted in the large volume of body water and then they are buffered. Buffering of hydrogen ions involves a reduction in the amount of body bases (bicarbonate ions, secondary phosphate ions, and a fall in the buffering capacity of proteins). Considering the bicarbonate-carbonic acid buffer system one might have a shift of HCO_3^-/H_2CO_3 from the normal ratio of 20/1 to 10/1. The condition of acidosis is characterized by a fall in the blood concentration of buffer base. This bicarbonate buffer system is of great importance in the body because each component of the system is under independent physiological control. The concentration of carbonic acid in blood is under the control of the respiratory centre of the medulla and the lungs and the bicarbonate concentration of blood is under renal control. Carbon dioxide dissolved in blood is a potent regulator of respiratory rate through the direct action on the respiratory centre of the medulla and the indirect action through the chemoreceptors of the aortic and carotid bodies. Under normal circumstances breathing is controlled by these means so as to maintain a partial pressure of CO_2 in the alveoli and arterial blood of 40 mm. Hg. An excess of CO_2 in the blood stimulates a rise in ventilation of the lungs to 'flush out' the excess CO_2 from the blood. A fall in pp. CO_2 in the arterial blood depresses ventilation so that less of the CO_2 generated in metabolism is excreted from the lungs and it accumulates in the blood until the normal concentration is achieved. Thus the denominator of the Henderson-Hasselbach equation, H_2CO_3 is maintained constant under normal conditions.

The chemoreceptors of the aortic and carotid glomi are also sensitive to the pH of the blood circulating through them. A fall up pH acting through them stimulates an increase in ventilation rate, and an increase in pH of the blood depresses ventilation. A change of 0.1 of a pH unit of the blood alters the ventilation two fold. Thus, after buffering of acids by the blood buffers and a fall in blood pH, modifications in ventilation occur which tend to restore the pH of the blood towards normal values. In the hypothetical example considered above, buffering of acid was associated with a change in the ratio of

HCO_3^-/H_2CO_3 from 20/1 to 10/1. This is associated with a fall in the pH of the blood. Acting via the chemoreceptors this stimulates ventilation so that more CO_2 and thus more H_2CO_3 is excrted from the blood, producing a further change in the ratio of buffers e.g. from 10/1 to 10/0.5.

This establishes a normal ratio of the two buffers and the pH of the blood returns to normal levels. Although the pH of the blood is quickly restored to normal by the change in ventilation of the lungs, the buffering capacity of the blood, which is determined by the total amount of substances rather than their ratio, has been greatly reduced – in the above example buffering capacity has been halved.

The restoration of buffer capacity is carried out by the kidney which restores bicarbonate ions to the plasma. This, however, is a slower process, taking hours and days contrasting with the immediate ventilatory response which may be accomplished within minutes.

By means of the process already described on the kidney tubule cells generate new hydrogen ions equivalent to those which have been buffered and these are excreted in the urine, and for each hydro ion excreted in the urine one new bicarbonate ion is returned to the plasma to restore the buffer capacity. The concentration of bicarbonate ions in the plasma gradually rises and the process of buffering is reversed.

1. The concentration of bicarbonate ions in the bicarbonate-carbonic acid system is returned to normal.
2. Primary phosphate is reconverted to the more basic secondary phosphate.

$$H_2PO_4^- + HCO_3^- - HPO_4^- + H_2CO_3$$
$$\downarrow$$
$$H_2O + CO_2 \text{ excreted via lungs.}$$

3. Hydrogen ions taken up by proteins are now given up.

$$HProt^{(n-1)-} + HCO_3^- - Prot^{n-} + H_2CO_3$$
$$\downarrow$$
$$H_2O + CO_2 \rightarrow \text{excreted via lungs.}$$

The reabsorption of bicarbonate ions from the glomerular filtrate was described. By the end of the proximal tubules of all the filtered bicarbonate ions have been destroyed by the tubular secretion of hydrogen ions, but each destroyed bicarbonate ion has been replaced by a new bicarbonate ion generated in the tubule cells. After the bicarbonate has been removed from the filtrate by this mechanism the secretion of hydrogen ions continues in the distal convoluted tubules and the bicarbonate ions, which pass into the blood from the tubule cells, are now in excess of the bicarbonate ions filtered off in the glomerular filtrate. It is this new additional bicarbonate which restores the buffer capacity of the blood.

Free hydrogen ions cannot be excreted into the urine for obvious reasons. The hydrogen ions secreted into the tubules are buffered by secondary phosphate ions which have been filtered off in the glomerular filtrate.

$$HPO_4^{--} + H^+ \rightarrow H_2PO_4^-$$

When all the phosphate has been used up in this way the pH of the urine is at its lowest of pH 4.3. However, further hydrogen ions can be excreted in the urine, and thus additional bicarbonate is restored to the plasma without further lowering the pH of the urine. This is because the kidney cells can produce the base ammonia, either from the amino acid glutamine or from the oxidative deamination of other amino acids. This ammonia is secreted into the tubular lumen with the hydrogen ions, as the very weak acid the ammonium ion.

$$\underset{\text{ammonia}}{NH_3} + H^+ \rightarrow \underset{\text{weak acid.}}{NH_4^+}$$

The amount of ammonium ions excreted by the kidney varies according to the state of the reaction of the body fluids, and large amounts can be excreted in acidosis.

The most important factor which determines how much of the bicarbonate base is restored to the blood passing through the kidney is the amount of bicarbonate which his filtered off in the glomerular filtrate which in turn depends upon the concentration of bicarbonate in the plasma. If this is low, e.g. because of buffering of acids by the blood, then less bicarbonate is filtered off in the kidneys and fewer of the hydrogen ions secreted by the tubule cells are used in 'reabsorbing' filtered bicarbonate ions and a greater quantity of 'new' bicarbonate ions are restored to the blood. If the concentration of bicarbonate ions in the plasma is high (alkalosis) more bicarbonate ions are filtered off in the kidneys, more hydrogen ions are used in 'reabsorbing' this bicarbonate and a reduced or zero quantity of 'new' bicarbonate ions is returned to the plasma. In fact all the hydrogen ions secreted by the tubule cells may be utilised in the 'reabsorption' of filtered bicarbonate ions, and these now begin to appear in the urine, making the urine alkaline.

Constituents of Urine and their Origin

Composition Of Human Urine in grammes per 24 hours.

Urea	12 - 34 G
Ammonia	0.6 - 1.2 G
Creatinine	0.8 - 2.0 G
Uric acid	0.3 - 0.8 G
Hippuric acid	0.7 G
NaCl	10 - 15 G
P	1.2 G
Na	2.5 G
S	1.2 G
K	1.5 - 2.0 G

Urea (Carbamide) $(NH_2)_2CO$

This the main nitrogenous excretory product found in mammals. It is very soluble in water and fairly toxic. Human blood has normally between 20-38 mg. urea/100 mls. Of blood. Urea is manufactured in the liver in mammals in a cyclic chemical reaction called the ornithine cycle. It was once thought to be produced directly from ammonia but it is now known to be produced from ammonia by way of the amino acid arginine. Under the influence of an enzyme arginase the arginine breaks down to urea and the amino acid ornithine. The ornithine is reconverted into arginine by taking up further ammonia derived from the breakdcwn of waste and excess amino-acids.

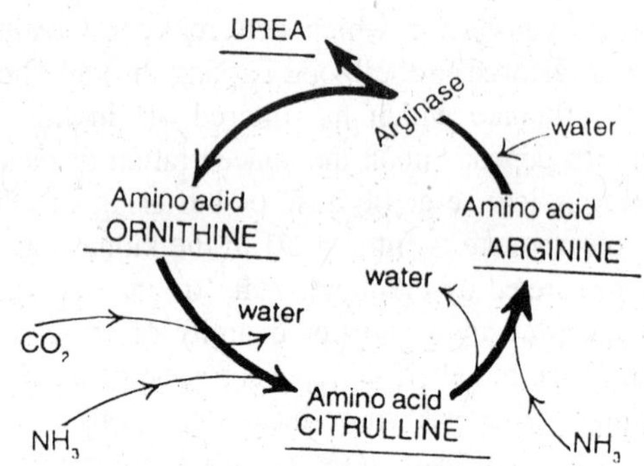

Fig. 14.11. The Ornithine cycle for the production of urea.

Ammonia

The ammonia found in the urine is formed in the kidney, as described on page 515. It is produced by the deamination of the amino acid glutamine in the cells of the renal tubules.

The process of deamination is an important one; it occurs also in the mammalian liver where the ammonia, however, is converted into urea by way of the ornithine cycle. Chemically, deamination is usually an oxidative process:

$$\underset{\text{amino acid}}{\mathrm{R.\overset{\displaystyle H}{\underset{\displaystyle COOH}{\overset{|}{\underset{|}{C}}}}{-}NH_2}} + \tfrac{1}{2}O_2 \rightarrow \underset{\text{keto-acid}}{\mathrm{R.\underset{\displaystyle COOH}{\underset{|}{C}}{=}O}} + \underset{\text{ammonia}}{NH_3}$$

The keto-acids produced are respired to produce energy and in the liver ammonia is incorporated into the amino acid arginine. Ammonia is a very toxic substance and as little as 5 mg. 100 ml. of blood will kill a rabbit. There is normally only about 0.001 mg. 100 ml. in mammalian blood.

Since ammonia is so poisonous it is not surprising to find that it is an important excretory product only in those animals that can quickly expel it from their bodies. In fresh water bony fishes where water is passing through the body in large quantities and there is a copious dilute urine, ammonia accounts for over 50% of the nitrogenous excretory products.

Uric acid

Uric acid is almost insoluble in water and therefore does not enter into the chemical reactions of the body. It is found only in small amounts (about 0.5 G/day) in man, but in reptiles and birds it forms the major excretory product. It is essential for an animal which lives for a long embryonic period in the egg to be able to produce an excretory product which will not poison the embryo, for there is no means of removing the excretory product when a shelled egg is laid on land. Reptiles and birds, which have large shelled eggs on land, have the ability to excrete most of their waste nitrogen as uric acid. In passing we may note that although the young mammal has a long embryonic period it can survive in spite of a relatively toxic soluble excretory product (urea) because the excretory waste is removed via the placenta and the kidney of the mother.

The small amount of uric acid produced in the urine of man comes from oxidative deamination of the purines adenine and guanine, which are produced when nucleic acids break down.

Nucleic acid → adenine → hypoxanthine

Nucleic acid → guanine → xanthine → (xanthine oxidase) → uric acid

An abnormality in the metabolism of uric acid in man may produce the disease gout, in which deposits of uric acid crystals may accumulate in certain joints and soft tissues; a diet excluding foods rich in purines is advised.

Phosphate and Sulphate. These are derived mainly from the breakdown of proteins containing phosphorus and sulphur, in addition to which there is a variable intake of sulphates and phosphates in foodstuffs.

Cholorides. The chlorides present in the urine come from the dietary salt, which is a varying quantity.

15

NERVOUS SYSTEM

Of all the cells in the body, those of which nervous tissue is composed have the properties of irritability and conductivity most highly developed. The detailed structure of nervous tissue has been discussed previously. It will be recalled that all nervous tissues are of ectoermal origin. They trace their beginning, for the most part, from the embryonic neural tube and neural crests. In a number of cases, *placodes*, or thickenings of the superficial ectoderm in certain restricted areas, also give rise to nervous elements.

In addition to nerve cells, or *neurons*, of various types, there are present in the central nervous system (brain and spinal cord) other components referred to as *neuroglia*. These very numerous ectodermal cells are interspersed among the nervous elements, providing support and some degree of protection since true connective tissue of mesodermal origin is not present. It is coming to be recognized that they play a very important role in the synthesis of certain proteins which are vital elements in the proper functioning of such higher nervous functions as thought and memory. Several types of neuroglia cells are recognized. They generally consist of very much-branched cells, the processes of which form an interlacing network between neurons. Other nonnervous, supporting, *ependymal cells* of an epithelial nature line the cavities of brain and spinal cord. They are often ciliated.

Gray Matter and White Matter

Nerve-cell bodies, their dendrites, the supporting neuroglia cells, and the proximal unmyelinated portions of the axons have a grayish appearance and form the greater part of the *gray matter* of the brain and spinal cord. Groups of nerve-cell bodies are called *ganglia*. The term ganglion is generally used to refer to an aggregation of never-cell bodies forming a mass or enlargement *outside* the central nervous system. Gray ganglionic masses *within* the brain and spinal cord are usually designated as nuclei, or nerve centers. White matter, on the other hand, is composed chiefly of bundles of glistening white myelinated, or medullated, fibers. Such bundles, which it is possible to trace in prepared sections of brain and spinal cord, are known as *fiber tracts*. In some regions gray and white matter are intermingled to varying degrees, such an arrangement being known as a *reticular formation*.

Primary Divisions

The nervous system is extremely complex, and all parts are structurally connected and functionally integrated. It is customary, nevertheless, to speak of two main divisions. These are (1) the *central nervous system*, composed of the brain, which lies within the cranial cavity of the skull, and the *spinal cord*, lying within the neural canal formed by the neural arches of the vertebrae, and (2) the *peripheral nervous system*, made up of nerves and ganglia which are connected to the brain and the spinal cord. Part of the peripheral nervous system is composed of autonomic fibers which are distributed to those parts of the body under involuntary control. These, in turn, consist of sympathetic and parasympathetic components which function in an opposite manner in mediating various involuntary activities of the body. The autonomic portion of the peripheral nervous system is sometimes spoken of as a third division of the nervous system, called the *autonomic nervous system*. However, since autonomic fibers are anatomically associated with spinal nerves and certain cranial nerves, it is most logical to include them under the category of the peripheral nervous system. In the following pages we shall refer to the autonomic components as the *peripheral autonomic system*.

Meninges

Both and spinal cord are surrounded by membranes, or menings (singular, meninx), the complexity in arrangement of which increases according to advance in the evolutionary scale. These membranes protect and give some support to the central nervous system.

In mammals the pia-arachnoid membrance differentiates into two layers, an inner, very vascular pia mater and an outer, nonvascular arachnoid membrane. A network of fine trabeculae composed of collagenic and elastic fibers joins the arachnoid membrane and pia meter. A subarachnoid space, filled with cerebrospinal fluid, makes

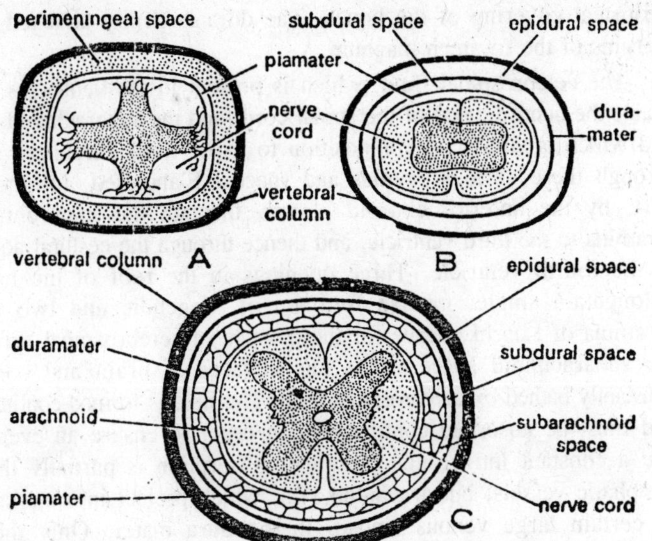

Fig. 15.1. T.S. of nerve cord showing meanings. A—Fish; B—Amphibian, reptile, and bird, C—Mammal.

its appearance between the two membranes. In the brain region the cranial dura mater fuses with the endorachis and the epidural space thus disappears. The subdural space between that dura and arachnoid membrane is shallow and contains only a miniumu of fluid which is not cerebrospinal fluid. Melanocytes are usually present in the portion of the pia mater of mammals on the ventral side of medulla oblongata.

About the spinal cord a fusion of spinal dura and endorachis does not occur, and the epidural space, which thus persists in this region, contains fatty and other connective tissues as well as a number of venis. The cranial dura is continuous with the spinal dura at the foramen magnum.

The cranial dura is a tough, dense membrane loosely attached to the cranium except at the sutures and the basal portion of the skill. The pia mater and arachnoid membrain, known collectively as the leptomeninges, are of a loose, delicate consistency. Those of the brain and cord are of similar structure and appearance.

Near the posterior end of the spinal core the subdural space disappears and the spinal dura forms a close-fitting sheath around the filum terminale. At the points where cranial and spinal nerves emerge, the dura forms sheaths around the nerves and their roots. The sheaths pass through the intervertebral foramina and are continuous with the periosteal covering of the bones. The dura is firmly attached to the margins of the foramen magnum.

The celebrospinal fluid, which is present in the ventricles of the brain, the central canal of the spinal cord, and the subarachnoid space, and which is similar in composition to tissue fluid, circulates slowly through these various cavities and spaces. Elaborated, for the most part, by the anterior choroid plexus, through the interventricular foramina to the third ventricle, and thence through the ceslbral aqueduct to the fourth ventricle. Three openings in the roof of the medulla oblongata-a single, median foramen of Magendie and two lateral foramina of Luschka-serve for the passage of cerebrospinal fluid into the subarachnoid space. Thus the surfaces of brain and cord are constantly bathed by cerebrospinal fluid. The fluid is formed continuously and must be absorbed continuously in order to ensure an even flow and a constant intracranial pressure. Absorption is partially through lymphatic vessels, but arachnoid villi, which project into the cavities of certain large venous sinuses in the dura mater. Only the thin mesothelial membranes of the arachnoid villi separate the cerebrospinal fluid in the subarachnoid space from the blood in the venous sinuses.

Certain modifications of the meninges are to be observed in a mammals. The cranial dura sends a process, or fold, the falx cerebri, down into the fissure separating the two cerebral hemispheres. A similar process, the tentorium, pushes down between the cerebral hemispheres and cerebellum. In some forms, as in the cat, the tentorium becomes ossified and fused to the parietal bones. On the ventral side of the brain, the dura encapsulates the pituitary gland but also forms a fold, the diapharagma sellae, over the sella turcis, or depression in the dorsal face of the sphenoid one, in which the pituitary gland lies. The stalk of the infudibulum, together with the pars tuberails, if present penetrates the diapharama sellae.

Brain

The brain, or encephalon, is that part of the central nervous system which is located in the cranial cavity. It is enclosed by the cranial meninges which are continuous at the level of the foramen magnum with the spinal meninges. The cranial dura mater consists of two layers, an outer endosteal layer and an inner meningeal layer. The meningeal layer undergoes reduplication to form a sickle-shaped partition, the falx cerebri, which is located in the great reduplicated between the cerebrum and the cerebellum so as to cover both surfaces of the bony tentorium cerebelli. The cranial arachnoid and cranial pia mater are similar to their spinal counterparts.

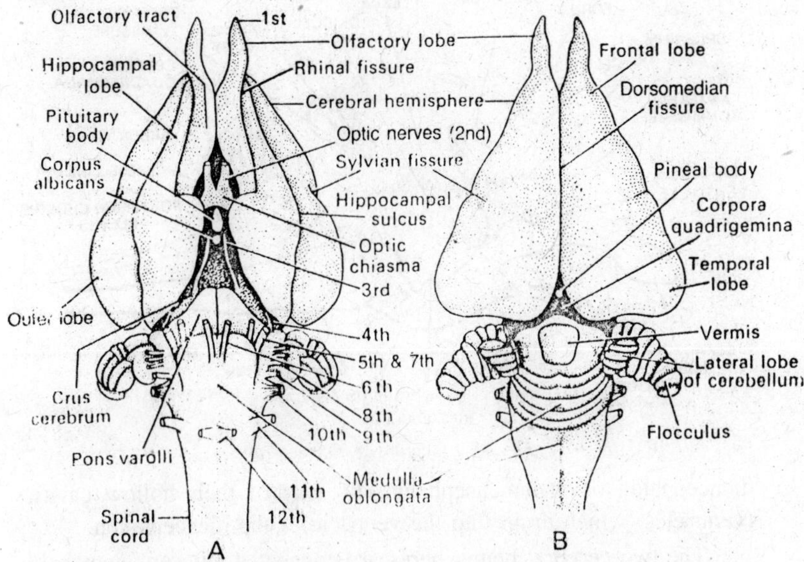

Fig. 15.2. Mammalian brain (A) ventral, (B) dorsal views; digits indicate cranial nerves.

The brain is composed of the following parts: the medulla oblongata, the pons Varolii, the midbrain, the diencephalons, the cerebellum, and the cerebral hemispheres. The first four parts constitute the brain stem. Each part is composed of gray and white matter, the arrangement of which varies with the part.

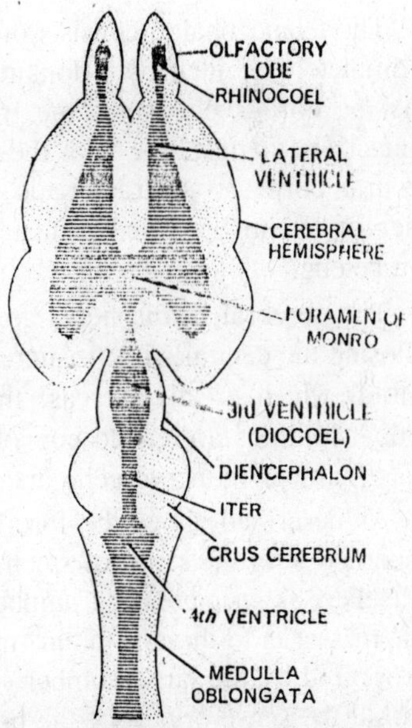

Fig. 15.3. Cavity of brain.

Fore Brain

This fore-brain is developed from the most anterior of the swellings of the hollow neural tube. There are two hollow lateral pouches of this swelling which develop into the cerebral hemispheres, the telencephalon. The cerebral hemispheres communicate with the unpaired portion of the fore brain, the

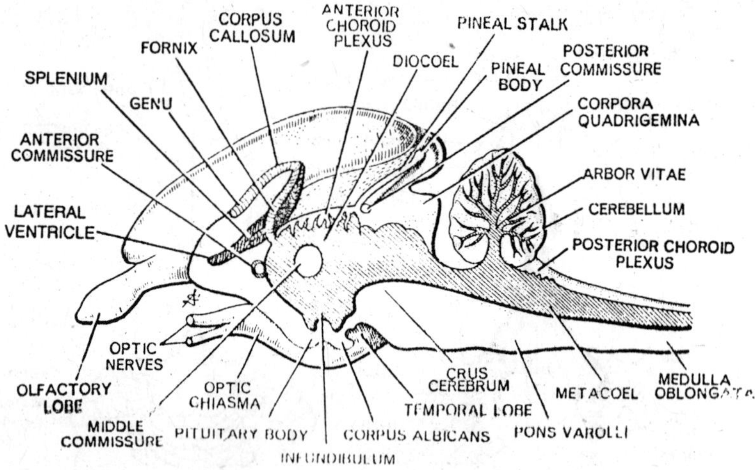

Fig. 15.4. Sagittal section of brain.

diencephalon or thalamencephalon, by way of their hollow cavities (ventricles) which drain into the ventricle of the diencephalon.

The two cerebral hemispheres are connected with one another by means of transversely running commissures of which the corpus callousm is the largest, a thick sheet of fibers running between the hemispheres.

Telecephalon

The telencephalon consists of the cerebral hemispheres which are incompletely divided by a longitudinal fissure in the floor of which runs the corpus collosum, the transversely running fibers of which connect one hemisphere with the other. The high development of the cerebral cortex is characteristic of mammals and especially of the order of mammals called primates, to which the monkeys, apes and man belong.

The cerebral hemispheres are covered by a layer of grey matter enclosing the central white matter. The cortex shows a complex pattern of folds which greatly increase the surface area of the cortex. These cortical foldings are called convolutions. The grey matter consists of a vast number of nerve cells arranged in several layers.

As described above the fore-brain in lower vertebrates is mainly concerned with the sense of smell but as one ascends the evolutionary scale there is an increasing number of other sense organs which send their messages to the fore-brain. In mammals the original 'small brain' is obscured by the large number of connections of the cerebral cortex with all the sense organs of the body. Passing through the dorsal wall of the diencephalons (the thalamus) to the cerebral hemispheres are fibers carrying information from the lower centres of the brain and from all the sense organs of the body.

This information, in the form of nerve impulses, passes to various arts of the cortex where it is synthesized with other sources of information, including that which arises from the memory of previous experience. Approximate action may be initiated by way of fibers which pass from the cortex to the lower centres of the brain and to the spinal cord.

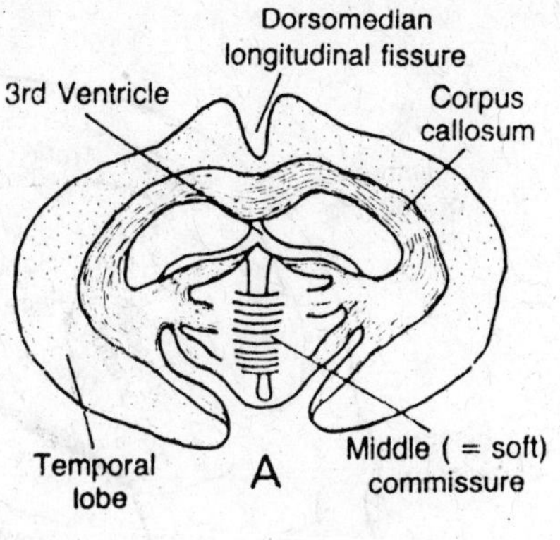

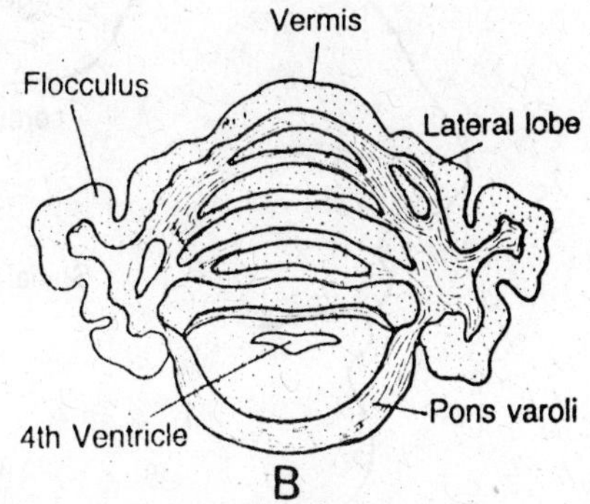

Fig. 15.5. Transverse sections of cerebrum (A) and cerebellum (B) of brain.

Localization of function in the cerebral cortex

In man, and other animals, many of the functions of the cortex have been localized to certain areas. These have been discovered in various ways, e.g. by noting the behaviour of animals and men with disease of certain areas of the cortex or in which certain areas of the cortex have been removed by operation. The brain itself is insensitive to pain and some brain operations may be performed on a conscious patient. In these operations electrical stimulation of the brain may also give rise to useful information. By stimulating small parts of the cortex concerned with movement, the movement of individual parts of the body may be noted. Stimulation of other areas may be followed by seeing remembered sights, or hearing noises or music. Although the functions of the cortex appear to be fairly localized each part of the cortex is connected to other parts by means of association fibers. It is interesting gives an indication of the amount of area of cortex devoted to each function gives an indication of the amount of information passing to and from that area. If we were to draw a picture of a human being as reflected in the sensory functions of the cortex we obtain in which the sensory areas of the fingers and lips appeal of great importance.

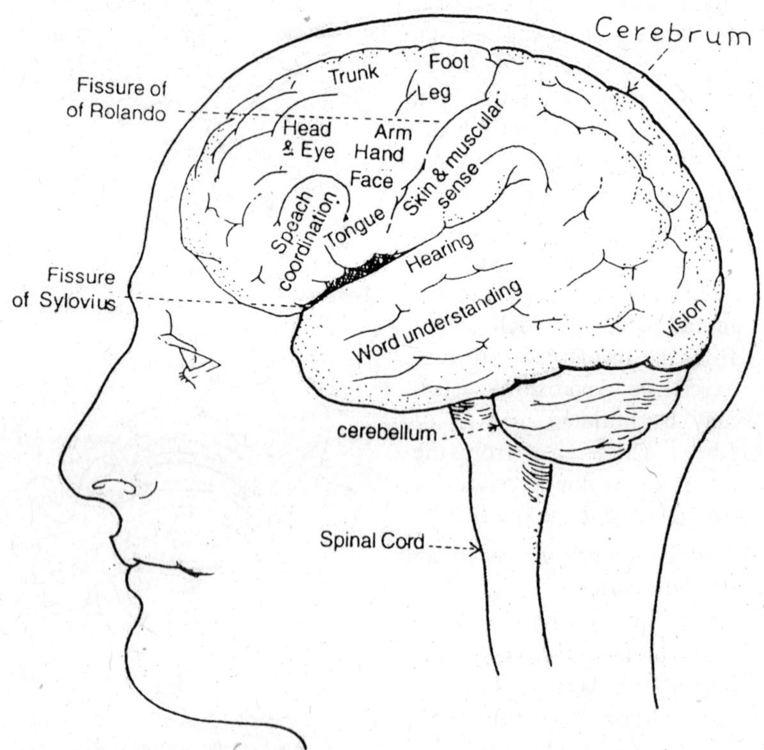

Fig. 15.6. Diagram of the association areas of the cerebral cortex.

Motor pathways from the cerebral cortex

From the motor areas of the cortex two sets of motor fibers pass inwards, the fibers of the pyramidal system and those of the extra-pyramidal system. The pyramidal fibers form a pair of thick bundles of fibers called the cerebral peduncles which pass in the floor of the mid-brain to the spinal cord where they terminate to synapse with the cells of the somatic motor fibers in the anterior (ventral) horn of the spinal cord.

The extra pyramidal fibers pass inwards from the cortex to collections of grey matter deeply placed within the hemisphere called the basal ganglia. These basal ganglia constitute the corpus striatum, which will be further discussed below.

These two motor systems, the pyramidal and extra pyramidal systems tend by to have opposing actions, and the interplay between the two permits normal coordinated movements and normal muscle

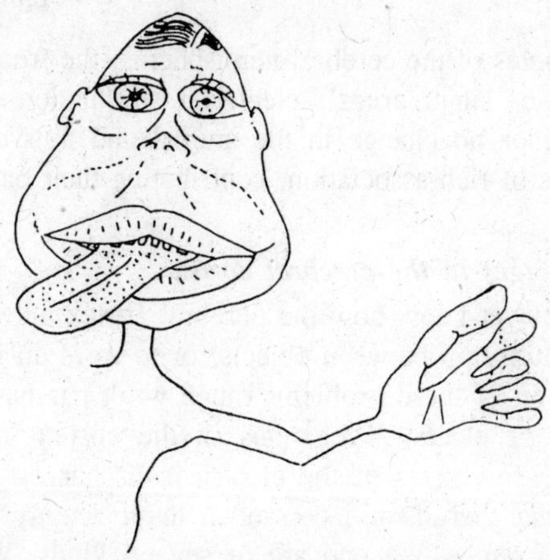

Fig. 15.7. Drawing of man, the size of part indicating the importance of the part as reflected in its sensory representation on the cerebral cortex.

tone. If one of this pair of opposing systems is damaged the function of the other will be seen exaggerated. Thus after a haemorrhage in the basal ganglia, a form of 'stroke', the relatively unopposed action of the pyramidal system is seen; the limbs stiffen as the tone of the muscles increases and the reflexes become exaggerated, because the normal antagonistic and inhibitory effects of the extra pyramidal system is not functioning properly. In cases of damage to the pyramidal system the muscles become flaccid as their tone is lost, due to the unopposed inhibitory effect of the extra pyramidal system. In the anterior horn of the spinal cord a great variety of influences act upon the somatic motor cells whose fibers innervate the voluntary muscles of the body, and the net effect of these influences depends upon the strength of the various factors which include impulses carried by pyramidal fibers, extra pyramidal fibers and the fibers relaying local segmental reflexes.

Intelligence

Intelligent behaviour is a striking feature of mammalian behaviour, particularly of the primates. In intelligent behaviour there is no fixed sequence of events but actions are performed appropriate to the task in hand, and if a usual pattern of events in the environment changes then the intelligent animal should be able to adjust its behaviour accordingly.

There is no fixed are of the cortex in which intelligence is localized but intelligence depends upon the associative activity of the entire cortex, the richer the association paths the greater the intelligence.

The anterior poles of the cerebral hemispheres, the frontal lobes, have been called the 'silent areas'; electrical stimulation of these areas produces little or no change in the animal and it is considered that these are areas of rich association, contributing their part to intelligent behaviour.

Electrical changes in the cerebral cortex

We do not yet know how the cerebral cortex allows us to think, to see the relationship between objects, or to store up information. In the investigation of these problems much work has been done on the measurement of electrical changes in the cortex in a variety of situations. Measurements of the electrical discharges from the brain in man show that there is an incessant rhythmic activity. These rhythms are electrical brain waves and are of several kinds. When we see a uniform visual field our brain waves are of a variety called the alpha type. When the visual field changes to a pattered type there is a change in the electrical activity to a beta type. When we sleep a different set of brain waves is recorded. These currents in the brain are apparently not affected by cutting the tracts to and from the cerebrum and they are thought to be a product of the activity of the cortex itself. Interesting models have been made using valves and electronic equipment which have been taught to remember certain things e.g. model toys which can learn to avoid obstacles. We are only at the very beginning of our search for useful scientific knowledge about the functioning of the brain.

Corpus striatum and instinct

The base of the telecephalon is also thickened in the mammal to form two very important swellings, one on each side, called the corpora striata. The corpus striatum is so called because it appears to be striped grey and white when it is sectioned. These striated bodies are important as they are the seat of the instinctive actions of the animal. Birds are par excellence the animals which behave instinctively and they are also the animal with the largest corpora striata. At first when the brain of a bird and mammal are seen side by side they present a superficially similar appearance for both of them have the large swellings in the bore-brain area. If the roof of the fore-brain is removed however in the case of the bird it is seen to be very thin whilst in the mammal it is thick. Under the thin roof of the cerebrum in a bird is the very big thickening of the floor – the corpus striatum, whilst in the mammal although the floor is thickened the striatum is not as big as it is in the bird. Thus in both these very highly advanced vertebrates there is a great development of the association centres in the telencephalon, in the bird the emphasis is on the corpus striatum and instinctive behaviour whilst in mammals the stress in on the cerebral cortex and intelligent behaviour. It is necessary to be quite clear at this stage what we mean by the terms instinct and intelligence. Instinctive behaviour can be very complex behaviour as for instance nest building and territory defence in birds. Birds seldom get to fighting each other, for when a cock bird enters the territory of another cock the latter makes a display of aggressiveness and the former eventually flees. This is a very sensible thing to do for it prevents birds of the same species fighting each other and also more and less guarantees that each bird will have a big enough territory and therefore enough food to feed the young when they are hatched. Now all this complex behaviour is not learned but is as it were built into the bird. He does not need to have seen any other bird or ever taken up a defensive posture and yet he can take up the exactly appropriate posture when required to do so. When he is threatened and he is in someone's territory he knows it is his turn to run away. The instinctive action is something that is of great biological importance to the bird, in this case it is concerned with the rearing of the young. The instinctive act is triggered off by what is called a sign stimulus, that is by some significant pattern in the environment. For instance a robin in the breeding season will fiercely attack an egg-shaped piece of wood on which is painted a red area if it is placed in his territory but he would take little notice of a stuffed robin provided that the red breast had been painted brown. Obviously the thing that causes a fighting reaction from a robin is a red area which does not run away when it is displayed at. The sign stimulus is red-circle-on-brown in this case. Thus we may think of an instinctive action as a complex action which is inborn in the animal and triggred off by a sign stimulus. It usually fulfils some important biological purpose for the animal and the action is usually accompanied by a fairly large amount of energy or 'drive'. The action cannot be modified to any great extent, for instance in the case of this robin it is obviously stupid to go on displaying at what to an intelligent animal is a piece of wood with a red circle on it, but the robin will go on and on and on for the bird cannot easily modify its instinctive behaviour. In the natural world it would be a very sensible action on the bird's part, it is only because some human has come along and artificially confused the issue by placing artificial sign stimuli about that the action seems stupid. Man has not many very obvious instinctive reactions, or if he has it is difficult to pick them out because he is always likely to apply his intelligence and inventiveness to any situation and modify the instinctive action. The rounded chubby features of a human baby's face or any other object which has a similar outline, especially if it has a large head and a relatively small body, will act as a sign stimulus to call out the pattern of events that we call maternal behaviour in a woman. This probably accounts for the popularity of the teddy bear and explains why human dolls are not shaped like snakes. It is obviously biologically useful if the form of the human infant elicits motherly love from the mother for then she will hold the baby to her breasts and allow it to suckle, an action which incidentally it can do without having to be taught, for it is instinctive. Small baby kittens push their noses into the mother's fur before their eyes open and when they feel the nipple on their lips they suck. They can do this a few hours after being born, what is more they push the mammary gland around the nipple with their paws and this helps to release the milk from the nipple. These activities associated with feeding of the new born are far too important to have to be learned for the baby needs feeding at once and time for learning can not be afforded, so these feeding reactions are instinctive.

Corpus striatum in man

Little is known for certain about the function of the corpus striatum in man, but we do know that it is important in the control of muscle tone and in steadying muscular movement. Patients who suffer damage to the corpus striatum sometimes have paralysis agitans or chorea.

Thalamencephalon or Diencephalon

The thalamencephalon is the more posterior part of the fore-brain and connects the cerebral hemispheres with the mid-brain. The roof of the thalamencephalon consists of a thin plate of non-nervous vascular tissue from which projects the pineal body or epiphysis. In lower vertebrates e.g., lamprey and some reptiles the pineal body has the structure of a third eye but in mammals its function is unknown although it has been suggested that it has an endocrine function.

Thalamus

The mammals there are large collections of grey matter in the lateral walls of the thalamencephalon constituting the thalamus. Through the thalamus large amounts of information pass from the lower centres of the brain to the cerebral cortex. But the thalamus is not merely a relaying station for the cortex and fibers pass from the cortex to the thalamus where some form of primitive integration is carried out.

This part of the brain is often referred to as the pain-pleasure-brain, since pain, extremes of temperature and rough contacts with the environment are appreciated in the thalamus. The thalamus gives us a crude sort of consciousness whereas the cerebral cortex gives a more sensitive impression. The cerebral cortex can inhibit the activity of the thalamus, and if this inhibitory influence is removed the animal then reacts in manner suggesting displeasure, even to stimuli which are only mildly irritating to an animal with an intact cortex. A cat with no cortex is often found snarling with claws extended.

Hypothalamus

The thickened floor of the thalamencephalon forms the hypothalamus. A projection from the floor of the hypothalamus, the infuncidbulum, is continuous with the posterior nervous lobe of the pituitary gland and is an important channel for the co-ordination of nervous and endocrine activity. The hypothalamus is the head of the autonomic nervous system;

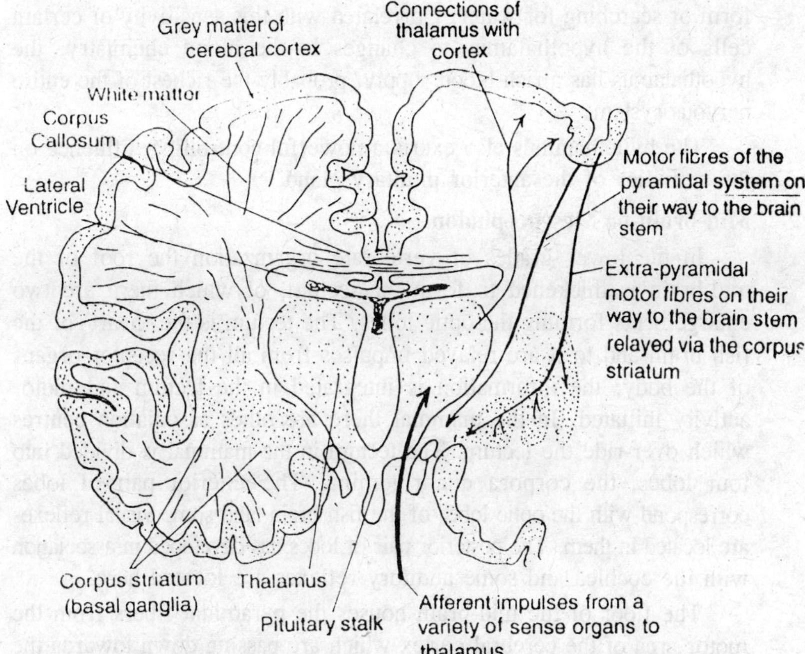

Fig. 15.8. Transverse section through human fore-brain showing cerebral hemispheres, corpu striatum and thalamus. The afferent (sensory) pathways to the cortex via the thalamus, and the two motor pathways (pyramidal and extra-pyramidal) from the cortex are indicated.

visceral afferent fibers reach the hypothalamus and after coordination with each other, efferent fibers carry impulses down the brain stem to lower subsidiary centres and eventually down the spinal cord and out along the spinal nerves. There seem to be two anatomically discrete centres in the hypothalamus one controlling the activity of the sympathetic nervous system, and the other controlling the parasympathetic nervous system.

In addition there are regions of the hypothalamus which initiate and control quite complex patterns of behaviour, e.g. sleep, feeding, aggression. The hypothalamus also contains certain sense organs, called osmoreceptors, which are sensitive to the sodium ion concentration of the blood. The axons of these cells pass down the infundibular stalk to the posterior pituitary gland, and granules of neurosecretory material migrate down the axons to the posterior pituitary where they are liberated. In addition to this effect of a rise in the salt concentration of the blood the hypothalamus initiates somatic motor activity in the form of searching for water. Correlated with this sensitivity of certain cells of the hypothalamus to changes in the blood chemistry, the hypothalamus has a rich blood supply, probably the richest of the entire nervous system.

The hypothalamus also exerts a powerful controlling influence on the activities of the anterior pituitary gland.

Mid-brain or Mesencephalon

In the lower grades of vertebrate organization the roof of the mid-brain is thickened to form the tectum, of which there are two enlargements forming the optic lobes. The tectum is the 'heart' of the fish brain and to it are relayed impulses from all the receptor organs of the body; the information is integrated in the tectum and motor activity initiated. In the mammal there are other association centres which over-ride the tectum. The tectum in the mammal is divided into four lobes, the corpora quadrigemina. The anterior pair of lobes correspond with the optic lobes of the fish brain land some visual reflexes are located in them. The posterior pair of lobes has developed in association with the cochlea and some auditory reflexes are located here.

The floor of the mid-brain houses the pyramidal fibers from the motor area of the cerebral cortex which are passing down towards the spinal cord. Two thick bundles of fibers run on either side of the floor of the brain stem. These thick bundles are called the cerebral peduncles (or crura cerebri) and they come to occupy the whole of the floor of the mid-brain.

In summary we can think of the mid-brain as being concerned with some visual and auditory reflexes and being an important link between the cortex and the rest of the brain.

Hind-brain or Metencephalon

The hind-brain consists of two important parts, the cerebellum and the medulla oblongata.

Cerebellum

The cerebellum is a large and complex association centre which dominates the appearance of the hind-brain in mammals. It receives large numbers of afferent fibers from the spinal cord and other parts of the nervous system, including the cerebral cortex, and from it efferent fibers pass to different parts of the brain stem. Many of the afferent fibers reaching the cerebellum are sensory, from a variety of sense organs, and a large number of these are from proprioreceptor organs. In spite of a large sensory input to the cerebellum none of these sensations reach consciousness and the activity of the cerebellum is unconscious. The sensory data supplied to the cerebellum is integrated there and used in the control of voluntary movement initiated by the cerebral cortex. The cerebellum does not initiate movement, rather it coordinates movement, ensuring that each muscle engaged in a particular movement contracts at the right time and to the right degree. Some parts of the cerebellum are specially concerned in the maintenance of equilibrium. Damage to the cerebellum in man may produce an unsteady gait, and jerky uncoordinated movements, and there may be also changes in muscle tone.

Medulla oblongata

The medulla oblongata is formed from the posterior end of the hind-brain swelling in the embryo. It is not sharply demarcated from the spinal cord into which it merges. The roof of the medulla is non-nervous and consists of a vascular choroids plexus under which lies the ventricle of the medulla, the fourth ventricle of the brain. In the roof of the fourth ventricle there are three pores, a central one and two lateral ones through which cerebro-spinal fluid passes from the ventricles of the brain into the subarachnoid space.

The medulla contains a number of nuclei (collections of nerve cells) of the more posterior cranial nerves, which supply the head and body (e.g. the vagus nerve). In addition there are large numbers of fibers passing through the medulla from the spinal cord on their way to higher centres of the brain, and from higher centres (e.g. pyramidal fibers from the cerebral cortex) on their way to the spinal cord.

The medulla is the centre of control for certain vital functions including the control of respiratory movements (for the activities of the medullary respiratory centre) and the control of the tone of blood vessels. The vagus nerve, the nucleus of which is located in the medulla, is concerned with reflexes of the heart and gastro-intestinal tract. An animal can live even when the brain is severed from its connections with the spinal cord at the level of the mid-brain. But if the medulla is damaged, even with the rest of the brain intact, the animal dies, either because of a cessation of breathing movements or a loss of tone in the blood vessels.

Cavities of Brain and Spinal Cord

With the differentiation of the neural tube into brain and spinal cord, its original cavity becomes modified to form the ventricles of the brain and the central canal of the cord. When the telecephalon develops into the paired cerebral hemispheres, the cavity extends into those structures as the *lateral ventricles*, or *ventricles* I and II. The cavity of the diencephalons is then known as the *third ventricle*. Each lateral ventricle communicates with the third ventricle by means of an opening, the *interventricular foramen*, or *foramen of Monro*. In higher vertebrates a narrow canal extends posteriorly from the third ventricle through the mesencephalon and is referred to as the *cerebral aqueduct*, or *aqueduct of Sylvius*. In lower forms the cerebral aqueduct may be considerably expanded on either side and the term *optic ventricles*, or *mesocoele*, is then applied. Posteriorly the cerebral aqueduct communicates with the enlarged cavity of the rhombencephalon, or the *fourth ventricle*. In numerous lower forms the anterior part of the fourth ventricle becomes modified into a chamber, the *metacoele*, extending into the cerebellum. The portion of the fourth ventricle within the medulla oblongata is then referred to as the *myelocoele*. It is continued posteriorly as the central canal of the spinal cord.

As development of the brain continues, its walls become thickened with nervous tissue except in two regions where only the epithelial ependymal layer is retained. These regions include the anterior part of the roof of the diencephalons and the roof of the medulla oblongata. In each region a fusion of the roof with a highly vascular membrane, the pia mater, forms a tela choroidea. Folds from the tela choroidea projects into the third and fourth ventricles and are known, respectively, as the anterior and posterior choroids plexuses. The ependymal layer lining the choroids plexus region is referred to as the choroid plexus epithelium.

Lymphlike cerebrospinal fluid is contained within the cavities of brain and spinal cord and circulates through the ventricles and central canal. It passes out of these cavities through openings, to be mentioned later, to bathe the entire central nervous system. The choroid plexures are important in forming the cerebrospinal fluid.

CENTRAL NERVOUS SYSTEM

Spinal Cord

The spinal cord or medulla spinalis, is a somewhat flattened cylindrical mass of nervous tissue which occupies the vertebral canal. At its cephalic end it is continuous with the medulla oblongata. The cord presents two enlargements, the cervical and the lumbar; and a conical termination, the conus medullaris. The enlargements correspond to the origin of the nerves which enter into the formation of the brachial and lumbosacral plexuses, and are due to the fact that a greater mass of tissue innervated (forelimb and hindlimb) requires a greater number of neurons.

The surface of the cord is marked by a number of longitudinal grooves. The ventral median fissure and the dorsal median sulcus incompletely divide it into right and left halves. Each lateral half exhibits a ventrolateral and a dorsolateral sulcus. The rootlets of a

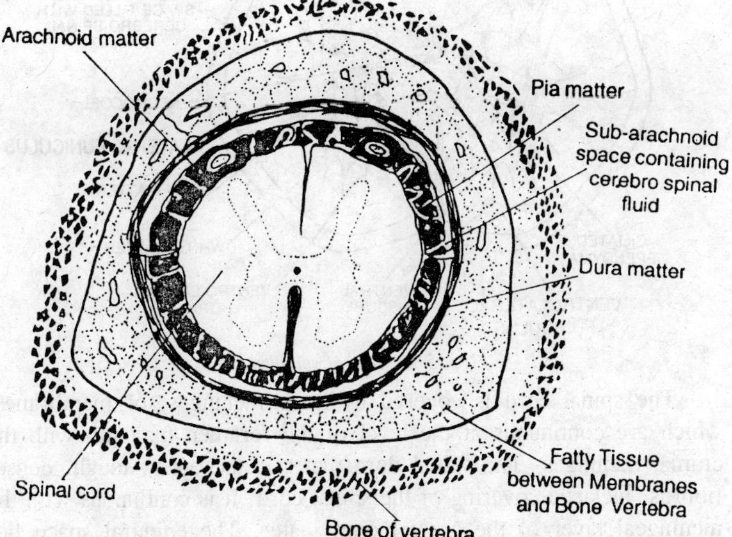

Fig. 15.9. Diagram of a transverse section through the spinal cord. The sub-archnoid space containing cerebro-spinal fluid is indicated in black.

dorsal root of a spinal nerve enter the cord via the dorsolateral sulcus, and the rootlets of the ventral root of the spinal nerve emerge from the ventrolateral sulcus. The space between the dorsal median sulcus and the dorsolateral sulcus is occupied by the dorsal funiculus; that between the dorsolateral sulcus and the ventrolateral sulcus by the lateral funiculus; and that between the ventrolateral sulcus and the ventral median fissure by the ventral funiculus.

The cord is divisible into cervical, thoracic, lumbar, sacral, and caudal or coccygeal portions. Each portion consists of a number of segments. A cord segment gives origin to a pair of spinal nerves. In the cat there are eight caudal segments must pass caudad to reach the foramina through which they exit from the vertebral canal since the cord does not occupy the entire canal. There is, therefore, in the caudal part of the vertebral canal a leash of nerves called the cauda equina.

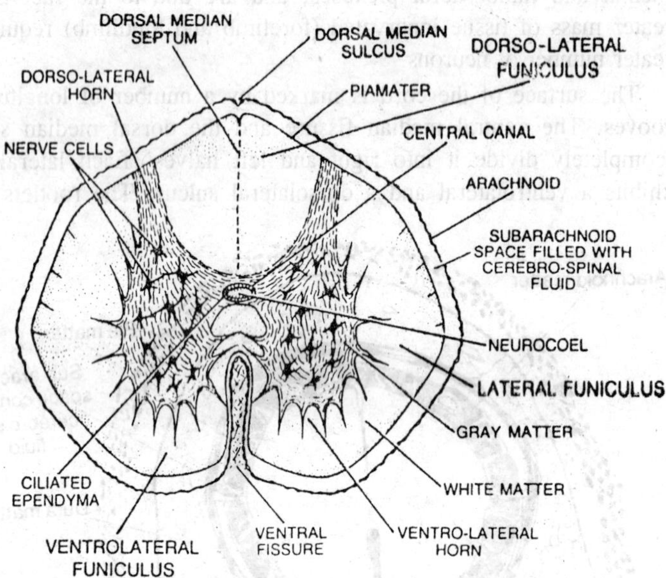

Fig. 15.10. T.S. of spinal cord.

The spinal cord is covered by three meninges, or membranes, which are continuous at the level of the foramen magnum with the cranial meninges. The spinal dura mater is the outer, tough, dense, fibrous, inelastic covering of the spinal cord. It is continuous with the meningeal layer of the cranial dura mater. The epidural space lies between it and the periosteum lining the walls of the vertebral canal.

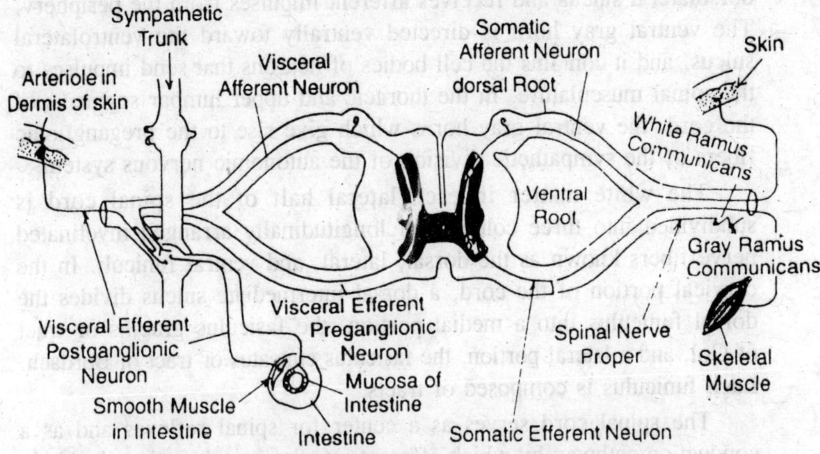

Fig. 15.11. Types of spinal reflex arcs.

The spinal arachnoid is a thin, delicate, gauze-like membrane lying beneath the dura. Like the dura mater, it does not dip into the grooves of the spinal cord. Between the spinal dura and the spinal arachnoid is the cleft-like potential subdural space. It contains a small amount of fluid to prevent friction between contiguous surfaces. The spinal pia mater is a thin, vascular membrane that closely envelops the spinal cord and is continued caudad beyond it as the filum terminale. Along the midlateral line of the cord the pia is thickened to form the ligamentum denticulatum, which sends toothlike processes or denticulations through the arachnoid to the dura mater to provide lateral anchorage for the spinal cord. The space between the arachnoid and the pia mater is called the subarachnoid space, and it contains cerebrospinal fluid.

A transverse section of the spinal cord at any level reveals the disposition of the gray and white matter. The gray matter is situated in the interior, and the white matter surrounds it. The gray matter in transverse section roughly has the outline of the letter "H." The cross member of the "H" is known as the gray commissure. It contains the central canal which ends blindly and caudally in the conus medullaris and communicates cranially with the fourth ventricle in the medulla oblongata. The white matter immediately dorsal to the gray commissure is called the dorsal white commissure, and that immediately ventral to it is known as the ventral white commisure. Each lateral half of the gray matter is subdivided into a dorsal gray horn and a ventral gray horn. The dorsal gray horn is directed dorsolaterally toward the dorsolateral sulcus and receives afferent impulses from the periphery. The ventral gray horn is directed ventrally toward the ventrolateral sulcus, and it contains the cell bodies of neurons that send impulses to the spinal musculature. In the thoracic and upper lumbar segments of the cord, the ventral gray horns which give rise to the preganglionic fibers of the sympathetic division of the autonomic nervous system.

The white matter in each lateral half of the spinal cord is subdivided into three columns of longitudinally arranged myelinated nerve fibers known as the dorsal, lateral, and ventral funiculi. In the cervical portion of the cord, a dorsal intermediate sulcus divides the dorsal funiculus into a medial portion, the fasiculus gracilis or tract of Gol, and a lateral portion, the fasiculus cuneatus or tract of Burdach. Each funiculus is composed of tracts.

The spinal cord serves as a center for spinal reflexes and as a conduction pathway by which afferent impulses reach various levels in the brain and by which efferent impulses pass from the various motor centers in the brain of efferent neurons which arise in the cord and supply effectors.

Peripheral Nervous System

Cranial Nerves

There are twelve pairs of cranial nerves, each of which will be considered subsequently from the aspects of superficial origin, course, and distribution.

Olfactory nerve

The filaments of the olfactory nerve spring from olfactory bulb and descend through the foramina of the cribiform plate the ethmoid bone for distribution to the olfactory mucosa.

Optic nerve

The optic nerve takes its superficial origin from the optic chiasma and enters the orbit by way of the optic foramen. In the orbit it described a sigma-shaped course dorsocraniad to the eyeball, finally penetrating the fibrous and vascular coats of the eye to reach the retina.

Oculomotor nerve

the oculomotor nerve emerges from the median aspect of the cerebral peduncle and leaves the cranium by way of the orbital fissure to enter the orbit between the superior rectus and lateral rectus muscles. Almost immediately after its entrance into the orbit, it sends a branch dorsad to the ventral surface of the superior rectus muscle. As it passes laterad of the optic nerve, it gives off a branch that courses ventrad of the optic nerve to terminate in the medial rectus muscle. After supplying a short branch to the inferior rectus muscle, it gives rise to a long branch that extends cranio-lateral over the dorsal surface of the interior rectus to reach the inferior oblique muscle. At the origin of the latter branch, there obtains a triangular enlargement, the ciliary ganglion, whence are derived the two short ciliary nerves that run on each side of the optic never to perforate the sclera of the eyeball for distribution to the ciliary muscle.

The oculomotor nerve also supplies branches to the levator palpebrae superioris and retractor oculi muscles.

Trochlear nerve

This nerve arises from the anterior medulllary velum and extends craniad of origin to make its way into the orbit by way of the orbital fissure. It then passes dorsad of the superior recturs muscle near its origin to terminate in the dorsal border of the belly of the superior oblique muscle.

Trigeminal nerve

The trigeminal nerve obtains its superficial original by two roots, a large sensory root and a small motor root, from the ventral surface of the pons near the lateral end of its caudal margin. On the sensory root near its origin from the brain is the semilunar or Gasserian ganglion from which arise the three great divisions of the trigeminal nerve: the ophthalmic, maxillary, and mandibular. The motor root, already mentioned, may be seen to join the mandibular division of the trigeminal nerve; and consequently the mandibular division contains both sensory and motor fibers.

Ophthalmic division

The ophthalmic division springs from the semilunar ganglion and enters the orbit through the orbital fissure. The branches of this division are as follows:

(a) *Frontal nerve*. As the ophthalmic division passes through the orbital fissure, it gives off the frontal nerve. The frontal nerve subsequently passes dorsocraniad near the lateral border of the superior oblique muscle. After coursing dorsad and laterad of the pulley of the superior oblique muscle, it finally emerges near the anterior end of the supraorbital crest for distribution to the medial portion of the integument of the upper eyelid and the contiguous region of the nose.

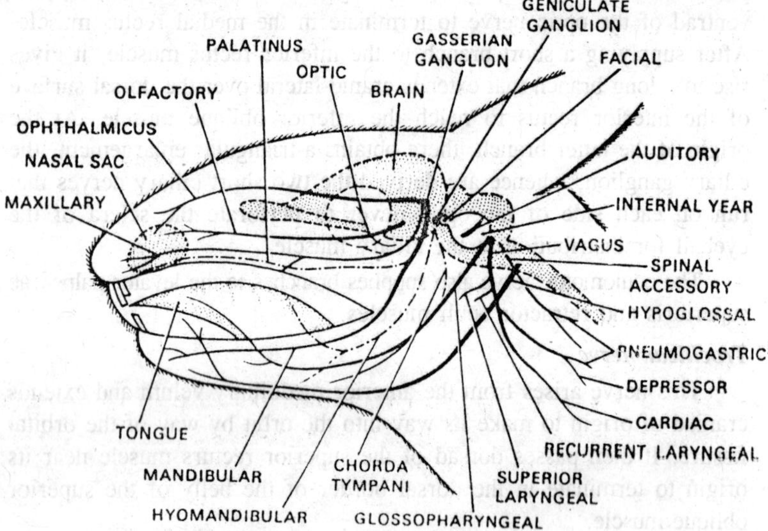

Fig. 15.12. Cranial nerves of rabbit.

(b) *Connection to the ciliary ganglion and long ciliary nerve*. The ophthalmic nerve leaves the orbital fissure, and, before passing dorsomediad of the optic nerve, it sends a short branch ventrad into the ciliary ganglion. In crossing the optic nerve, the ophthalmic division gives rise to the long ciliary nerve which proceeds craniad along the medial surface of the optic nerve for final distribution in the fibrous coat of the eye. As the ophthalmic division comes to lie between the superior rectus and medial rectus muscles, it divides to form ethmoidalis and infractrochlearis nerves.

(c) *Ethmosidalis nerve*. The ethmoidalis nerve descenas ventro-mediad between the contiguous surfaces of the superior oblique and medial rectus muscles to enter the ethmoidal foramen with the ethmoidal artery for distribution to the nasal mucosa.

(d) *Infratrochlearis nerve*. The infratrochlearis nerve proceeds craniad between superior rectus and superior oblique muscles, finally passing ventrad of the pulley of the superior oblique for distribution to the integument near the medial angle of the upper eyelid.

Maxillary division

The maxillary division takes its origin from the semilunar ganglion and emerges from the cranium through the foramen rotundum. Immediately after is transmission through the latter foramen, the maxillary division divided to form the zygomatic, lacrimal, infraorbital, and sphenopalatine nerves.

(a) *Zygomatic nerve*. This nerve arises from the maxillary division at the foramen rotundum and courses craniodorsad, lying mediad of the temporalis muscle, to the medial surface of the frontal process of the malar bone through which it is transmitted by the malar canal or foramen. After leaving the malar canal, the zygomatic nerve divides into at least three branches, two of which terminate in the skin around the general vicinity of the frontal process of the malar bone. The third branch of the zygomatic nerve, relatively long, passes craniad above the dorsal border of the malar bone to distribute itself in the skin that lies ventrad of the lower eyelid.

(b) *Lacrimal nerve*. This nerve springs from the maxillary division along with the zygomatic nerve and courses dorsocraniad, paralleling the course of the zygomatic nerve for a short distance and ultimately coming to lie in close relation to the caudal surface of the periorbita. As it ascends dorsad along the periorbita, it sends a couple of slender branches to the lacrimal gland. Continuing dorsad of the lacrimal gland, it divides into two branches: one enters into an anastomosis with the zygomatic branch of the dorsal ramus of the facial nerve and sends a branch caudad for a short distance to terminate nerve; the other branch loops over the zygomatic branch of the facial nerve to run mediad for a sort distance before distributing itself to the skin between the eye and the external ear.

(c) *Infraorbital nerves*. The infraorbital nerves – initially in the form of a single, thick, fused nerve trunk and later about midway in its course diverging to form at least two stout nerves – pass directly craniad over the dorsal surface of external pterygoid muscle and thence onto the orbital plate of the maxillary bone where they give off filaments to the posterior dental foramina for the innervation of the upper molar teeth. As the infraorbitals begin their passage over the orbital plate of the maxillary bone, there is seen to arise from the ventral surface of the infraorbital nerves a slender branch that enters the anterior dental foramen to innervate the incisor, canine, and premolar teeth of the upper jaw. The infraorbital nerves leave the orbital plate of the maxilla through the infraorbital foramen and again divide to form a number of branches that distribute themselves in the skin just lateral to the nose, to the skin in front of the infraorbital foramen, and to the vibrissae.

(d) *Sphenopalatine nerve*. The sphenopalatine nerve accompanies the infraorbital nerves for a short distance over the dorsal surface of the external pterygoid muscle and abruptly diverges mediad toward the sphenopalatine ganglion – a triangular ganglion lying on the craniodorsal surface of the external pterygoid muscle, just laterad of the sphenopalatine foramen. As the sphenopalatine nerve approaches the latter ganglion, it generally divides into two branches: one enters the sphenopalatine ganglion, and the other branch, the greater palatine nerve, continues craniad to enter the posterior palatine canal for distribution to the mucosa of the hard palate.

Nerves communicating with the sphenopalatine ganglion:

(i) *Lesser palatine nerve*. This nerve leaves the lateral side of the sphenopalatine ganglion and course directly laterad over the craniodorsal surface of the external pterygoid muscle, passing ventrad of the greater palatine and infraorbital nerves in its lateral passage and receiving also a long slender branch from the maxillary division. This small nerve then curves ventrad over the craniolateral border of the external pterygoid muscle and subsequently follows a caudomedial course along the ventral surface of the external pterygoid to terminate finally in the soft palate.

(ii) *Posterior nasal nerve*. The posterior nasal nerve, on the other hand, arises from medial side of the sphenopalatine ganglion and immediately enters the sphenopalatine foramen for distribution to the middle and ventral portions of the mucosa in the nasal cavity.

(iii) *Vidian nerve*. The vidian nerve may be seen at this point to enter the caudal end of the sphenopalatine ganglion. Its origin and course will be fully discussed under the facial nerve.

Mandibular division

The mandibular division of the trigeminal nerve is transmitted through the foramen ovale and immediately thereafter gives rise to the following nerves:

(a) *Lingual nerve*. The lingual nerve, a few millimeters after its origin from the mandibular division, receives the chorda tympani nerve from the facial nerve; then it passes across the dorsal surface of the external pterygoid muscle to turn ventrad over its lateral border so as to continue dorsad or beneath the mylohyoid muscle. In its course beneath the mylohyoid muscle it crosses, at right angles, the combined ducts of the submaxillary and sublingual glands. In traversing the latter ducts, the lingual nerve gives off a slender branch that runs caudad along the ducts to both the submaxillary and sublingual glands. The lingual nerve itself finally enters the lateral surface of the tongue.

(b) *Inferior alveolar nerve*. The inferior alveolar nerve leaves the mandibular division and crosses dorsad of the chorda tympani and thence along the dorsal surface of the external pterygoid muscle at the dorsolateral border of which it enters the mandibular foramen along with the artery and vein of the same name. Shortly before the inferior alveolar nerve enters the foramen, it gives rise to a small branch that bifurcates ventrad of origin to enter the internal surface of the digastric muscle and the external surface of the mylohyoid muscle. The inferior alveolar nerve itself courses through the mandibular canal to send filaments (inferior dental nerves) into the teeth of the mandible. As the inferior alveolar nerve reaches the cranial end of the mandibular canal, it sends several branches through the mental foramina to the skin adjacent to the foramina.

(c) *Buccinator nerve*. This nerve passes craniolaterad from the mandibular division across the dorsal surface of the external pterygoid muscle to that part of the maxilla where the masseter muscle takes origin. Here it divides to supply branches to the masseter, to the lips, and to the mucous membrane of the mouth.

(d) *Pterygoideus nerve*. This nerve arises from the mandibular division along with the lingual nerve. The nerve descends ventrad for about a millimeter and gives off three branches: one extends craniolaterad, paralleling the course of the buccinator nerve, to terminate in the dorsal surface of the external pterygoid muscle; the second branch, slender and about a millimeter in length, penetrates the dorsal surface of the internal pterygoid muscle; and the third branch, also rather slender, proceeds caudad to enter the tympanic cavity for distribution to the tensor tympani muscle.

(e) *Auriculotemporal nerve*. The auriculotemporal nerve, shortly after its origin from the mandibular division, curves dorsad around the zygomatic process of the temporal bone in the region of the mandibular fossa; and then ascends further dorsal to its point of emergence at the caudal border of the masseter muscle where it divides into a temporal and an auricular branch. The temporal branch first receives an intercommunicating branch from the dorsal ramus of the facial nerve; subsequently it courses craniad along the zygomatic arch, finally descending ventrad to anastomose with the superior buccal nerve from the ventral ramus of the facial nerve. In its course along the zygomatic arch, the temporal branch supplies several branches to the skin. The auricular branch, on the other hand, ascends dorsad in a concavity between the temporalis muscle and the cartilaginous portion of the external ear, lying mediad of the temporal branch of the dorsal ramus of the facial nerve. The nerve finally terminates in the skin of the external ear.

(f) *Massetericus nerve*. This nerve takes its origin from the cranial end of the mandibular division and ascends dorsad to turn laterad through the sigmoid or mandibular notch. After passing through the sigmoid notch, it turns ventrocraniad into the fibers of the masseter muscle, which it innervates.

(g) *Deep temporal*. The deep temporal arises from the mandibular division along with massetericus nerve and immediately passes dorsocraniad into the temporalis muscle. Frequently there may be two or more branches coming from the mandibular division in this area – these collectively constitute the deep temporal nerve.

Abducens nerve

The abducens nerve takes its superficial origin from the ventral aspect of the medulla oblongata between the pyramid and the trapezoid body and enters the orbit through the orbital fissure. In the orbit it proceeds cranioventrad to the internal surface of the lateral rectus muscle along the medial surface of which it extends for a short distance before actually entering the muscle.

Facial nerve

The facial nerve obtains its superficial origin from the lateral potion of the trapezoid body near its cranial end and enters the internal auditory meatus to course through the facial canal wherein it enlarges to form the geniculate ganglion. During its course through the facial canal, the facial nerve gives rise to three branches:

Stapedial nerve

This nerve enters the stapedial fossa to innervate the stapedial muscle.

Superficial petrosal nerve

This nerve leaves the facial canal through the hiatus facialis to enter the tympanic cavity wherein it combines with the deep petrosal, which comes into the tympanic cavity from the caudal end of the tympanic bulla. The confluence of the superficial and deep petrosal nerves culminates in the formation of the Vidian nerve. The Vidian nerve leaves the tympanic bulla on the dorsolateral surface of the Eustachian tube. Diverging from the Eustachian tube after its exit from the bulla, it passes craniad for a short distance to enter the foramen for the Vidian nerve at the caudomedial border of the alisphenoid bone. The Vidian nerve then continues craniad through the pterygoid canal to pass through the orbital fissure and thence onto the dorsal surface of the external pterygoid muscle, which it traverses to enter the caudal angle of the sphenopalatine ganglion.

It might be interesting to note at this point that the superficial petrosal nerve contains preganglionic fibers (parasympathetic) from the facial nerve, which synapse in the sphenopalatine ganglion. On the other hand, the deep petrosal nerve embodies postganglionic fibers (sympathetic) from the superior cervical ganglion, which pass through the sphenopalatine ganglion without synapsing. For a more detailed consideration of the latter nerves, refer to the autonomic system.

Chorda tympani

This nerve leaves the facial nerve a short distance before the facial nerve makes its exist through the stylomastoid foramen and enters the tympanic cavity by way of a small opening in the anterior wall of the facial canal. It then crosses the tympanic cavity to leave it through the canal of Hugier. From the canal of Hugier, the chorda tympani passes ventrad of the zygomatic process of the temporal bone and then craniad to unite with the lingual nerve three or four millimeters laterad of the lingual nerve's origin from the mandibualr division of the trigeminal nerve.

The facial nerve proper leaves the facial through the stylomastoid foramen. Just after its emergence from the latter foramen, it gives off the posterior auricular nerve which shortly divides into at least four branches: three branches penetrate the cartilage of the external ear; and the fourth branch, a rather stout nerve, ascends further dorsad to innervate the muscles caudal to the external ear (levator auris longus, etc.). The facial nerve then passes over the posterior auricular artery to wind craniad around the more proximal portion of the auricular concha. Just before passing over the posterior auricular artery, the facial nerve sends a branch ventrocaudad into the anterior border of the digastric muscle. In the course of its curving craniad around the auricular concha, the facial nerve divides into a dorsal and a ventral ramus.

The ventral ramus of the facial nerve proceeds ventrocraniad around the caudal surface of the masseter muscle. As it turns craniad in its course, it divides to form the superior and inferior buccal nerves. The superior buccal nerve innervates the musculature of the upper lip and the adjacent area and also anastomoses with the temporal branch of the auriculotemporal nerve from the mandibular division of the trigeminal nerve. The inferior buccal nerve innervates the musculature of the lower lip.

The dorsal ramus of the facial nerve rises dorsad between the auricular concha and caudal border of the masseter muscle to the ventral level of the zygomatic process of the temporal bone. Here it sends an intercommunicating branch to the temporal branch of the auriculotemporal nerve and subsequently divides into the zygomatic and temporal nerves. The zygomatic nerve courses craniad over the zygomatic arch to approach the caudal angle of the eye. Here it sends branches into the upper and lower eyelid. The main part of the zygomatic nerve then continues dorsad of the eye, finally descending entromediad into the musculature along the lateral side of the nose. On the other hand, the temporal nerve ascends directly dorsad from the dorsal ramus of the facial nerve for a distance to distribute itself in the musculature laying craniad of the external ear.

Branches of the dorsal ramus perforate the substance of the parotid gland to innervate the overlying cutaneous musculature. Some small branches to the cutaneous musculature in this region may also arise from the facial nerve proper before it bifurcates into its dorsal and ventral rami.

Acoustic nerve

The acoustic or auditory nerve arises from the medulla oblongata caudad of the facial nerve. It passes into the internal acoustic meatus and divides into the vestibular and cochlear nerves.

The fibers of the vestibular nerve are the central processes of neurons whose cell bodies constitute the vestibular ganglion of Scarpa. This ganglion lies in the depths of the internal acoustic meatus. The peripheral processes of these neurons pass through minute openings in the medial wall of the vestibule and terminate in the cristae in the ampullae of the membranous semicircular canals and in the maculae in the sacculus and utriculus.

The fibers of the cochlear nerve are the central processes of bipolar neurons whose cell bodies constitute the spiral ganglion of the cochlea. The peripheral processes of these cells terminate about the hair cells of the spiral organ of Corti.

At this point, it is interesting to note that the vestibular nerve conveys proprioceptive impulses concerned with the maintenance of body posture into the vestibular nuclei in the brain stem and to the cerebellum. On the other hand, the fibers of the cochlear nerve form a part of the hearing pathway to the temporal cortex. They also form the afferent limb of reflex arcs which have to do with reflexes to sound waves.

Glossopharyngeal nerve

The glossopharyngeal nerve arises by a number of rootlets from the lateral surface of the medulla oblongata caudad of the acoustic nerve, and almost immediately after its origin from the medulla oblongata it is characterized by a small swelling, the superior ganglion. It emerges from the cranium through the jugular foramen and subsequently enlarges to present another ganglionic swelling, the ganglion petrosum. From the ganglion petrosum the glossopharyngeal nerve turns craniad over the external surface of the tympanic bulla. After having crossed the tympanic bulla, the glossopharyngeal nerve sends a branch (nerve of Hering) into the carotid body and sinus at the cranial end of the common carotid artery. Shortly thereafter the glossopharyngeal nerve contributes a branch to the stylopharyngeus muscle and subsequently approximates the cranial border of the constrictor pharynges medius to the internal surface of which it sends a branch. The glossopharyngeus nerve then passes mediad of the cranial cornu of the hyoid bone and, in so doing, divides into at least three branches: two pass into the wall of the pharynx; the other branch continues craniad into the root of the tongue.

Vagus nerve

The vagus nerve derives its superficial origin from the area ovalis of the medulla oblongata. Just before leaving the cranium through the jugular foramen, or in the course of its transmission through the foramen, the vagus enlarges to form the jugular ganglion. The vagus then continues caudad from the foramen in close association with the spinal accessory and hypoglossal nerves and then presents another ganglionic enlargements, the ganglion nodosum. From the ganglion nodosum the vagus courses caudad, in intimate association with the sympathetic trunk, along the dorsal surface of the common carotid artery. Near the level of the first rib, the vagus and sympathetic trunks diverge, with the vagus lying relatively more ventrad.

At the level of the first rib, on the left side, the vague will be seen to give off a branch, the left cardiac nerve, which almost at once splits into two filaments. One, the thicker filament, runs along the lateral surface of the left common and innominate arteries to

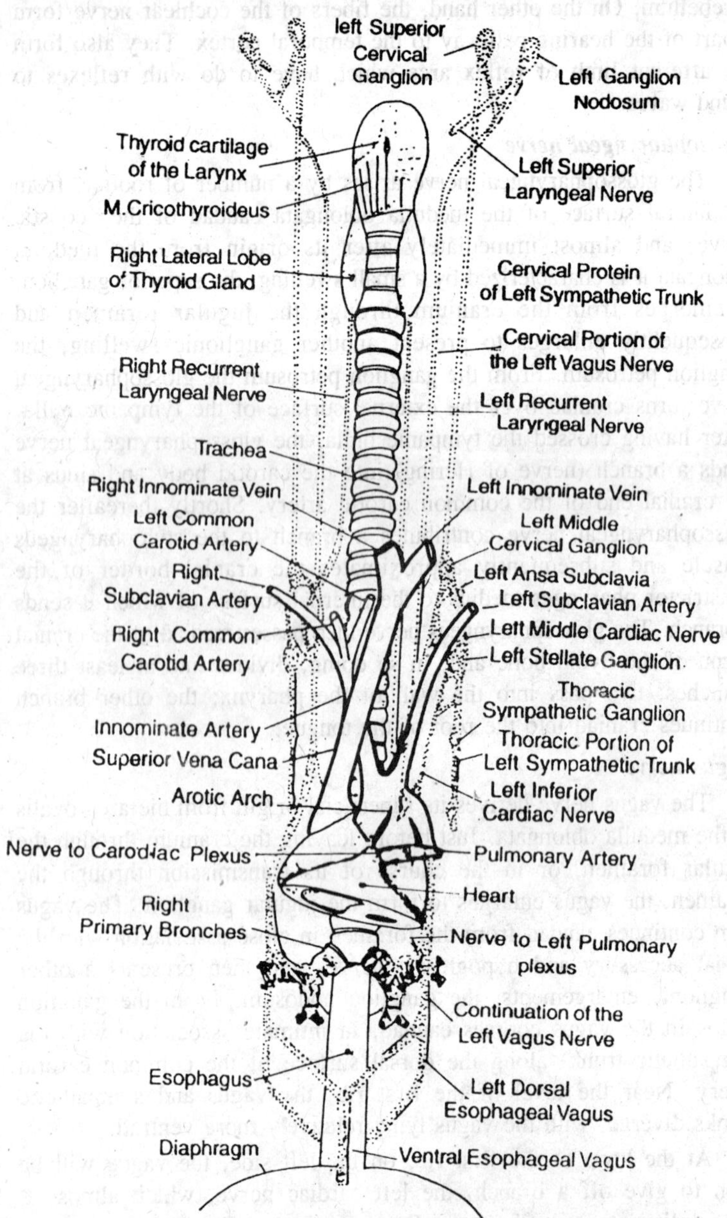

Fig. 15.13. Cervical and thoracic portion of the vagi and sympathetic trunks.

cross the aortic arch and to wind mediad around the latter in order to combine, just caudad of the arch or beneath its dorsal surface, with the cardiac nerve from the vagus on the right side. From this confluence there arises a fairly well-sized nerve which runs under the dorsal surface of the pulmonary artery to join the course of the descending ramus of the coronary artery, finally terminating in the ventral surface of the heart. The other filamentous branch of the left cardiac nerve passes along the dorsal surface of left common carotid and innominate arteries to reach the aortic arch in which it terminates. This filament is, apparently, the depressor nerve of the aorta.

The left vagus itself follows the course of the left subclavian artery in close approximation to the lateral surface of the esophagus to reach the aortic arch. After crossing the ventral surface of the aortic arch, it gives off the left recurrent laryngeal nerve which winds around the aortic arch to run back along the lateral surface of the trachea to larynx. The left recurrent laryngeal nerve enters the larynx along the lateral surface of the cricoid cartilage to supply the muscles of the larynx. The left vagus then follows a caudal course to the root of the lungs, laying ventrad of the dorsal aorta. At the root of the lungs, the left vagus gives off numerous branches which contribute to the formation of the pulmonary plexus and continue onto the heart to form the cardiac plexus. Branches from these plexuses supply the lungs, heart, pericardium, pulmonary artery, etc.

On the other hand, the right vagus gives rise to its recurrent laryngeal nerve just caudad of the right subclavian artery. The recurrent laryngeal nerve on this side winds around the right subclavian artery to attain the lateral surface of the trachea along which it continues craniad to the larynx. Just caudad of the right subclavian artery, also, the right vagus will be observed to give off the right cardiac nerve which subsequently runs along the medial surface of the superior vena cava and then dorsad of the aortic arch to combine with a branch of the left cardiac nerve previously described. The right vagus resumes its caudal course along the lateral surface of the trachea, and at the root of the lungs it becomes involved in the formation of the pulmonary and cardiac plexuses.

A short distance caudad of the root of the lungs, each vagus divides into a dorsal and a ventral branch. The ventral branches from the vagi shortly become confluent on the ventral surface of esophagus to form the ventral esophageal vagus. The ventral esophageal vagus passes caudad along the ventral surface of the esophagus to penetrate the diaphragm so as to attain the ventral surface of the cranial end of the stomach. This continuation of the vagus into the abdomen is known as the ventral gastric vagus. The ventral gastric vagus passes to the lesser curvature of the stomach where it forms the ventral gastric plexus, branches of which continue dorsad over the lesser curvature to unite with similar branches from the dorsal gastric plexus. The two dorsal esophageal vagi also become confluent a short distance craniad of the diaphragm to form a single dorsal esophageal vagus which subsequently penetrates the diaphragm to attain the dorsal surface of the stomach along the greater curvature. This continuation of the vagus into abdomen is known as the dosal gatric vagus, and almost immediately breaks up into numerous branches, some of which contribute to the formation of the dorsal gastric plexus and others of which contribute to the formation of the dorsal gastric plexus and others of which pass further caudad to become interconnected with celiac plexus.

Other branches of the vagus:

Auricular nerve

This nerve arises from the jugular ganglion of the vagus, and passes through the occipital and temporal bones by means of a minute canal to enter the facial canal. It leaves the facial canal through the sytlomastoid foramen for distribution to the external ear.

Pharyngeal nerve

This nerve originates by numerous rootlets from the vagus craniad of the ganglion nodosum. The nerve then proceeds mediad; and, after sending a communicating branch to the superior laryngeal nerve and receiving a filament from the glossopharyngeal nerve, it breaks up into numerous branches which terminate in the muscles of the pharynx.

Superior laryngeal nerve

The superior laryngeal nerve has its origin from the ganglion nodosum; and runs ventrad, passing mediad of the common carotid artery. It finally perforates the lateral surface of the thyroid cartilage at the lateral border of the thyrohyoideus muscle to supply the laryngeal mucosa.

Spinal Accessory Nerve

The spinal accessory nerve has to roots of origin: one, the cranial root, arises by many rootlets from lateral surface of the medulla oblongata; the other, the spinal root, is formed by the coalescene of rootlets that are derived from first five or six cervical segments of the spinal cord. The spinal root passes craniad through the foramen magnum to combine with the cranial root so as to form the spinal accessory nerve. The spinal accessory nerve leaves the cranium by way of the jugular foramen together with the vagus and glossopharyngeal nerves. After its emergence from the latter foramen, the spinal accessory nerve becomes interconnected with the hypoglossal, vagus, and sympathetic by numerous minute fibers (the pharyngeal plexus). The spinal accessory then passes caudad for a short distance, turning lateral to pierce the cleidomastoid muscle which it innervates. As the spinal accessory emerges from perforating the cleidomastoid, it receives an intercommunicating branch from the ventral ramus of the second cervical nerve, and at the same time sends a branch to the sternomastoid muscle. The spinal accessory subsequently cuts a branch to the sternomastoid muscle. The spinal accessory subsequently cuts dorsad across the ventral ramus of the second cervical nerve, and in doing so sends a couple of branches to the clavotrapezius muscle. The spinal accessory now continues caudad along the dorsal border of the levator scapulae ventralis, and at the cranial border of the acromiotrapezius muscle it receives an intercommunicating branch from the ventral ramus of the fourth cervical nerve. (This branch from the ventral ramus of the fourth cervical nerve loops around the lateral surface of the levator scapulae ventralis to attain its connection with the spinal accessory nerve). The spinal accessory nerve sends delicate fibers into the acromiotrapezius across the internal surface of which it continues caudad to enter the external surface of the spinotrapezius near its origin.

Hypoglossal nerve

The hypoglossal nerve takes its superficial origin from the ventrolateral sulcus of the medulla oblongata, and leaves the cranium through the hypoglossal canal. In the initial part of its course, the hypoglossal nerve lies between vagus and spinal accessory nerves, and thus passes through the pharyngeal plexus. It then turns abruptly ventrad a short distance craniad of the ganglion nodosum to cross the lateral surface of the common carotid artery. Subsequently it follows the course of the lingual artery, finally traversing the lateral surface of the artery to attain the lateral surface of the hyoglossus muscle at the medial border of which the hypoglossal nerve enters the musculature of the tongue. Before it passes into the intrinsic musculature of the tongue, it supplies slender branches to the hyoglossus, geniohyoid, and genioglossus muscles in the area.

About five millimeters before the hypoglossal nerve crosses common carotid artery, it gives rise to the ramus descendens hypoglossi which runs caudad in close approximation to the ganglion nodosum. Five or six millimeters caudad of its origin from the hypoglossal nerve, the ramus descendens receives a branch from the ventral ramus of the first cervical nerve. The ramus descendens thereafter crosses the common carotid artery to course caudad along its medial surface. In its course the ramus descendens gives off two branches: one innervating the cranial end of the stemothyroid and sternohyoid muscles; the other terminating in the caudal ends of the same two muscles. There also arises from the hypoglossal nerve along with ramus descendens a slender nerve which extends ventrad across the lateral surface of the common carotid artery to terminate in the thyrohyoid muscle and that portion of the stemphyoid muscle immediately ventrad of thyrohyoid muscle.

Spinal Nerves

A typical spinal nerve, insofar as its origin, components, and autonomic connections are concerned, has been discussed. The peripheral distribution of the spinal nerves in the cat will now be considered in some detail.

Cervical nerves

There are eight pairs of cervical nerves. Each cervical nerve divides to form a dorsal and a ventral ramus.

First cervical nerve

The first cervical nerve emerges from the vertebral canal by way of the atlantal foramen.

The dorsal ramus (suboccipital nerve), almost after its formation, supplies short branches that enter the internal surfaces of the complexus, obliquus capitis superior, and obliquus capitis inferior muscles. The dorsal ramus, as it begins to pass dorsal, sends a short filamentous branch into the dorsal surface of the rectus capitis posterior minor; and subsequently dividies, with one branch entering the lateral border of the rectus capitis posterior medius and the other branch terminating in the dorsal surface of the rectus capitis posterior major.

The ventral ramus descends directly ventrad from the atlantal foramen, passing mediad of the notch of the atlas and across the lateral surface of the longus capitis muscle to receive an interconnecting branch from the ventral ramus of the second cervical nerve and subsequently to approximate rather closely the ganglion nodosum. Here it sends a branch to the ganglion nodosum and to the superior cervical ganglion, after which the ventral ramus proceeds to cross the common carotid artery to join the descending ramus of the hypoglossal nerve. The

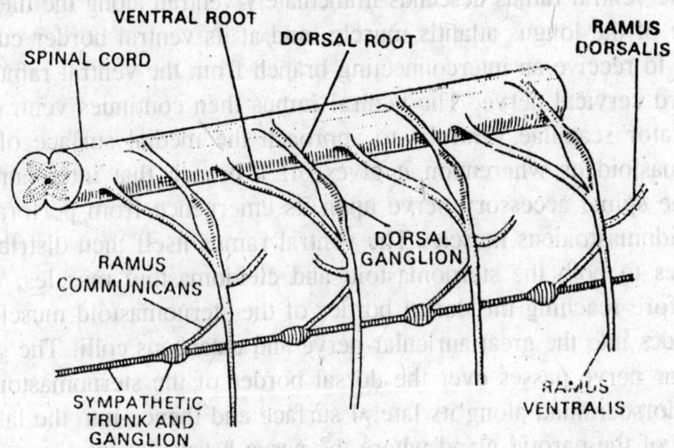

Fig. 15.14. Pattern of spinal nerves.

latter nerves, thus united, follow a course along the lateral border of the sternothyroid muscle, distributing branches to the cranial and caudal ends of both the sternohyoid and stemothyroid muscles.

Second cervical nerve

The second cervical nerve leaves the vertebral canal through an area below the dorsal arch of the atlas and lamina of the axis, with the cranial extension of the spinous process of the axis forming the dorsal boundary of tlie area. Directly laterad of the area of exit for the second cervicalnerve will be seen the dorsal root ganglion, the external surface of which lies in close relation to the inner surface of the obliquus capitis inferior. The second cervical nerve then divides into a dorsal and a ventral ramus.

The dorsal ramus (the greater occipital nerve) subdivides into a lateral branch and a dorsal branch. The lateral branch emerges ventrad of the obliqus inferior muscle to enter the inner surfaces of both the longissimus capitis and splenius. The dorsal branch, relatively longer, first proceeds along the ventral border of the obliquus inferior; it then curves dorsad over its lateral surface, sending an offshoot into the internal surface of the complexus as it does. Rising further dorsal, the dorsal branch perforates the biventer cervicis over the dorsal surface of which it travels craniad. From the cranial end of the biventer cervicis, it continues beneath the caudal portion of the levator auris longus to issue forth finally between the adjacent borders of the cranial and caudal portions of the levator auris longus for ultimate distribution to the skin and cutaneous muscle between the external ears.

The ventral ramus descends immediately ventrad along the internal surface of the longus atlantis muscle, and at its ventral border curves laterad to réceive an interconnecting branch from the ventral ramus of the third cervical nerve. The ventral ramus then continues ventrad of the levator scapulae ventralis to approach the medial surface of the cleidomastoideus whereupon it gives off a branch that interconnects with the spinal accessory nerve upon its emergence from perforating the cleidomastoideus muscle. The ventral ramus itself then distributes branches to both the sternomastoid and cleidomastoid muscles, and, just before reaching the dorsal border of the sternomastoid muscle, it subdivides into the great auricular nerve and cutaneous colli. The great auricular nerve passes over the dorsal border of the sternomastoid to travel dorsocraniad along its lateral surface and thence into the lateral surface of the parotid gland where the nerve breaks up into numerous filaments, some of which supply the skin around the caudal and lateral surfaces of the external ear and the others of which terminate in the skin over the parotid gland. The cutaneous colli nerve, slender in contrast to the great auricular nerve, passes ventrad along the lateral surface of the sternomastoid to ramify in the skin just ventral to the masseter muscle.

Third cervical nerve

The third cervical nerve is transmitted from the vertebral canal through the intervertral foramen between the axis and the third cervical vertebra. The dorsal ramus divides near origin into at least three branches. One courses laterad through the fibers of the longus atlantis

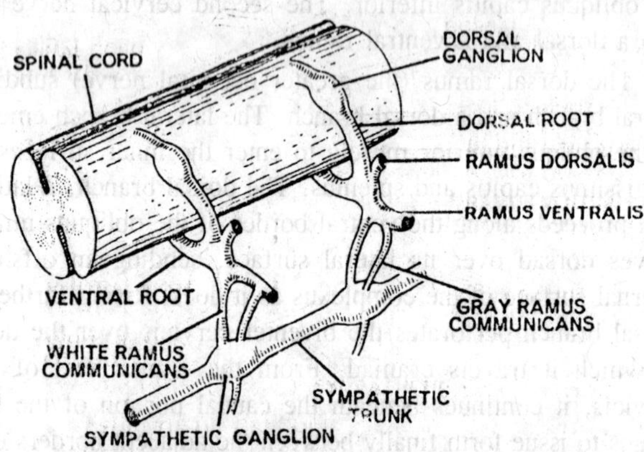

Fig. 15.15. Pattern of branching of thoracic, first three pairs of lumbar and second to fourth pairs of sacral nerves in rabbit.

muscle, innervating it in its course and finally emanating from its lateral surface to distribute slender branches to both the longissimus capitis and splenius muscles. The second branch ascends dorsad for a short distance along the ventrocaudal border of the obliquus inferior to supply the complexus and, thereafter, rises further dorsad to enter the ventral surface of the biventer cervicis muscle. The third branch curves around the ventrocaudal border of the obliquus inferior to ascend dorsad along its lateral surface for distribution to the biventer cervicis.

The ventral ramus descends directly ventrad to pass mediad of the transverse process of the second cervical vertebra. It then turns laterad, running between the cervical portion of the scalene muscle above and the longus capitis below. As it passes between the later muscles, it gives rise to four branches: one, fairly large, interconnects with the ventral ramus of the second cervical nerve; the second branch goes to the sternomastoid and cleidomastoid muscles and also generally contributes to the formation of the cutaneous colli nerve; the third branch, a slender filament, terminates in the internal surface of the clavotrapezius muscle; and the fourth branch is distributed to the levator scapulae ventralis muscle.

Fourth cervical nerve

The fourth cervical nerve makes its exit from the vertebral canal by way of the intervertebral foramen between the third and fourth cervical vertebrae.

The dorsal ramus loops around the ventral border of the semispinalis cervicis into the which it sends a branch, and then continues dorsocaudad along the lateral surface of the semispinalis cervicis to distribute itself finally to the caudal third of the biventer cervicis.

The ventral ramus passes mediad of the origin of the levator scapulae from the transverse process of the third cervical vertebrae and, subsequently, courses between the bundles of the cervical portion of the scalene muscle to appear on its lateral surface whence it distributes itself to the clavotrapezius, levator scapulae ventralis, and occipito-scapularis muscles.

Fifth cervical nerve

The fifth cervical nerve leaves the vertebral canal through the invertebral foramen between the fourth and fifth cervical vertebrae.

The dorsal ramus, after sending a branch into the ventral border of the semispinalis cervicis muscle, ascends dorsad over its lateral surface to supply the caudal end of the biventer cervicis muscle.

The ventral ramus, near its origin, divides into three branches. One branch enters the levator scapulae muscle directly. The other two branches pass laterad between two bundles of the cervical scalene muscle. At the lateral border of the cervical scalene muscle, the more slender branch turns caudad to join a branch from the ventral ramus of the sixth cervical nerve in the formation of the phrenic nerve. The other larger branch passes laterad and breaks up into four branches: one interconnects with the spinal accessory nerve near the cranial border of the acromiotrapezius muscle; the second branch enters the levator scapulae ventralis muscle; the third branch passes beneath the ventral border of the levator scapulae ventralis near the scapula, and thence continues along the caudal border of the clavotrapezius and clavobrachialis to innervate the skin over the acromiodeltoideus muscle; the fourth branch, after interconnecting with a small branch from the ventral ramus of the fourth cervical, pierces the clavotrapezius just above the raphe to innervate the skin over the clavobrachialis and clavotrapezius on either side of the raphe.

The sixth, seventh, and eighth cervical together with part of the fifth cervical, are concerned in the formation of the brachial plexus.

Brachial plexus

The ventral rami of the fifth, sixth, seventh, and eighth cervical nerves and that of the first thoracic, by virtue of interconnecting branches, contribute to the formation of the brachial plexus. The plexus is situated in the axilla together with the axillary artery and vein and furnishes innervation for the shoulder and the forelimb.

The disposition and various interrelations of the ventral rami in the constitution of plexus, although subject to some variation, present the following general pattern of arrangement. The confluence of a long slender branch from the fifth cervical and a short branch from the sixth cervical nerve results in the formation of the phrenic nerve. From the sixth cervical is derived the supracapular nerve. From the sixth and seventh cervical nerves spring the cranial subscapular nerves. The seventh cervical nerve gives rise to the middle subscapular, axillary, musculocutaneous, and posterior thoracic nerves. From the union of branches from the seventh and eighth cervical nerves issues forth the caudal subscapular nerve. The anterior thoracic nerves obtain their origins from a junction of branches from the eight cervical and first thoracic as well as from the seventh cervical nerve alone. The seventh and eighth cervical nerves and the first thoracic are involved in the formation of the radial nerve. The ulnar nerve obtains its origin from a union of branches from the eight cervical and first thoracic nerves. The seventh and eighth cervicals and first thoracic nerve contribute to the formation of the median nerve in a manner subsequently described. From the caudal surface of the first thoracic nerve arises the medial cutaneous nerve.

The nerves that arise from the brachial plexus will now be considered in some detail from the aspects of origin, course, and distribution:

Suprascapular nerve

This nerve is derived from the sixth cervical nerve, and pursues a lateral course to coracoid border of the scapula where it enters the supraspinatous fossa through the suprascapular notch. In the supraspinatous fossa, it gives a branch to the supraspinatus muscle; and then descends, mediad of the acromion, to the infraspinatus fossa to terminate in the infraspinatus muscle.

Cranial subscapular nerves

The cranial subscapular nerves, two in number, spring from the sixth and seventh cervical nerves; and pass laterad to enter the medial surface of the subscapularis muscle.

Middle subscapular nerve

The middle subscapular nerve takes origin from the seventh cervical, and passes laterad to the medial surface of the scapula. In its course it divides to send a branch to the subscapularis muscle and another to the teres major muscle.

Caudal subscapular

The caudal subscapular nerve derives its origin from the confluence of branches from the seventh and eighth cervical nerves. It then follows a caudolateral course to enter the medial surface of the latissimus dorsi muscle.

Axillary nerve

The axillary nerve emanates from the seventh cervical nerve and extends ventrolaterad to the axilla through which it courses to emerge laterally between the teres minor and the ventral border of the proximal end of the head of the triceps brachii. Here it sends a slender branch to the teres minor and then gives off a number of branches to the acromiodeltoideus and spinodeltoideus before passing beneath the acromiodeltoideus to its final point of innervation in the proximal portion of the clavobrachialis muscle.

Musculocutaneous nerve

The musculocutaneous nerve obtains its origin from the ventral surface of the seventh cervical nerve and continues laterad to the

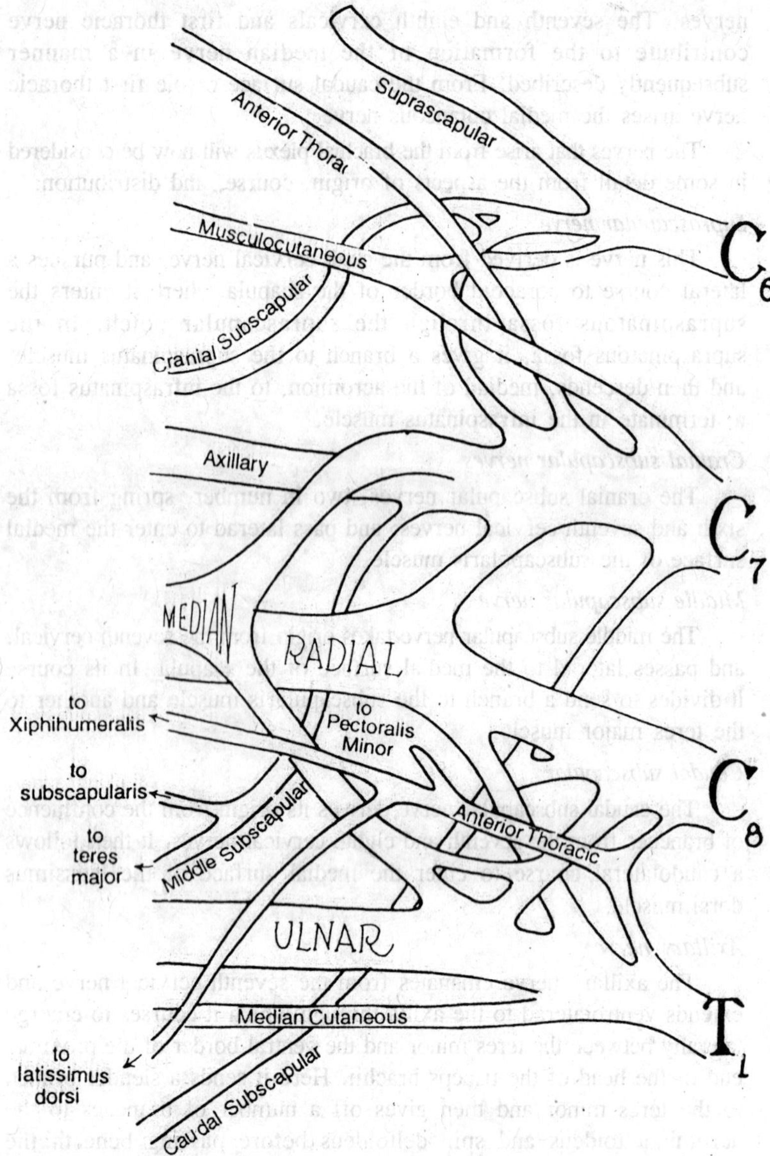

Fig. 15.16. Diagrammatic presentation of the brachial plexus of the cat from aspect of its origin.

shoulder Joint. After sending a slender branch to the coracobrachialis muscle, it approximates the proximal end of the biceps brachii into which it sends a branch. The nerve subsequently descends along the dorsal border of the biceps almost to its distal end where it abruptly curves ventrad beneath its internal surface to reach the brachialis muscle. After innervating the brachialis muscle, the nerve again passes slightly ventrad between the distal ends of the pectoantibrachialis and the clavobrachialis muscles to enter the forearm for distribution to the skin along its ventral border.

Phrenic nerve

The phrenic nerve originates from the confluence of a long slender branch from the fifth cervical nerve and a short branch from the ventral surface of the sixth cervical nerve. The nerve then follows a caudal course into the thorax, passing ventrad of the thyrocervical and subclavian arteries to parallel course of the vagus to the level of the heart where it enters the middle mediastinum and, ultimately, adheres to the side of the posterior vena cava in the final phase of its journey to the diaphragm. Incidentally, it might be well to indicate here that the phrenic nerve is not only motor to the diaphragm, but that it is also sensory to the diaphragmatic pleura and sensory to the mediastinal pleura.

Anterior thoracic nerves

The anterior thoracic nerves are two in number. One springs from a confluence of short branches from the eighth cervical and first thoracic nerves. This nerve passes laterad with the axillary artery which it leaves to accompany the anterior thoracic artery to the pectoralis muscles. At the point where the latter nerve has its origin, there arises a slender branch which proceeds directly to the xiphihumeralis muscle. The second anterior thoracic nerve comes from the seventh cervical only, and runs laterad to embrace the cranial surface of the anterior thoracid artery; in its course it interconnects with the first anterior thoracic nerve by means of a filamentous branch before termination in the pectoralis musculature.

Radial nerve

The radial nerve arises from a point of confluence between the eighth cervical and the first thoracic, after which it receives a rather thick contributory branch from the seventh cervical nerve. Almost immediately after its origin from the branches aforementioned, it gives off a muscular branch which in turn supplies a slender branch to the epitrochlearis muscle and a stouter branch to the proximal third of the long head of the triceps brachii. The radial nerve proper then passes through the intermediate and long portions of the middle head of the triceps so as to emerge, on the lateral surface of the arm, along the dorsal border of the brachio-radialis near its origin. As the radial nerve issues forth from between the intermediate and long portions of the middle head of the triceps, it sends branches to the lateral head of the triceps and to the anconeus muscle. In the region of the elbow, the radial nerve divides into a superficial radial and a deep radial nerve (dorsal interosseous nerve).

The superficial radial nerve, (cutaneous or sensory branch of the radial nerve) passes between the lateral head of the triceps and the brachialis muscle; upon emerging from between the latter muscles, it parallels the course of the vena cephalica and supplies the overlying integument. It finally enters the hand along its dorso-medial surface and ultimately breaks into seven terminal filaments: one to each side of the thumb; one to the integument on the medial side of the second digit; one to the integument on each adjacent side of the second and third digit; and one to each adjacent side of the third and fourth digit.

The deep radial nerve sometimes referred to as the dorsal interosseous nerve, descends down into the forearm between the distal end of the brachialis and the proximal end of the extensor carpi radialis, shortly passing beneath the internal surface of the latter to which it contributes two short branches. The deep radial then runs underneath the extensor communis digitorum and, at the same time, distributes a fairly large branch to the latter muscle and a slender branch to the supinator muscle. The deep radial now overrides the outer surface of the supinator muscle and supplies branches to the overlying extensor lateralis digitorum and to the upper third of the adjacent extensor carpi ulnaris. The deep radial nerve continues distad along the supinator muscle, sending several filamentous branches to the extensor indicis and finally terminating in the proximal end of the extensor brevis pollicis.

Ulnar nerve

The ulnar issues forth from a confluence between the eighth cervical and first thoracic nerves to embrace the dorsal surface of the brachial artery from which it diverges near the supracondyloid foramen. The ulnar nerve then swerves dorsad over the dorsal surface of the medial epicondyle of the humerus and thence passes beneath the short division of the middle head of the triceps to enter the forearm between the two heads of the flexor carpi ulnaris, which it also innervates. The ulnar nerve now descends along the dorsal border of the second head of the flexor pro-fundus digitorum and, in the initial part of its course, sends a branch to the first head of the flexor profundus digitorum. Midway along the dorsal border of the second head of the flexor profundus digitorum, the ulnar nerve divides to form the dorsal cutaneous nerve and the palmar nerve.

The superficial dorsal cutaneous nerve traces a distal course along the dorsal border of the first head of the flexor profundus digitorum to traverse the transverse ligament, after which it proceeds to the lateral side of the dorsum of the hand. Here it finally terminates in three delicate filaments: one to the lateral side of the fifth digit; and the other two to the adjacent sides of the fourth and fifth digits.

The palmar nerve continues distad along the dorsal border of the second head of the flexor profundus digitorum to pass beneath the transverse ligament, dividing shortly thereafter into a superficial palmar branch and a deep palmar branch, both of which enter the palm.

The superficial palmar branch, after sending a slender twig into the conical pad in the region of the pisiform bone, extends along the ventral surface of the hand to terminate in delicate filaments in the integument on the ventrolateral (ulnar) side of the fifth digit and the ventral integument on the adjacent sides of the fourth and fifth digits.

The deep palmar branch, on entering the palm, proceeds dorsad of the ligament of the pisiform bone; thereafter it swerves to the medial side of the palm where it ends in a number of filaments to the interossei (short muscles occupying the intervals between the metacarpal bones) and the musculature of the fifth digit.

Median nerve

The median nerve is embodied by the union of three branches: the initial confluence of a slender branch from the eighth cervical nerve and a relatively stouter branch from the seventh cervical nerve, subsequently joined by a rather thick branch that springs from the junction of the eighth cervical nerve and the first thoracic nerve. After its formation, the median nerve runs along the medial surface of the brachial artery to accompany the artery in its passage through the supracondyloid foramen. Thereafter, it enters the forearm between the pronator and distal end of the biceps brachii, supplying a branch to the pronator teres in the course of its passage. After sending a slender branch into the internal surface of the flexor carpi radialis, it gives off, a short distance distad, two branches: one to the proximal end of the third head of the flexor profundus digitorum; the other branch to the second head of the flexor profundus digitorum, the fibers of which it also perforates to gain access to the internal surface of the overlying palmaris longus muscle. A little further distad, again, the median sends branches into the ventral border of the fourth head and into the medial surface of the fifth head of the flexor profundus digitorum. The median nerve now descends along the medial surface of the fifth head of the flexor profundus digitorum and emerges superficially on the medial surface of the forearm between the tendon of the flexor carpi radialis and the distal end of the third head of the flexor profundus digitorum. As it reaches the superficial surface, it sends a small branch (posterior interosseous nerve) dorsad to supply both divisions of the flexor sublimes digitorum as well as the pronator quadratus muscle. The median then courses dorsad of the common tendon of the flexor profundus digitorum to enter the palm where it gives rise to three branches: the medial branch to the integument of the thumb and medial side of the second digit; the intermediate branch, after sending a filament to the trilobed pad, continues to the adjacent sides of the second and third digits; the lateral branch also sends a filament to the trilobed pad before terminating in the integument on the adjacent sides of the third and fourth digits.

Medial cutaneous nerve

The medial cutaneous nerve arises from the caudal surface of the first thoracic nerve and extends laterad along the medial surface of the median nerve. Approximately midway along the arm, it leaves its association with the median to emerge to superficial surface between the apposing borders of the pectoantibrachialis and the epitrochlearis muscles. Shortly after is emergence from between the latter muscles, it divides into two branches which continue distad into the forearm. One branch distributes itself to the skin on the medial surface of the forearm while the other branch turns dorsad for distribution to the skin along the dorsal border of the forearm.

Posterior thoracic nerve

This nerve arises from the ventral ramus of the seventh cervical nerve. Thence it passes laterad beneath the ventral border of the cervical scalene near its origin and then turns abruptly dorsad between the apposing surfaces of the dorsal and cervical scalene muscles to emerge from beneath the dorsal border of the dorsal scalene muscle to emerge from beneath the dorsal border of the dorsal scalene muscle whence it continues caudad along the external surface of the digitations of the serratus anterior muscle. In its course it distributes filamentous branches to the levator scapulae and serratus anterior muscles.

Thoracic nerves

There are thirteen pairs of thoracic nerves, each of which leaves the vertebral canal through an intervertebral foramen to divide subsequently into a dorsal and a ventral ramus. Since the majority of the thoracic nerves, with the exception of those noted, possess essentially the same peripheral distribution, the following general description of the thoracic nerves will suffice for most practical purposes.

The dorsal ramus of a thoracic nerve divides a short distance from origin into two branches. One branch courses lateral between the opposing surfaces of the lognissimus dorsi and iliocostalis. In its course it supplies the iliocostalis, at the dorsalateral border of which it curves directly dorsad over the lateral surface of the longissimus dorsi to pierce the overlying musculature, either the rhomboideus or latissimus dorsi as the case may be, so as to reach the skin along the dorsal surface of the back. The other branch ascends directly dorsad into the fibers of the longissimus dorsi, supplying it, the spinalis dorsi, and the adjacent vertebral musculature.

The ventral ramus (intercostals nerve) descends ventrad along the caudal surface of each rib, supplying the intercostal musculature in its course. A short distance from its origin, the intercostals nerve gives off a slender branch which accompanies it for about a centimeter ventrad and which then pierces the intercostal musculature to continue laterad along the ventral surface of the iliocostalis for distribution to the internal surface of the serratus posterior or inferior as the case may be. Approximately midway in its course along the caudal surface of the rib, the intercostals nerve gives rise to a lateral cutaneous branch which penetrates the intercostals musculature to pass laterad through the intervals of the serratus anterior, finally perforating the latissimus dorsi to reach the skin and subcutaneous tissue in the area. However, in the region of the digitations of origin of the external abdominal oblique, the lateral branches of the intercostals nerve are not only distributed to the skin but also to muscle. Accordingly, it will be observed that the lateral branch of the intercostals in this region pierces the external abdominal oblique near its origin on the rib and subsequently divides into a muscular branch which enters the lateral surface of the external abdominal oblique and a cutaneous branch which terminates in the skin lying over the external abdominal oblique muscle. Finally as the intercostal nerve nears the completion of its course along the caudal surface of each rib, it gives off, about five millimeters from the midventral line, a branch which perforates the internal intercostals muscle in the area to enter the internal surface of the overlying rectus abdominis.

The transverses costarum muscle in innervated by a branch of the ventral ramus of the first thoracic nerve, which, you will recall, contribution to the formation of the branchial plexus. This branch to the transverses costarum leaves the first thoracic nerve just craniad of the first rib and proceeds directly ventrad to terminate in the outer surface of the muscle.

The ninth, tenth, eleventh, and twelfth intercostals nerves leave the caudal border of each rib in the region of the costal arch and follow a ventrocaudal course across the fibers of the transverses abdominis to enter the dorsal surface of the rectus abdominis muscle. As these intercostals nerves pass ventrocaudad over the fibers of the transversus abdominis, they send delicate twigs into the muscle.

The ventral ramus of the thirteenth thoracic nerve, the caudal costal nerve, divides a short distance from origin into two branches. One branch emerges between the third and fourth head of the serratus posterior inferior muscle and then passes ventrad under the external abdominal oblique, finally piercing the latter muscle. Shortly after perforating the external abdominal oblique, this branch subdivides, with one branch entering the outer surface of the external abdominal oblique and the other continuing ventrocaudad for about nine centimeters to terminate in the integument over the ventral surface of the abdomen. The other main branch of the caudal costal nerve courses ventrocaudad between the internal abdominal oblique and the transverses abdominis for about ten centimeters, finally entering the dorsal surface of the rectus abdominis about nine and a half centimeters craniad of the pubis.

Lumbar Nerves

There are seven pairs of lumbar nerves. They leave the vertebral canal by way of the intervertebral foramina of the lumbar vertebrae, and subsequently each of them divides to form a dorsal and ventral ramus.

Each dorsal ramus almost immediately subdivides into two branches. One branch ascends dorsad to innervate the longissimus dorsi and adjacent vertebral musculature while the other branch proceeds laterad through the fibers of the longissimus dorsi to its lateral border whence it continues into the adjacent integument.

The ventral rami of the four lumbar nerves interconnect with one another to form the lumbar portion of the lumbosacral plexus. On the other hand, the ventral rami of the first three lumbar nerves are separate from each other and, consequently, will be considered individually.

(a) *The ventral ramus of the first lumbar nerve,* after sending a branch to the crus of the diaphragm, passes dorsad of the lateral component of the iliposoas muscle to its lateral border where it divides to form a lateral and a medial division. The lateral division almost immediately penetrates the transverses abdominis and thence courses laterad between the transverses abdominis and the internal abdominal oblique for about three centimeters, after which it perces the internal abdominal oblique and external abdominal oblique to proceed along a caudolateral course to its termination in the skin on the lateral surface of the abdomen. The medial division also perforates the transverses abdominis and then passes ventrocaudad between the transverses abdominis and internal abdominal oblique for about eight centimeters to enter the dorsal surface of the rectus abdominis.

(b) *The ventral ramus of the second lumbar gives off,* near its origin, branches to the quadratus lumborum, lateral portion of the ilipsoas, and psoas minor; then it continues dorsad of the iliposoas to its lateral border where it bifurcates into a lateral and a medial branch. The lateral branch ascends directly dorsad through the muscles of the abdominal wall and then continues ventrocaudad on the outer surface of the external abdominal oblique, ending in the skin at about the level of the seventh lumbar vertebra. The medial branch passes caudad along the medial surface of the transverses abdominis for about three centimeters and pierces the transverses abdominis to pass ventrocaudad between the transverses abdominis and internal abdominal oblique to terminate in the dorsal surface of the rectus abdominis about four centimeters craniad of its origin. In its course to the rectus adbnominis, the medial branch sends a slender branch to the internal abdominal oblique muscle.

(c) *The ventral ramus of the third lumbar nerve* also branches to the quadratus lumborum, to the lateral portion of the iliopoas, and to the psoas minor; then it continues dorsad of the iliopsoas to emerge at its lateral border where it divides into a lateral and medial branch. The lateral branch promptly turns dorsad through the musculature of the abdominal wall and passes ventrocaudad along the outer surface of the external abdominal oblique, giving off several branches to the skin in its course and finally terminating in the skin of the lateral part of the abdomen about one centimeter craniad of the thigh. The medial branch curves abruptly caudad along the dorsalateral surface of the iliopsoas, emerging from the dorsal surface of the latter muscle just craniad of the iliolumbar artery. The medial branch then passes ventrad of the lateral femoral cutaneous nerve and finally ventrocaudad to innervate the caudal ends of the transversus abdominis and rectus abdominis.

Lumbar portion of the Lumbosacral plexus

As has already been indicted, the ventral rami of the last four lumbar nerves contribute to the formation of the lumbar portion of the lumbosacral plexus. From the ventral ramus of the fourth lumbar nerve originate the medial genitofemoralis as well as the lateral genitofemoralis whenever it is present. The lateral femoral cutaneous nerve arises from an interconnecting band between the fourth and fifth lumbar nerves. The femoral nerve obtains its origin from the confluence of two roots, one from the fifth lumbar nerve and the other from the sixth lumbar nerve. The obturator nerve takes its origin in a similar fashion from the sixth lumbar and seventh lumbar nerves. The sixth and seventh lumbar nerves together with the first sacral nerve, are involved in the formation of the lumbosacral cord from which the sciatic nerve has its origin.

The nerves that arise from the lumbar portion of the lumbosacral plexus will now be considered in some detail from the aspects of origin, course, and distribution.

Medial genitofemoralis

The medial genitofemoralis arises from the ventral ramus of the fourth lumbar nerve whence it courses ventrocaudad between the psoas minor and the iliopsoas near origin. It shortly emerges to view on the ventromedial surface of the iliopsoas over which it continues caudad, passing dorsad of the iliolumbalis artery and vein to the ventral surface of the external iliac artery and ultimately terminates in the skin over the pelvic region of the thigh.

Lateral genitofemoralis

The lateral genitofemoralis, if present, also takes origin from the ventral ramus of the fourth lumbar nerve and courses through the fibers of the iliopsoas to appear on its ventral surface at the level of the fifth lumbar vertebra. It then runs caudolaterad across the ventral surface of the iliopoas muscle so as to pass through the abdominal wall to its distribution in the skin on the craniomedial surface of the thigh.

In many cats, the lateral genitofemoralis nerve is not found.

Lateral femoral cutaneous

The lateral femoral cutaneous nerve originates from an interconnecting band between the ventral rami of the fourth and fifth lumbar nerves. It then proceeds ventrad between the psoas minor and the lateral component of the iliopsoas to which it sends a small bifurcating branch, finally coming into view at the level of the fifth lumbar vertebra. The nerve then travels caudolaterad over the ventral surface of the ilioposoas to join the clurse of the iliolumbalis artery for ultimate distribution to the skin along the dorsolateral and lateral surfaces of the thigh.

Femoral nerve

The fermoral nerve arises from the confluence of a long descending root from the fifth lumbar nerve and a rather short root from the sixth lumbar nerve. It then travels caudad through the fibers of the iliopsoas muscle to which it gives off several short branches in its course. The femoral nerve finally rises to the ventral surface of the iliopsoas just as the muscle makes its exit from the abdominal cavity and forthwith divides into lateral, intermediate, and medial branches. The lateral branch almost immediately penetrates the ventral border of the sartoris muscle. The intermediate branch passes between the rectus femoris and the vastus medialis and subsequently divides into the three branches: one to the rectus femoris, another to the vastus lateralis, and a third to the vastus medialis. Each of the branches to the vastus lateralis and vastus medialis sends a slender branch to the vastus intermedius before entering the muscle to which it is directed. The medial branch joins the course of the femoral artery and is known as the long saphenous nerve. The long saphenous nerve travels along with the femoral artery for a short distance, finally leaving it to accompany the saphenous artery for ultimate distribution to the skin along the medial surface of the lower leg and the dorsal surface of the foot.

Obturator nerve

The obturator nerve originates from the confluence of a long root from the sixth lumbar and a short ascending root from the seventh lumbar nerve. It then passes dorsad of the common iliac vein near the origin of the hypogastric and then directly caudad through the obturator foramen to the cranial border of the external obturator muscle to which it contributes slender branches. The nerve itself then divides into at least four branches: one branch courses through the fibers of the adductor femoris muscle to enter the cranial border of the gracilis muscle; the other three branches penetrate the adductor femoris after sending branches to the adjacent adductor longus and pectineus muscles.

Sacral Nerves

There are three sacral nerves. The dorsal rami of these nerves are transmitted through the dorsal foramina of the sacrum and through the foramina between the caudal end of the sacrum and the first caudal vertebra. These dorsal rami then ascend abruptly dorsad to innervate the skin and musculature over the sacrum. On the other hand, the ventral rami leave the sacral canal through the ventral foramina of the sacrum and through the notches at its ventro caudal border. These ventral rami enter into the formation of the sacral portion of the lumbosacral plexus.

Sacral portion of the lumbosacral plexus

The large ventral ramus from the seventh lumbar nerve together with its interconnection from the sixth lumbar nerve converges dorsocaudad with the thick ventral ramus from the first sacral nerve to form the lumbosacral cord from which arises the sciatic nerve. Between the ventral rami of the first and second sacral nerves, there is an interconnecting band. It is thus possible that the ventral ramus of the second sacral nerve may contribute some fibers to the lumbosacral cord by way of this interconnection. The second and third sacral nerves unite into a complex confluence of fibers from which the pudendal, posterior femoral cutaneous, and inferior hemorrhoidal nerves take origin.

The nerves that arise from the sacral portion of the lumbosacral plexus are now to be considered from the aspects of origin, course, and distribution.

Sciatic nerve

The sciatic nerve originates from the lumbosacral cord and courses dorsolaterad to leave the pelvis by passing through the great sciatic notch. The sciatic nerve then swerves slightly caudolateral over the dorsal surface of the gemellus superior muscle and subsequently passes under the pyriformis muscle where it communicates, by a short slender branch, with the pudendal nerve. In this area the sciatic nerve also gives off a branch that abruptly turns ventrad near the caudal border of the gemellus superior so as to continue caudad beneath the gemellus inferior muscle to which it supplies innervation before terminating in the internal surface of the quadratus femoris muscle. The sciatic nerve now resumes its course in continuing some what laterad over the tendon of the obturator internus to the cranial border of the biceps femoris where it gives off a muscular branch that accompanies the sciatic nerve proper under the biceps femoris. The muscular branch divides shortly thereafter into three branches which distributed to the biceps femoris, semimembranosus, and semitendinosus muscles. The sciatic nerve itself proceeds down along the lateral surface of the adductor femoris and semimembranosus and, in its course, sends a slender branch into the tenuissimus muscle. Just above the level of the knee joint, the sciatic nerve bifurcates into a medial branch, the tibialis nerve, and a lateral branch, the peroneus communis nerve.

The tibialis nerve passes between the heads of the gastrocenemius and at once gives off slender branches to the gastrocnemius and to the plantaris muscle. From the tibialis nerve at this point, there also arises a large muscular branch which arborizes, at the level of the origin of the soleus muscle, into a number of branches which innervate the soleus, flexor hallucis longus, tibialis posterior, and popliteus muscles. The tibialis nerve then continues distad along the medial edge of the plantaris muscle and thence through a groove-like area between the medial malleolus and the calcaneus to the plantar surface of the heel. Here the tibialis nerve divides into the lateral and medial plantar nerves.

The medial plantar nerve, after passing mediad of the flexor brevis digitorum, splits to send branches to the second and third digits.

The medial plantar nerve, after coursing laterad across the tendon of the flexor longus digitorum, forks into two branches, one of which goes laterad to the fifth digit and the other of which terminates in a number of filamentous branches to the interossei and the other small muscles in the sole of the foot.

The lateral division of the sciatic nerve, the peroneus communis, extends down between the distal portion of the biceps femoris and the lateral head of the gastrocnemius and then courses beneath the fascia that gives origin to the lateral head of the gastrocnemius. The peroneus communis then passes between the extensor longus digitorum and the peroneus longus where it divides into the superficial peroneal nerve and deep peroneal nerve and sends some slender branches to the proximal end of the extensor longus digitorum.

The superficial peroneal nerve continues distad along the internal surface of the peroneus longus and, in its course, sends off a muscular branch which subdivides to innervate the personeal muscles. The superficial peroneal nerve then emerges, on the dorsolateral surface of the lower leg, between the distal ends of the tibialis anterior and extensor longus digitorum; it proceeds superficially over the transverse ligament on the tendons of the tibialis anterior and extensor longus digitorum to the dorsum of the foot. On the dorsum of the root, the superficial personeal nerve divides into four slender branches, each of which bifurcates distally to pass to both sides of each digit of the foot.

The deep peroneal nerve, on the other hand, passes beneath the proximal end of the extensor lognus digitorum to the internal surface of the tibialis to which it sends a short branch. The deep peroneal then descends along the internal surface of the tibialis anterior to emerge finally to superficial surface between the tendons of tibialis anterior and the extensor lognus digitorum. Thereafter it courses along the dorsomedial surface of the root, passing under the medial tendon of the extensor longus digitorum and ultimately dividing to send branches to the adjacent surfaces of the fourth and fifth digits as well as to the skin overlying these digits.

Pudendal nerve

The pudendal nerve derives its fibers from the confluence of the ventral rami of the second and third sacral nerves and proceeds caudolaterad of its origin to receive a small communicating branch from the sciatic nerve as it passes beneath the pyriformis muscle. The nerve then the makes its way along the dorsal surface of the gemellus superior to the medial surface of the obturator internus muscle. Here it divides into two branches: one, called the middle hemorrhoidal nerve, extends directly caudad toward the anus, and in its course, sends several branches ventrad to the muscle around the caudal end of the rectum; the second branch, the dorsal penile nerve, turns sharply caudoventrad along the medial surface of the obturator internus muscle to to attain the dorsal surface of the penis for distribution to the glans penis and the compressor urethrae muscle.

In connection with the pudendal nerve, it may be noted at this point that the innervation to the obturator interns muscle obtains its origin from the intercommunicating branch between the pudendal and sciatic nerves. This nerve passes caudad, generally along with the pudendal, to enter the medial surface of the obturator internus muscle.

Posterior femoral cutaneous nerve

The posterior femoral cutaneous nerve also derives its fibers from the confluence of the ventral rami of the second and third sacral nerves. It joins the course of the pudendal nerve to a point just caudad of the pyrifonnis muscle. Here it divides into two main branches, one of which passes directly caudad to the skin that lies laterad of the anus and the other of which goes caudolaterad to the skin over the proximal ends of the biceps femoris muscle.

Inferior hemorrhoidal nerve

This nerve takes origin mainly from the ventral ramus of the second sacral nerve. It may receive a delicate branch from the ventral ramus of the third sacral nerve. The nerve passes directly ventrad from origin and joins the course of the middle hemorrhoidal artery. On the lateral surface of the rectum, it splits into two branches, one of which extends craniad to the lateral surface of the bladder and the other of which courses caudad along the ventrolateral surface of the rectum.

Superior gluteal nerve

The superior gluteal nerve arises from the lumbosacral cord at its point of formation. The nerve passes dorsolaterad over the dorsal border of ilium and then divides into at least three branches. One branch, after sending a slender filament into the pyriformis, penetrates directly into the overlying gluteus medius; the second branch terminates in the gemellus superior; and the third branch, after sending a small subdivision to gluteus minimus, passes between the cranial border of the gemellus superior and gluteus minimus to emerge ventrad of the latter muscle in its course to the medial surface of the tensor fasciae latae.

Inferior gluteal nerve

The inferior gluteal nerve originates from the lumbosacral cord about a centimeter caudad of the origin of the superior gluteal nerve. It passes directly caudad to supply innervation to the gluteus maximus and caudofemoralis muscles.

AUTONOMIC NERVOUS SYSTEM

General Functions of the Autonomic Nervous System

The autonomic nervous system is a part of the nervous system which is concerned with the regulation of visceral functions, glands and smooth muscles, contrasting with the somatic nervous system concerned as it is with the functions of voluntary muscle, and with receiving information from the external environment. The somatic nervous system integrates and adapts the animal to the external environment whilst the autonomic system is concerned with the maintenance and modifications of the internal environment. This division

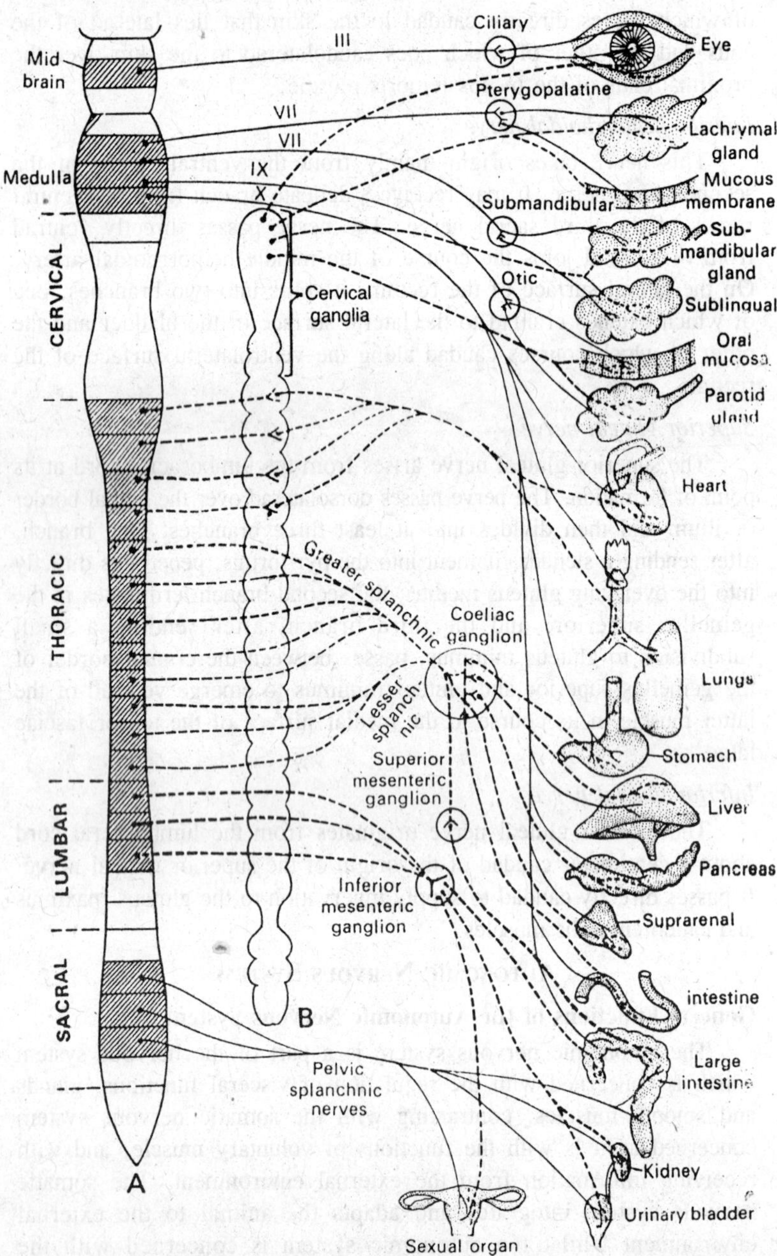

Fig. 15.17. Sympathetic (dotted) and parasympathetic (continuous lines) nervous systems.

of function is not as clear cut as it might seem for both systems may be integrated and act at the same time. Thus whilst the somatic nervous system may be receiving information from the external environment, by way of the sense organs, to which the animal responds by rapid movement (e.g. hunting prey) initiated through the somatic motor system, the autonomic nervous system, by means of modifications of breathing, heart beat, the tone of smooth muscle in walls of arterioles, is initiating those changes of the internal environment which make the response of the somatic motor system possible. Because these functions are involuntary and not normally under conscious control the autonomic nervous system is sometimes called the involuntary nervous system. Because of the connections of the autonomic with the somatic motor system at many points this is not always strictly true. Thus in micturition there is an element of conscious control, although the bladder muscle and internal bladder sphincter are under the control of the autonomic system.

Anatomy of the Autonomic Nervous System

The centres which integrate and control the activities of the autonomic nervous system lie within the central nervous system where there are many connections between the autonomic (visceral) and somatic functions. Centres of integration of autonomic activity occur at various levels in the central nervous system and include the cerebral cortex (although the autonomic functions of this area are not well understood), the hypothalamus, medulla oblongata and spinal cord. The higher centres of integration are connected to the cell bodies of neurones lying within the brain and the lateral grey column of the spinal cord. Axons from these neurones pass out of the central nervous system to terminate in ganglia (containing collections of nerve cells) which lie at varying distances from the central nervous system. Some of these ganglia lie close to the spinal cord, others are actually situated in the organs innervated. The preganglionic fibers arising within the central nervous system synapse with post-ganglionic fibers then leave the ganglia and pass to the organs innervated. The length of the post-ganglionic fiber depends on the central nervous system, short if it is situated in the wall of the organ innervated. Thus whilst there is only one neurone between the central nervous system and the effector organ in the somatic motor system, in the autonomic system there are two neurones which synapse with one another in a special structure called a ganglion. The preganglionic fiber is a medullated fiber, the post-ganglionic fiber nonmedullated.

So far as we have described the autonomic nervous system it consists of higher controlling centres, with subsidiary centres connected motor pre-ganglionic fibers to a series of autonomic ganglia from which post-ganglionic fibers pass to the various visceral structures. This is the classical description of the autonomic system and it will be seen that it is entirely motor in function, i.e. a visceral efferent system. The classical description does not include a visceral afferent system. The classical description does not include a visceral afferent or sensory system. This afferent system must obviously exist to supply the higher autonomic centres with information on the basis of which modifications in the pattern of visceral activity can occur. The rectum and bladder contract only when full; this would be impossible without a system of visceral afferents from these organs. These visceral afferents are not anatomically separated from the somatic afferents and pass into the spinal cord in the dorsal roots of the spinal nerves, from which they are projected upwards in the cord to the higher autonomic centres.

Sympathetic and the Parasympathetic Divisions of the Autonomic Nervous System

The autonomic nervous system is divided into two major divisions, the sympathetic and the parasympathetic systems. These are anatomically distinct and each system opposes the other in its functions.

Sympathetic Nervous System

The sympathetic nerves consist of those autonomic nerve fibers which leave the spinal cord in the thoracic and lumbar regions. These groups of medullated pre-ganglionic fibers leave the thoraco-lumbar region of the cord by way of the ventral root of the spinal nerve and leave the ventral root in a bundle of fibers called the white ramus communicants. The white ramus leads to a ganglion where the pre-ganglionic fibre synapses with the post-ganglionic fiber. In the thoracic and lumbar regions of the cord these ganglia form a chain on either side of the anterior (ventral) surface of the vertebrae, each ganglion being connected to the one in front and behind it. The non-medullated post-ganglionic fibers leave the ganglion and rejoin the ventral root of the spinal cord by way of the grey ramus communicants. The post-ganglionic fibers are then distributed to the effector organs by way of the spinal nerves.

However, not all the pre-ganglionic fibers have their synapse in the para-vertebral sympathetic ganglia, and some pass through the ganglia on their way to more peripherally situated ganglia.

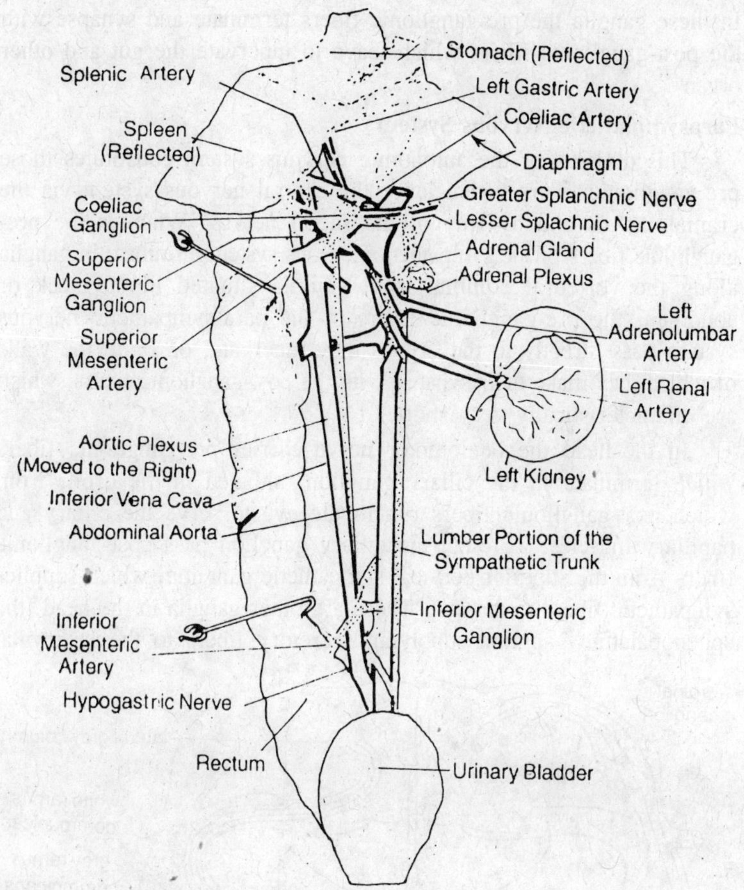

Fig. 15.18. Abdominal portion of the sympathetic nervous system: coeliac, superior, mesenteric, and inferior mesenteric ganglia.

From the upper end of the thoracic sympathetic chain pass pre-ganglionic fibers which germinate in ganglia in the neck, the superior, middle and inferior cervical ganglia, which supply post-ganglionic fibers to the structures of the head and the heart. The cervical ganglia have no direct connection with the spinal cord.

From the lower parts of thoracic chain of paravertebral ganglia pass pre-ganglionic fibers which go to the abdomen where a complex network of ganglia is situated around the major branches of the aorta in the abdomen, forming the celiac plexus, situated around the celiac artery, with subsidiary ganglia around the other arteries of the abdomen.

In these ganglia the pre-ganglionic fibers terminate and synapse with the post-ganglionic fibers which leave to innervate the gut and other organs.

Parasympathetic Nervous System

This division of the autonomic nervous system constitutes those pre-ganglionic fibers which leave the central nervous system via the cranial nerves and by way of the sacral nerves. Whereas the pre-ganglionic fibers of the sympathetic nervous system terminate in ganglia along the vertebral column or in ganglia situated in the neck or abdomen, the pre-ganglionic fibers of the para-sympathetic nervous system pass directly to the organs they innervate, on or in the walls of which they make their synapse with the post-ganglionic fibers, which are characteristically very short.

In the head the oculomotor nerve carries pre-ganglionic fibers which terminate in the ciliary ganglion, situated in the orbit, from which post-ganglionic fibers run to the eye to serve the ciliary and papillary muscles. Through the ciliary ganglion pass post-ganglionic fibers from the superior cervical sympathetic ganglion, which supplies sympathetic fibers to the eye. There are other ganglia in the head, the sphenopalatine ganglion supplying secretory fibers to the lachrymal

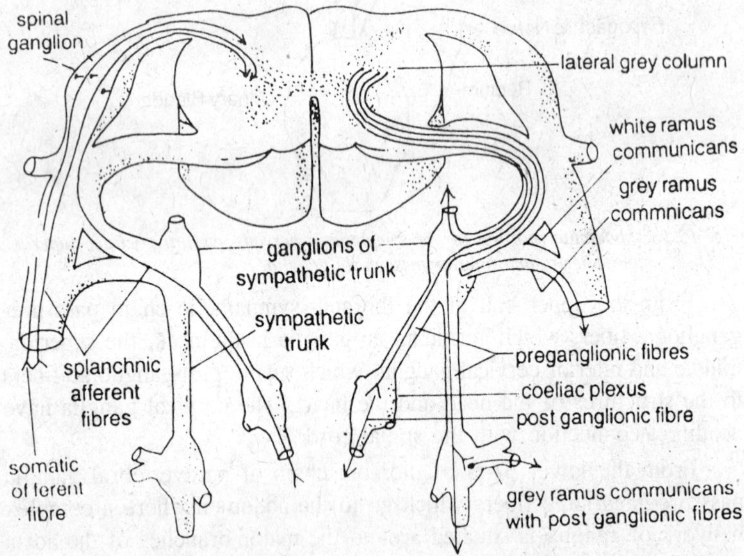

Fig. 15.19. Diagram showing the course of sympathetic afferent fibres and sympathetic efferent fibres.

gland, the otic ganglion supplying fibers to the parotid salivary gland, and the submandibular ganglion supplying fibers to the submandibular and sublingual salivary glands. All these ganglia receive pre-ganglionic parasympathetic neurones by way of the cranial nerves. In addition the vagus nerve arises in the head which carries pre-ganglionic parasympathetic fibers supplying the lungs, heart and gastro-intestinal tract. The pre-ganglionic fibers of the vagus have their synapses with the post-ganglionic fibers on or near the walls of the structure they innervate. The post-ganglionic neurones of the gut actually lie in the wall of the gut.

The remaining outflow of pre-ganglionic parasympathetic fibers comes from the sacral region of the spine. These fibers terminate in a series of ganglia around the organs of the pelvis, rectum, bladder and internal genital organs, and some fibers pass up into the abdomen to supply the lower reaches of the gastro-intestinal tract with a parasympathetic nerve supply.

Functions of the autonomic nervous system

The two divisions of the autonomic nervous system, the sympathetic and the parasympathetic, have been described as having opposing actions on the structures they innervate. Thus stimulation of the sympathetic nerves supplying the muscle of the iris causes the pupil to dilate, whereas stimulation of the parasympathetic nerves causes the pupil to constrict. The heart rate is quickened in response to sympathetic stimulation, slowed by parasympathetic (vagal) stimulation. In the bladder the muscles of the bladder wall (the detrusor muscle) relax and the sphincter of the bladder contracts in response to sympathetic stimulation whereas the detrusor contracts and the sphincter relaxes, thus achieving micturition, when the parasympathetic is stimulated. The effects of sympathetic and parasympathetic stimulation on a variety of structures is shown in table 15.1. In some organs, e.g. salivary glands, the two systems reinforce each other. Some structure e.g. sweat glands are innervated by only one system.

But the functional differences between the two divisions of the autonomic system are wider than this dual control of various visceral structures. The sympathetic nervous system, under certain circumstances, tends to be activated as a whole, producing widespread effects upon the organism. This is in part due to the fact that the sympathetic fibers ramify to a far greater extent than those of the parasympathetic system. And in the sympathetic ganglia a single pre-ganglionic fiber may be linked to as many as twenty or thirty post-ganglionic fibers; one pre-ganglionic fiber when stimulated can give rise to widespread effects. But the spread of sympathetic stimulation is also due to the fact that the hormones adrenaline and nor-adrenaline produced by the medulla of the adrenal gland produce effects on visceral structures which mimic those of sympathetic stimulation. The adrenal medulla is innervated by pre-ganglionic fibers of the sympathetic system from the celiac plexus, and these fibers terminate around the secretory cells of the medulla. These medullary cells are considered to be modified post-ganglionic neurones which arise in development from the tissues of the spinal cord, from where they migrate to the abdomen where many of them become surrounded by adrenocortical tissue to form the adrenal medulla. The terminals of the post-ganglionic fibers of the sympathetic nervous system are thought to produce their effects on visceral structures by the release of the hormones nor-adrenaline and adrenaline. The post-ganglionic fibers which come to form the secretory cells of the adrenal medulla are specialized in this respect and pour out large amounts of these hormones into the circulation when the sympathetic nervous system is activated.

Table 15.1. The effect of sympathetic and parasympathetic stimulation on various structures.

Structure	*Sympathetic Stimulation*	*Parasympathetic stimulation*
Heart	Quickening of heart rate	Slowing of heart rate
Blood vessels of skin and gut	Constriction	Dilatation
Coronary arteries	Dilatation	Constriction
Iris of eye	Dilatation of pupil	Constriction of pupil
Bladder	Relaxation of detrusor muscle and contraction of bladder sphincter	Contraction of detrusor muscle and relaxation of sphincter
Bronchial muscle	Relaxation	Constriction

The function of widespread sympathetic activity is to prepare the animal for 'fight or flight' and this function is discussed. We can summarize here the effects of a generalized sympathetic discharge.

1. There is a constriction of the arterioles of the skin and gut so diverting blood into those regions which may need a greatly increased blood supply during the effort of fight or flight – the voluntary muscles and the lungs. This widespread vasoconstriction leads to a rise in blood pressure which is available to force more blood through the dilated arterioles of the active muscles.

2. The pyloric sphincter contracts so that food does not pass into the lower digestive tract which has had its blood supply reduced.
3. The coronary arteries, supplying the heart muscle with blood dilate. This permits a greater cardiac output of blood by supplying the heart muscle with increased amounts of oxygen and glucose. Adrenaline increases the rate of conduction in the heart and increases the ability of the heart muscle to utilize glucose as a source of energy.
4. The glycogen stores in the liver, under the influence of adrenaline, are converted into glucose, providing the muscles and brain with a readily available source of energy.
5. In some species of mammal the spleen contracts, pouring out reserves of red blood cells into the circulation.

These responses of the organism to sympathetic stimulation prepare the organism for a sudden burst of activity.

The parasympathetic nervous system, with its post-ganglionic fibers located on or in the walls of the structures which it innervates permits a much more precise response to stimulation. Its activities, concerned as they are with such things as movements of the bowel, evacuation of the rectum and bladder, the secretory activity of the digestive glands, and its restraining influence on the heart, are more concerned with maintaining the status quo and conserving the organism. The terminals of the post-ganglionic fibers of the parasympathetic liberate the chemical substance acetyl - choline at their endings and this substance acts upon the effector organ. Acetyl-choline is rapidly destroyed in the body by enzymes called cholinesterases, and even if acetyl - choline is injected intravenously there are minimal effects because it is so rapidly destroyed. Thus the parasympathetic activity is rendered precise (anatomically limited) not only by the organization of its fibers but also in its chemical mediator acetyl-choline. Adrenaline enjoys a far longer life when released into the general circulation from the adrenal gland, and it is a dangerous substance to inject intravenously because of its potent effects on the activity of the heart.

Eye

The eye is one of the major special sense organs in most mammals, especially in man. It is light sensitive and capable of supplying information to the brain about the size, shape, and in many cases the colour of objects in the environment, and also about the direction and intensity of light.

Structure of the Eye

It is complex organ. The eye is a spherical hollow organ in which is suspended a lens system which focuses light onto the back of the eye. The lens system separates the eye into anterior and posterior chambers. (in medical texts the small space between the lens and the iris is often called the posterior chamber, but here the term is used meaning the whole of the chamber posterior to the lens). The anterior chamber is full of a watery fluid called the aqueous humour while the posterior chamber contains a more viscous vitreous humour. The wall of the anterior chamber is transparent and is called the cornea, which is covered by a layer of squamous epithelium on its outer surface. At the margin of the cornea this layer of epithelium is continuous with the conjunctiva, a loose layer of epithelium covering the visible portion of sclera and reflected forwards to form the inner layer of the eyelids. The cornea contains a great number of sense endings, so that one becomes painfully aware of the smallest particle resting on the surface of the cornea. The posterior chamber has a three-layered wall, the outer layer of which is a tough membrane called the sclera; it contains many white fibers and the external eye muscles which are responsible

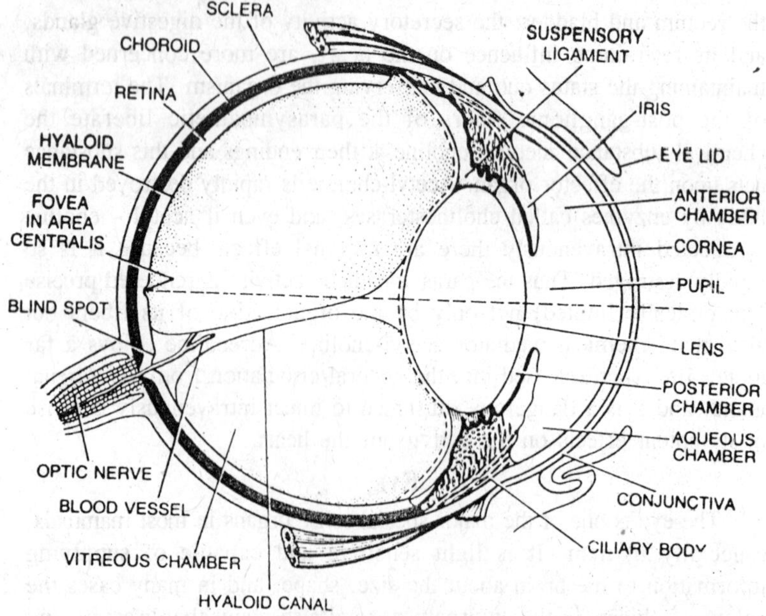

Fig. 15.20. Diagrammatic V.S. of mammalian eye.

for moving the eye within the bony orbit of the skull are attached to it. The visible portion of the sclera forms the white of the eye. The inner most layer of the wall of the posterior chamber is the retina, which is the light sensitive layer of the eye and corresponds to the film in a camera. Between the retina and the sclera is a soft brown vascular tissue called the choroids, responsible for the nutrition of the retina. The brown colour is due to pigment cells called melanophores which absorb the light after it has passed through the retina thus preventing internal refection within the eye. At the junction of the sclera and cornea is a ring of tissue called the ciliary body which forms the anterior edge of the brown choroids. Within the ciliary body are the ciliary muscles which are attached to the sclera near the corneo-sclerotic junction. The fibers of the muscle run in an anterior-posterior direction. The posterior part of the ciliary body forms the ciliary processes from which the suspensory ligaments arise. These ligaments form a circular disc of tissue in the centre of which the crystalline lens is held. The suspensory fibers in this circular ligament run radially and in the centre are continuous with the outer capsule of the lens. The lens is a cellular structure and the columnar cells at the centre of the lens are called lens fibers. Their protoplasm is fluid so that the lens behaves like a water filled balloon and when the suspensory ligaments are not putting the lens under tension it assumes a more or less spherical shape. The anterior margin of the ciliary body continues in front of the lens as the iris, which is a pigmented disc with a central aperture, the pupil. The iris has muscles which allows the pupil to vary in size and shape. The colour is due to pigment cells in the iris, and it is this structure which makes the human eye brown or blue.

The eye is protected by the eyelids. In man there are only two of these but in some mammals e.g. cat, there is a third eyelid which moves from the nasal to the temporal side of the eye. This is called the nictitating membrane and is very thin and transparent and can best be seen when a cat is just waking from a sleep. The eyes are kept free from dust by the lubricating action of the tears which are formed by the lachrymal gland, the duct of which opens under the upper eyelid. The well-known blink reflex also serves to protect the eye by causing the eye lids to close.

Passage of Light through the Eye

The cornea is curved and has a higher refractive index than the surrounding air (air R.I. = 1.0, cornea R.I. = 1.38) so that parallel light entering the eye converges onto the retina. The effect of the high refractive index of the aqueous humour lens and vitreous humour is to cause the light to continue to converge towards the retina, where the light stimulus is converted into electrical impulses which pass along the optic tract to the brain. The nerve fibers run along the inner surface of the retina, adjacent to the vitreous humour, and converge to a point where they penetrate the capsule of the posterior chamber to reach the brain; at this point there are no light sensitive cells and therefore it is called the blind spot. In some animals e.g. the squirrel the blind area forms a narrow band in the retina. This is an advantage to the squirrel because if the image of a branch to which it is jumping, falls on the blind area, it can, by a very small movement of the eye in a vertical plane, move the image onto a light sensitive area. If the blind spot were circular, as in man, and the image fell on the centre of the circle the eye would have to move through a greater distance in order to reach a light sensitive area.

Structure of the Retina

The retina is the inner light sensitive layer of the posterior chamber of the eye, and under the high power of an ordinary microscope it is seen to consist of several layers. This layered appearance is produced because there are three main kinds of cells in the retina; the visual elements, called rods and cones, the bipolar cells and the ganglion cells. The nuclei of these cells and the connections of one type of cell with another cause the layered appearance. It will be that light has to pass through the nerve fibers, ganglion cells and bipolar cells before it reaches the rods and cones. This is very surprising when it is realized that the rods and cones are the light sensitive cells and the bipolar and ganglion cells are merely conducting elements. The vertebrate retina is said to be inverted because it seems to be back to front. The reason for this inversion is explained later.

The rod

The *rod* is the most common light receptor in the human retina and it has been estimated that there are 130,000,000 of them in a pair of human eyes. There are some mammals e.g. the opossum and some bats which have pure rod retinas. In nocturnal mammals the rod is the most important light receptor for it is extremely sensitive and sends out nerve impulses to the bipolar cells even when light of very low intensity falls upon it. Since there are only about 450,000 nerve fibers in each optic tract this means that each nerve fiber must supply about 150 rods. It is seen that several rods are connected to each bipolar

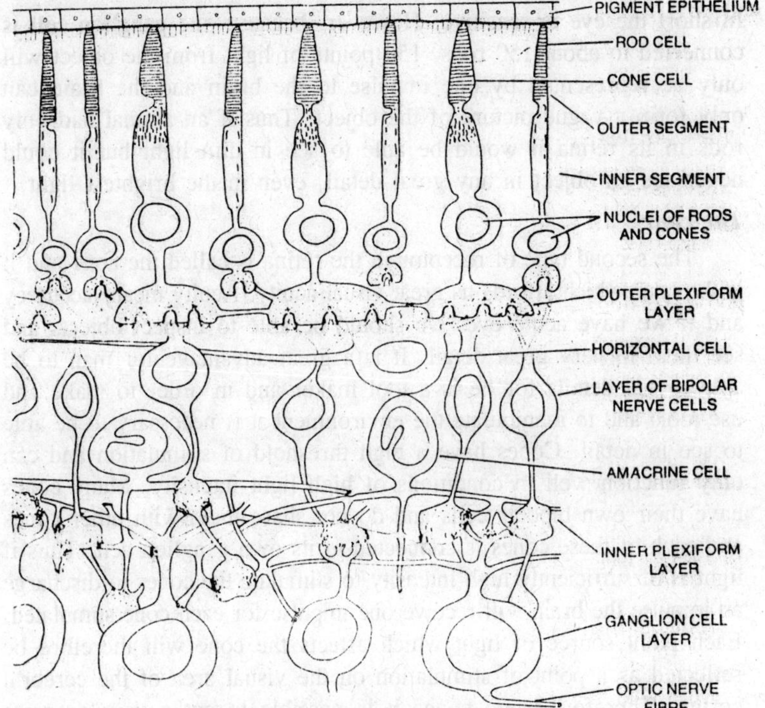

Fig. 15.21. Ultrastructure of optic retina.

cell and several bipolar cells to each ganglion cells, whose axon is the nerve fiber of the optic tract. Experiments have shown that when a lot of rods are stimulated it is possible to cause the ganglion cell which units these rods to discharge an impulse to the brain, whereas if only a few rods are stimulated the ganglion cell may not discharge. In technical language we say that there is summative interaction between the rods stimulated. The rods are interconnected to a greater degree around the periphery of the cup of the retina in the human eye than at the base of the cup near the back of the eye and this is perhaps one of the reasons why the periphery of the cup of the retina is very sensitive to dim light. It is well known that you can see better in dim light if you do not look directly at an object but look a little to one side so that light falls on the rod rich area of the retina. We have seen that rods discharge impulses when light of low intensity falls upon them and that these impulses are summated i.e. added together, so that they can overcome the threshold of a ganglion cell, ensuring that the latter discharges an impulse to the brain. Thus the brain is informed of an object even when it is lit by very dim light.

In short the eye is sensitive. However, because one ganglion cell is connected to about 150 rods, 150 points of light from the object will only be represented by one impulse to the brain and the brain can only form a vague picture of the object. Thus if an animal had only rods in its retina it would be able to see in dim light but it could never see an object in any great detail, even in the brightest light.

The cone

The second type of receptor in the retina is called the cone and it is the cone which affords us great visual acuity. Acuity means accuracy and if we have acute eyes we should be able to inspect objects and see them in very great detail. It is a great advantage for man to be able to see details for he is a tool maker and in order to make and use tools and to manipulate the environment it is necessary to be able to see in detail. Cones have a high threshold of stimulation and can only function well in conditions of high light intensity. Many cones have their own bipolar cells and do not share them with other cones and each of these cones is connected to its own ganglion cell. Thus if light is of sufficiently high intensity to stimulate the cones to discharge an impulse the brain will receive one impulse for each cone stimulated. Each point source of light which affects the cone will therefore be reflected as a point of stimulation on the visual area of the cerebral cortex. Therefore using cones it is possible to get a very accurate image of the object in the brain, i.e. the cones provide us with acute vision. The disadvantage is that high light intensity is needed for this acute vision. You should be reading this book in good illumination otherwise the cones cannot operate and you will be attempting to read using the more sensitive rods which cannot provide you with a sharp image. Since you get a blurred image you try to use your eye muscles to focus the eyes and still you cannot see clearly. You are obviously not using these very delicate instruments under the proper conditions and perpetual misuse of your eyes like this will probably give you a headache.

In order to explain the principles of acute vision we have described how each cone has its own ganglion cell but this is only so at certain special points in the retina e.g. at the yellow spot (fovea). There are about 8 cones in the retina for every neurone in the optic tract since each eye has about 3,500,000 cones and only 450,000 neurones in each optic tract. It is obvious therefore that each cone cannot have its own fiber in the optic tract and there must be some connections between the cones as their dendrites join the bipolar cells. It is known that at some places in the retina many cones do share one bipolar cell, and there are in fact some cells called horizontal cells which link rods and cones together. The situation is obviously more complicated than the simple case which has previously been put forward. It is clear that the old idea that each cone had an individual nerve supply is inaccurate but since there are many fewer cones than rods the general principle already stated holds true that cones are capable of allowing much more acute vision than are rods. The fovea is obviously an area in which very precise images can be formed, and passed on to the brain. The position of the fovea is so arranged in man that when an object is held in the hand at about one foot from the nose the image falls on the fovea. We automatically hold things in about this position for close examination for we have learned by trial and error that this is how we are able to see them best.

We have seen that the human eye has two kinds of receptor, the rod suitable for night vision, and the cone suited for day vision; the rod gives us a sensitive eye and the cone an acute eye. The rod and cone are so called because it is sometimes possible to distinguish them by their appearance under the microscope. We must not imagine that all retinas have the same distribution of rods and cones as we have, the retinas of different species of animals are as different as are their bodies.

Biochemical Basis of Colour Vision

The presence of light sensitive pigments within the cells of the retina allows the retina to respond to light energy and ultimately to send nerve impulses to the brain. Electron micrographs of the retina show that the ends of both rods and cones are filled with thin membraneous sacs which contain the light-absorbing pigments. These light sensitive pigments are called retinal or chromophores. These chromophores are combined with fatty protein molecules called opsins. In the rods the chromophore-opsin complex is called rhodopsin and this is not associated with colour vision in man and is sensitive to white light. The cones are responsible, in man, for colour vision and they chromophore-opsin complexes called iodopsins. There are 3 different proteins associated with the chromophore in human cones and hence there are 3 different iodopsins, but each cone only contains one kind of iodopsin. These 3 kinds of iodopsin are sensitive to blue, green and yellow light respectively. Colour vision in man is thus said to be trichromatic as it depends upon 3 populations of cones each containing a pigment sensitive to either blue, green or yellow light.

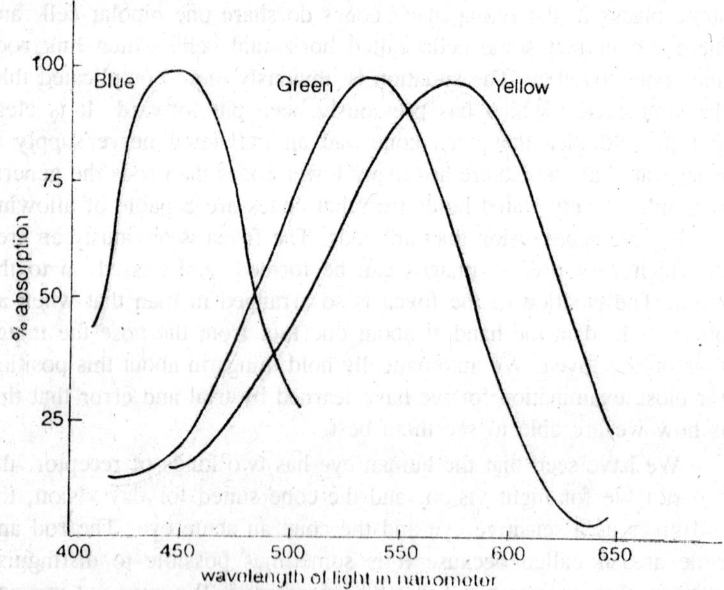

Fig. 15.22. Colour vision in man depends upon the presence of three different opsins combined with the chromophore retinal in the cones.

Study of purified chromophore-opsin combinations has shown that each kind of combination has a different affinity for light of a particular wavelength. The 3 opsins in human cones show peak absorption of light at 450, 525 and 550 nanometers – which are the blue, green and yellow regions of the visible spectrum respectively. When a pattern of coloured light meets the retina, this pattern is reflected in the nerve impulses which leave the retina because of the presence in the retina of these 3 different types of cone cells, each sensitive to light of a particular wavelength.

Molecular Changes Induced by Light in the Chromophore

On excitation by light all the 4 human opsins change in the same way. They only differ in the kind of light to which they respond. The molecule of the chromophore (retinal) is characterized by the presence of alternating single and double bonds, in chemical terms a conjugated system. Such molecules are structurally fairly stable because the groups of atoms which are attached by double bonds cannot rotate around these bonds. However, the molecules, if adequately extended, have a low energy state and can be excited by light. The effect of excitation by light is to 'relax' the double bond character so that the molecule can change from a cis configuration to a trans structure.

This change in the molecular configuration of the chromophore (retinal) is the only one which light is involved, and it occurs in an unbelievably small fraction of a second. Following this cis-trans transformation in the pigment are many complex chemical events in the associated protein opsin but these are not light dependent and can occur, as it were, in the dark. The change of the chromophore from cis to trans configuration sets in train this complex series of changes in the structure of the associated protein opsin and these changes finally lead to the release of the chromophore from the opsin.

The released all-trans chromophore is later reduced to the alcohol form and then oxidized back to the aldehyde form to regenerate all-cis chromophore which is ready to respond again.

The study of the rapid events that take place when light impinges upon chromophores has been revolutionized by low temperature chemistry. Solutions of chromophores can be held at the temperature of liquid nitrogen (– 140°C) and their responses to irradiation by light of different wavelengths and intensifies can be studied at rates which are far slower than they actually occur at body temperature.

Colour Blindness

Normal human vision utilizes 4 different opsins, 1 in the rods to make rhodopsin and 3 in the cones for the colour vision pigments. Each of these must be specified by a different gene. Normal colour vision is trichromatic, depending as it does on the presence of 3 kinds of opsins in the cones. Most colour blindness is dichromatic, resulting from an absence or deficiency of one kind of opsin in the cones, producing blue-, green-, or red-blindness. Red- and green-blindness is caused by a sex-linked recessive mutation. About 1% of men are red colour blind, about 2% green colour blind, whereas both conditions are usually very rare in women. Blue-blindness is very rare, not sex-linked and affecting only 1 in 20,000 persons.

Inverted retina

One of the very surprising things about the retina is that the rods and cones seem to be pointing the wrong way, as it were, for the light has of pass through the nerve layer then the bipolar cell layer and only finally does it reach the receptor cell. Such a retina is said to be inverted. The reason for this peculiar design is to be found in a study of the way in which the eye develops. One of the important ways in which vertebrate animals differ from those without backbones is that the vertebrate animals have a central nervous system which is hollow and dorsal whereas the invertebrate has a ventral solid nerve cord. The vertebrate nerve cord is a tube which is formed by a rolling up of the skin. Note how in the diagram the outer layer of the skin becomes the inner lining layer of the nerve tube. How the fore-brain bulges out to form buds which develop into the optic cup. If the layers are carefully followed on the diagram you will see how cells which were once on the outer skin of the animal come to lie in the position of the rods and cones in an inverted retina. The outer wall of the optic cup becomes pigmented and forms the pigment layer which backs the retina. This black layer in man serves to absorb the light after it has passed through the receptor layer. If the layer behind the receptors was shiny then light would be reflected back through the receptor cell layer and the cones might be restimulated and confusion could arise because of this internal dazzle effect. Some nocturnal animals e.g.

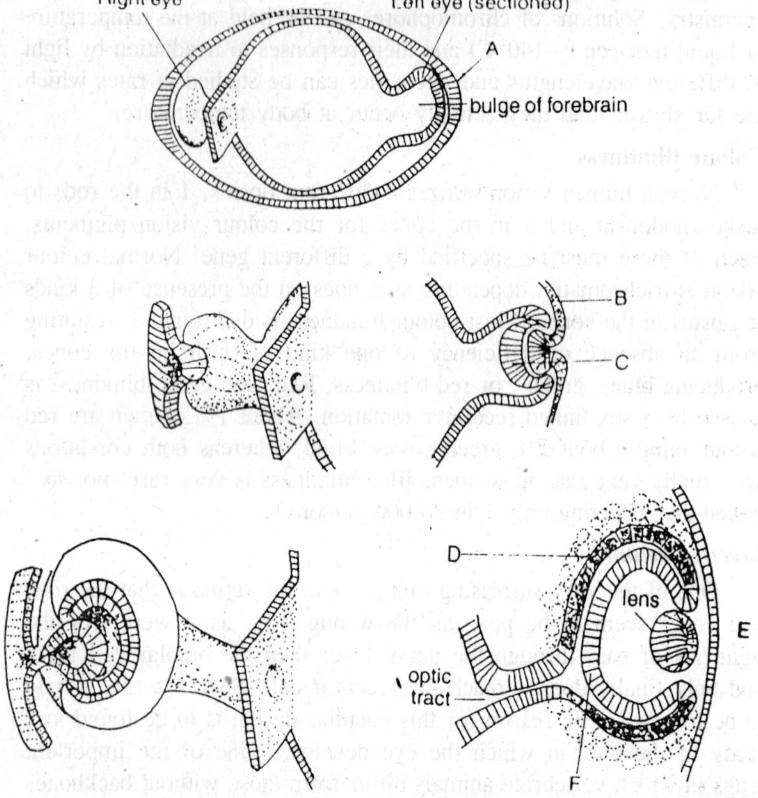

Fig. 15.23. Diagrams to show how the optic cup is formed.

cats do have a shiny backing to the retina called the tapetum. The shiny tapetum does reflect the light back into the receptors and the rods are restimulated, there is no dazzle of course for the cat is a nocturnal animal and is only using this tapetum in dim light on a rod rich retina. If such a dark adapted animal is caught in the beam of a car headlight the reflected light can easily be seen. The unfortunate animal will probably be dazzled and unable to see anything for the internal reflection inside the eye.

Before leaving the developmental story of the eye it must be mentioned that the lens is formed in a very interesting way. The epidermis overlying the developing optic cup is said to be organized to form the lens. Chemical substances are made by the developing optic cup and as these diffuse outwards they effect a change in the surrounding tissues, the epidermis folds in to form a lens and the new outer layers become the transpartent cornea. Mesodermal cells congregate around the cup bringing blood to the developing retina. Incidentally here is a physical reason why the inverted retina may have been so successful, for by inverting the retina the receptors are brought very near to the blood supply. Other connective tissue cells form the fibrous protecting sclerotic around the outside of the eye. This story of eye development was one of the early examples of organization that was discovered. We now know that the direction which the cells of our body take as they differentiate is governed by the position that these cells have on a particular diffusion gradient. A differentiating organism is a very complex nexus of diffusion gradients.

Accommodation

Ciliary muscles have their origin at the junction of the cornea and sclerotic and are inserted into the ciliary body, which is attached to the choroids layer. The suspensory ligaments are stretched from the processes of the ciliary body to the capsule of the lens, with which they are continuous. When the ciliary muscle contracts the ciliary body and the choroids are pulled forward to a point at which the diameter of the eye is smaller. Because the suspensory ligaments are now stretched over a smaller distance there is less tension in them and therefore there is less tension on the lens, which because of its elasticity, becomes more spherical. As it assumes a spherical shape its focal distance shortens and it becomes capable of turning light through a greater angle; it is adjusted for seeing near objects, from which the light rays are divergent. When the ciliary muscles relax, the tension in the stretched choroids is transmitted to the suspensory

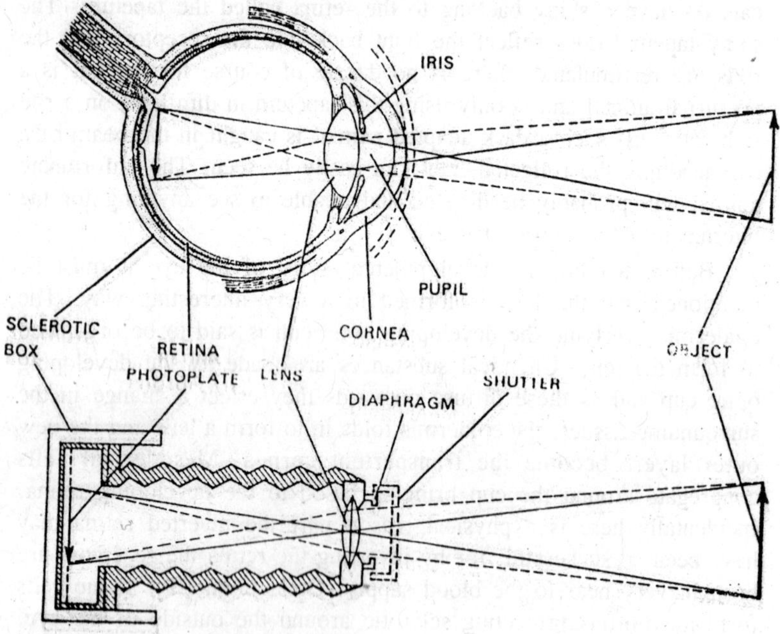

Fig. 15.24. Analogy in working of an eye and photo camera.

ligament and the lens assumes a less spherical shape, its normal resting position. It can now only bend light through a small angle and is suited for dealing with light from far distant objects, since light from these objects is parallel rather than divergent. It is a surprising fact, but a true one, that when the ciliary muscle contracts, tension falls in the suspensory ligament and vice versa. This is the method of accommodation in the mammal, but it must not be thought that all animals accommodate in this way. Many fishes accommodate by moving the lens backward and forward in the eye by means of a special muscle called the retractor lentis. Birds and some reptiles accommodate by the ciliary muscle actually squeezing the lens and making it more convex.

Accommodation, visual acuity and size of eye

Accommodation is the method used to focus the eye accurately so that objects can be seen with greater precision, but precise focusing is worthless unless the retina is capable of recording a detailed image. An image can only be seen with great precision in a cone containing retina for reasons already explained. The image must also be a large one so that it falls on many cones. Photographers know that if they want to make negatives which have to be enlarged many times they have to have a fine grain emulsion on their film, since the negative must be very precise if it is to be enlarged. It is the same with the retina; the image if it is to be precisely seen must fall on a fine grained retina, that is one containing many cones. For the light to fall on many cones the retina must be large. Therefore acuity of vision demands large cone rich retinas, and a lens system which is capable of accurate focusing and able to throw a large image on the retina. If the eye is small as in a mouse or any small rodent the image formed will also be small and the number of cones stimulated will be small. Therefore in small rodents acuity of vision is not possible and it is interesting to note that these small animals cannot accommodate. They are nocturnal creatures and their eyes are used only to indicate the

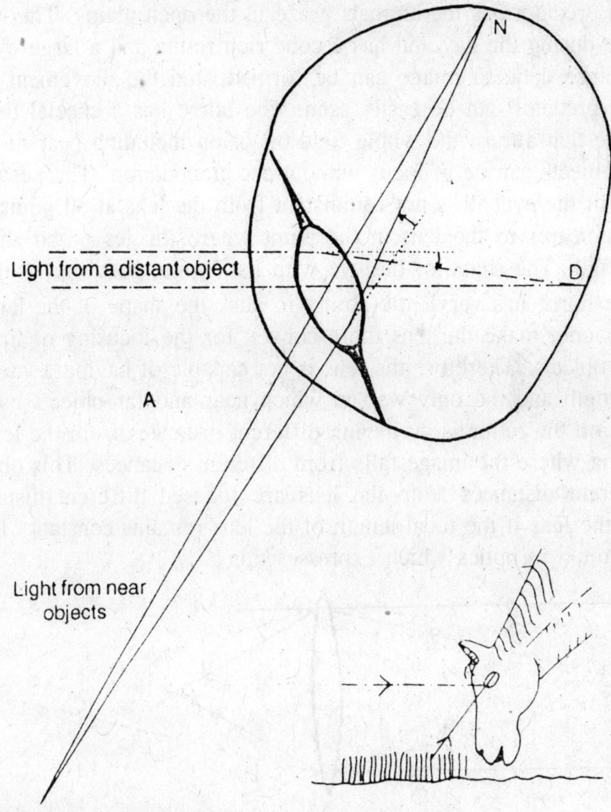

Fig. 15.25. The ramped retina of the horse.

direction and intensity of light, not to see in accurate detail. To these small nocturnal rodents the sense of smell is more important than sight. In the primates, a group which includes monkeys, apes and man, the eyes are relatively large and capable of throwing a large image onto the retina. Accommodation is now worth while and we have seen something of the method in man.

In mammals the ciliary muscles are poorly developed in rodents and show an increasing development in herbivores, carnivores, until they reach a maximum development in the primates. In general the extent of accommodation is low in mammals except in primates.

Herbivore Eye

The eye is important in these animals for safety. In the typical case of the horse the eye is useful for detection of the movement of possible predators as the animals graze in the open plains. The horse is active during the day and has a cone rich retina and a large eye so that a large detailed image can be formed, and the movement of a stalking predator can be easily seen. The horse has a special device to ensure that almost the whole field of vision including near and far distant objects can be in focus without accommodation. The posterior surface of the eyeball is not equidistant from the lens at all points but is much nearer to the lens at the point where images of far distant objects fall. The shape of the eye with its sloping posterior wall.

The horse has very little ability to alter the shape of the lens so that it cannot make the lens more convex for the focusing of images of near objects. Therefore the lens is not capable of having a variable focal length and the only way in which near and far objects can be focused on the retina is by having different distances from the lens to the retina where the image falls from different distances. This objects at different distances from the lens are focused different distances behind the lens if the focal length of the lens remains constant. There is a formula in optics which expresses this:

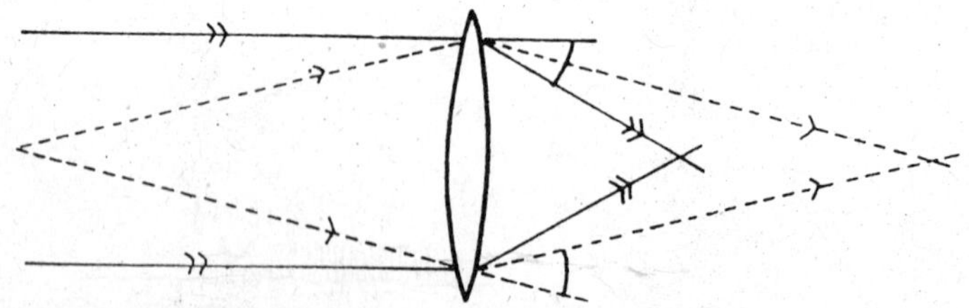

Fig. 15.26. Diagram illustrating the relationship of the position of the object and the position of the image with a convex lens of fixed focal length.

$$\frac{1}{v} - \frac{1}{u} = \frac{1}{f}$$

u – distance of objects from lens

v – distance of image from lens

f – focal length of lens

When the horse's head is bent down as it eats, the image of the grass focuses in the upper part of the retina at point N, whilst distant objects focus at point D in the lower part of the retina. The light from near and distant objects is bent through the same angle but both objects are in focus simultaneously because of the sloping retina. Such a retina is called a ramped retina. The ramped retina replaces the ciliary muscles functionally.

The eyes of the horse are large and set high in the front of the head, with the result that the horse can see what is happening all

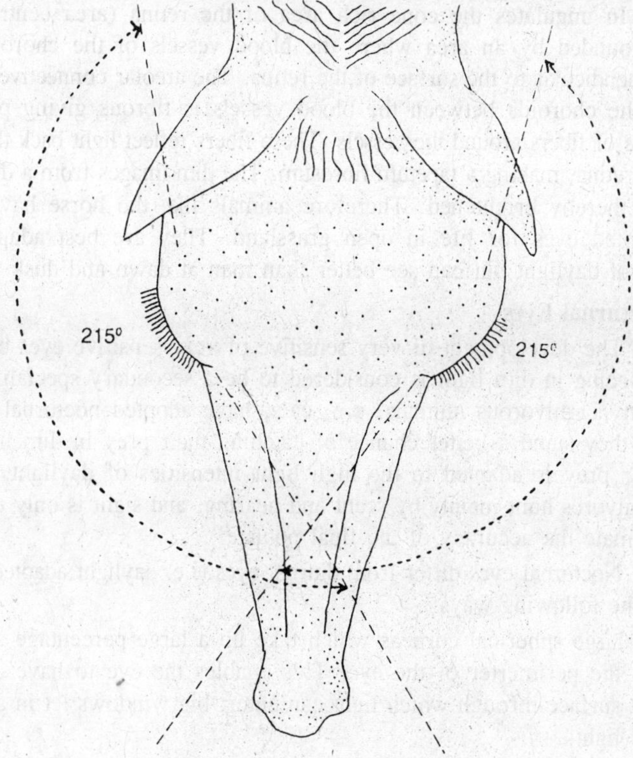

Fig. 15.27. The visual field of the horse.

round it except for a small area the width of its head behind its body. The visual field of each eye overlap in front of the head. The horse has a broad, horizontal oblong pupil, so that it can take advantage of the possibility of all round vision, even when the pupil is small in very bright light.

The horse's has a very small anterior chamber and a large posterior one, thus giving a large image. Movement of the object is detected by movement of the image on the retina. The image will move further in a large eye than in a small one. Thus in the large eye of ungulates movement is easily detected, so that the animal easily sees the predator coming and can flee.

It is necessary for distant vision to have sensitive cells in the retina, as the intensity of light from a distant object may be small. The ungulate eye is large enough to pack in rods, which are sensitive, between the cones.

In ungulates the cone rich area of the retina (area centralis is surrounded by an area where the blood vessels of the choroid run perpendicular to the surface of the retina. The areolar connective tissue of the choroids between the blood vessels is fibrous giving parallel rows of fibers around the vessels. These fibers reflect light back through the retina, making a tapetum fibrosum. The dim images from a distance are thereby brightened. Therefore animals like the horse have well adapted eyes for life in open grassland. They are best adapted to bright daylight but can see better than man at dawn and dusk.

Nocturnal Eyes

The development of very sensitive of very sensitive eyes capable of seeing in dim light is considered to be a secondary specialization. Many carnivorous animals, e.g. cats, have adopted nocturnal habits for they stand a better chance of catching their prey in dim light, if their prey is adapted to the high light intensities of daylight. Many carnivores hunt mainly by scent and hearing, and sight is only used to estimate the accuracy of the final pounce.

Nocturnal eyes differ from diurnal eyes (i.e. daylight adapted eyes) in the following ways:

(*a*) large spherical corneas which take up a large percentage area of the perimerter of the eye. This enables the eye to have a large surface through which light can enter; big windows let in a lot of light.

(*b*) the lens is spherical and the posterior chamber is small relative to the anterior chamber. The nocturnal eye must not be merely large, it must be disproportionately so. It needs to be large to let in a lot of light but it must not have large distance from lens to retina, for the intensity of light diminishes as the square of the distance; that is if you double the length of the eye you double the distance light has to travel from lens to retina and the intensity is diminished by four times. If the eye becomes three times longer, the illumination intensity is nine times smaller. Now the eye capable of seeing in dim light must make the image as bright as possible so it must have a short posterior chamber. If the posterior chamber is small then the lens must be powerful enough to bend the light through a large angle in order to focus light on the retina. The spherical lens of the nocturnal mammal is capable of this.

(*c*) The retina of a nocturnal animal must be sensitive and contain many rods whose effects are summated. But which fly at night have pure rod retinas.

(*d*) Tapetum. In many nocturnal animals there are layers of cells either in the retina, or more usually in the choroids which reflect light back through the light sensitive retina. Such layers are called tapetal layers. Cats eyes shine green at night, as the light is reflected back through the eye by a tapetal layer in the choroids. As the white light is reflected through the choroids and retina, red light is absorbed by the red pigment in blood and so the light appears green, since green is the complementary colour to red. In the cat the epithelium proliferates around the blood vessels in the choroids to form thin tiers of cells. The joints between these coincide and blood vessels run straight between the piles of cells. The piles of cells reflect the light and form what is known as a tapetum cellulosum. The cells in these piles contain crystalline threads in the protoplasm. These threads are very fine and form bundles running in varying directions, and it is these refractile threads which reflect the light.

Twenty-four Hour Eye

We have seen that some eyes are adapted for use in strong light or in dim light. Since animals live half their lives in light and half in dim light or darkness it is not surprising to find that most animals can see under a great variety of light intensities although they often have a very good performance under only a small range of light conditions. The opossum is incapable of seeing if he is disturbed in bright light – he can only see in the dark, but animals like the opossum are exceptions. Man can see well in bright light and fairly well in the dark and he manages this by having a special area of the retina called the fovea (yellow spot) which is very rich in cones for daylight use and other areas of the retina which are rich in rods for nighttime use. Many mammals are like man in that they possess dual purpose retinas.

The iris disphragm is a most useful organ for regulating the amount of light which enters the eye. It is a circular sheet of muscle with a central aperture, the pupil, size and shape of which can be altered to control the passage of light into the eye. The eye of a cat at night has a large circular pupil as the iris is allowing all the available light to enter. In bright light the cat's pupil becomes a very narrow vertical slit so that only a little light can enter, and the sensitive rods of the retina are protected. The iris thus allows the cat to live successfully during all light conditions.

Eye Defects

Far Sightedness

As we get older the lens becomes progressively more inelastic which means that as the ciliary muscles contract during accommodation for near vision the lens assumes a less spherical shape, and so is less able to converge the diverging rays of light from near objects. The closer the object to the eye the more divergent are the rays of light from it and the more converging the lens has to perform. There is a maximal amount of convergence that can be brought about by the lens system and there comes a point when an approaching object can no longer be focused properly on the retina. The point of maximum capacity of convergence is called the near point, and with increasing age the near point recedes progressively. A youth should be able to focus adequately up to about ten inches from his nose end but in older persons the near point is often three feet away. This defect may be easily corrected by fortifying the eye's powers of convergence of light by supplying spectacles containing biconvex (converging) lenses. People suffering from this defect can see distant objects well with the unaided eye and the defect is thus called far sightedness.

Long Sightedness

Persons with long sightedness can only see distant object clearly. This is not due to a loss of elasticity in the lens, as in the above case, but is due to a defect in tne shape of the eye. The eye of a long sighted person is too short, the retina being too near the lens, and the maximal powers of accommodation are unable to converge the light from near objects to focus them on the retina. The rays of light are in

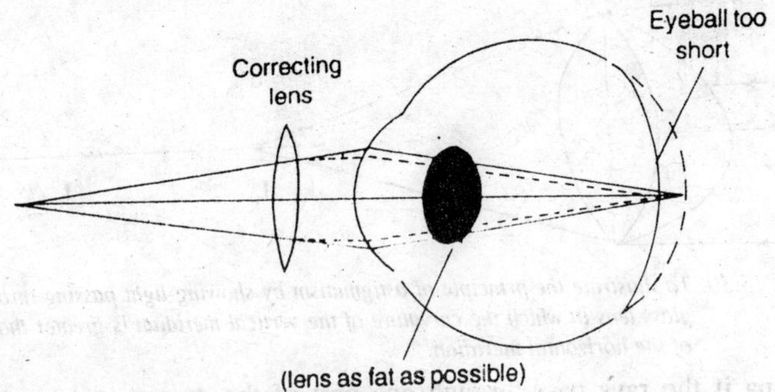

Fig. 15.28. The correction of long sightedness.

focus behind the retina and thus the image is blurred. The defect may be corrected by spectacles containing biconvex (converging) lenses.

Short Sightedness

In this condition the eyeball is too long so that the retina is further away from the lens than in a normal eye. This places for less strain on the powers of accommodation and the light from near objects can be focused onto the retina with less activity of the ciliary muscles. A short sighted person can see objects clearly when they are only a few inches from the eye. However the images of distant objects are focused in front of the retina even when the accommodation processes are completely at rest, because of the increased length of the eyeball. The person with short sightedness can be made to see distant objects clearly by the use of spectacles with divergent lenses (biconcave).

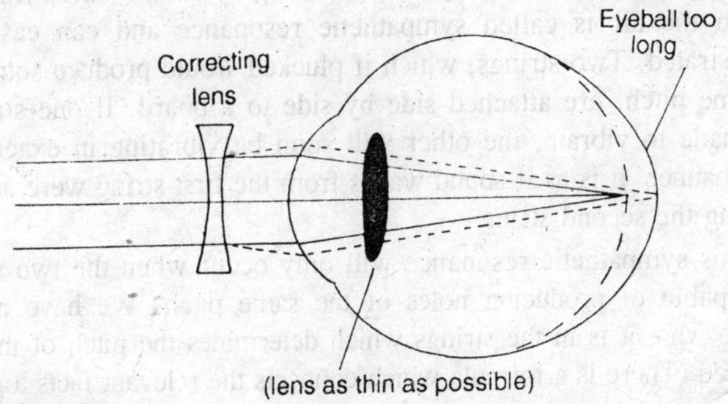

Fig. 15.29. The correction of short sightedness.

Astigmatism

This is a defect of the eye in which there is an irregularity of the corneal surface so that a point source of light may be focused on the

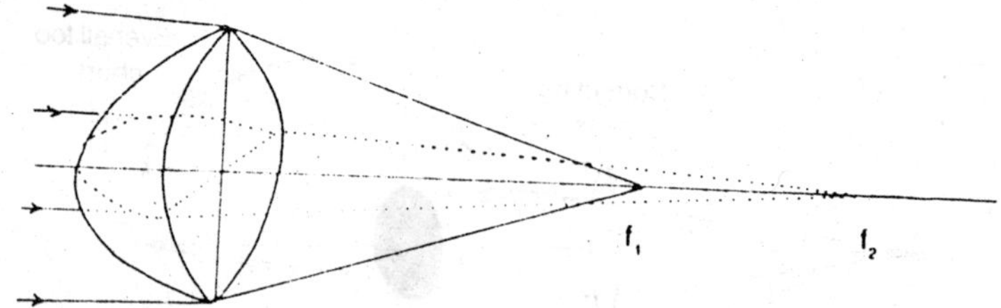

Fig. 15.30. To illustrate the principle of astigmatism by showing light passing through a glass lens in which the curvature of the vertical meridian is greater than that of the horizontal meridian.

retina if the rays pass through one part of the cornea, yet be out of focus if the rays pass through another part of the cornea. Irregularity of the corneal surface, which is normally spherical, can also lead to a distortion of the image on the retina. Provided that the irregularity is a regular one e.g. a flattening of the surface in one meridian the defect can be corrected by using a cylindrical lens so that the lens and cornea have the same effect as a normally shaped cornea would have.

The Ear

Principles of the Mechanism of Hearing

We can perhaps best understand the complex process of hearing by comparing the sensory apparatus of the ear to a piano sounding board. Whereas the strings of the piano are made to vibrate when they are struck by hammers the strings in the ear apparatus are made to vibrate indirectly by means of sound vibrations in the air.

The production of vibrations in a string by sound waves traveling through the air is called sympathetic resonance and can easily be demonstrated. Two strings, which if plucked would produce sounds of the same pitch, are attached side by side to a board. If one string is now made to vibrate, the other will soon be vibrating in exactly the same manner. It is as if sound waves from the first string were actually plucking the second string.

This sympathetic resonance will only occur when the two strings are capable of producing notes of the same pitch. We have now to enquire what it is in the strings which determines the pitch of the note produced. There is a formula which connects the relevant facts together:

$$n = \frac{1}{2\ell}\sqrt{\frac{t}{m}}$$

where

l – length of string

t - tension on string

m - mass of unit length of string

n - number of vibrations per second.

The pitch of the note is determined by the number of vibrations per second (n); the greater the number of vibrations per second the higher the pitch and vice versa.

Now the number of vibrations per second in the string is determined by various qualities of the string, its length, tension and mass. From the formula which connects these factors it is evident that to produce a note of high pitch the string should be relatively short, taut and light. To produce a note of lower pitch the string should be longer, slacker or heavier.

Thus in the piano the strings on the right hand side, producing notes of high pitch, are relatively short, taut and fine. The strings on the left hand side are longer, less taut and are bound round with copper wire to make them heavier.

The structure in the ear which we have compared to the piano sounding board is called the cochlea. Its essential structure is that of a fine membrane, surrounded by fluid and embedded deep in bone. This membrane may be likened to a lot of strings closely applied side by side. As in the piano sounding board the qualities of the strings vary from one end of the membrane to the other. At one end the fibers are shorter, more taut and are probably less heavily loaded with fluid (i.e. less mass) than at the other end.

We can now obtain a picture of how this membrane functions. Sound waves passing in the air reach the ear apparatus and set up vibrations in the fluid surrounding the membrane of the cochlea. Parts of the membrane now begin to vibrate, but only those parts that vibrate with the same frequency as the sound waves reaching the ear. Nerve fibers attached to the part of the membrane which is actually vibrating are now stimulated and electrical impulses pass to the auditory areas of the brain.

Structure of the Hearing Apparatus

We can now discuss how the ear apparatus is designed to carry out these activities. For descriptive purposes the ear apparatus may be divided into three parts, the outer, middle and inner ears.

The outer ear

The outer ear consists of the pinna, the 'ear' of everyday speech and a tube, the external auditory meatus which leads into the skull.

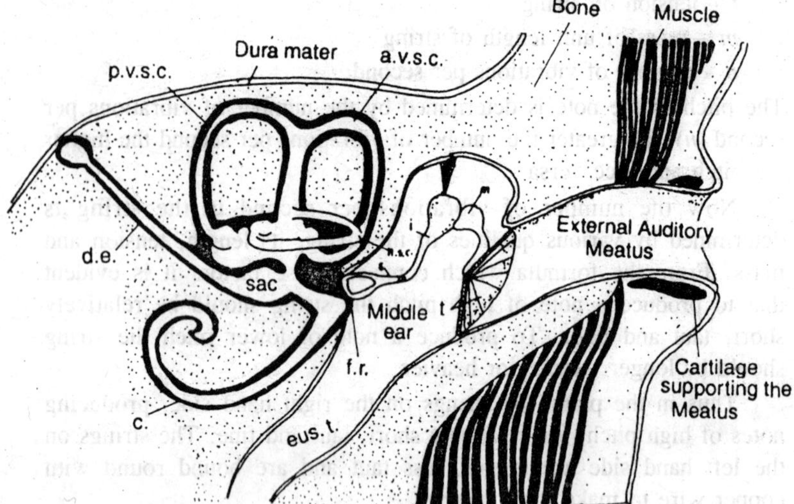

Fig. 15.31. Diagram of the structure of the hearing apparatus.

The pinna deflects waves of sound into the external auditory meatus. In some animals e.g. dog and rabbit, the pinna is moveable and can be directed independently from the head towards the source of particular sound waves. In man the pinna faces more or less forwards and is immovable and thus he is compelled to look and hear, as it were, in the same direction. Animals without a pinna e.g. amphibian are at a distinct disadvantage.

The external auditory meatus conducts sound vibrations inwards to the middle ear. It is separated from the middle ear by a taut membrane, the ear drum or tympanum, which is set in vibration by the sound waves.

The middle ear

The middle ear is an air filled cavity in bone, closed from the outside by the tympanum and leading into the inner ear through an oval aperture in the bony wall, the fenestra ovalis. Three bones, the ear ossicles, are suspended across the air filled cavity of the middle ear; the malleus (hammer), incus (anvil) and stapes (stirrup). These three bones are articulated to one another, with synovial joints between them. The arm of the malleus is attached to the tympanum and the stapes abuts onto the fenestra ovalis.

The bones serve to transmit the vibrations from the air in the external auditory meatus to the fluid which bathes the delicate apparatus of the inner ear.

The ear ossicles are suspended from the wall of the middle ear cavity by ligaments and the tension across the bones is regulated by means of two small muscles.

The area of the tympanum is large compared with that of the fenestra ovalis so that the pressure (i.e. force per unit area) acting upon the fenestra ovalis is greater than that on the tympanum. Thus increase in force is available for transmitting vibrations from the air of the middle ear to the more dense medium (i.e. fluid) of the inner ear.

The air in the middle ear is kept at atmospheric pressure by means of a connection to the back of the throat, the eustachian tube. Thus the pressure on the two sides of the tympanum is kept equal, at atmospheric pressure. The aperture of the eustachian tube in the throat is opened during swallowing. If the tube is blocked, as it often is by mucus plugs when one has a cold or catarrh, oxygen is absorbed from the air of the middle ear by the blood; the tympanum now bulges inwards because of the lower pressure in the middle ear, producing the familiar ringing and buzzing sounds.

The inner ear

The inner ear consists of a delicate hollow membranous structure, the membranous labyrinth filled with fluid called endolymph, bathed externally by the fluid perilymph and embedded deep in the temporal bone. The membranous labyrinth performs two functions, that of balance and orientation in space, and that of hearing. The part of the labyrinth called the cochlea is responsible for the sense of hearing.

The cochlea is a membranous structure would spirally like a snail shell inside a bony canal. Running along the whole length of the bony canal and stretched from one side wall of the canal to the other are two membranes. The part of the canal enclosed by the two membranes, the scala media, contains the fluid endolymph which is continuous with the endolymph of the rest of the membranous labyrinth. It is virtually a closed cavity. Of the outer two cavities the upper one is called the scala vestibuli because it communicates with the fluid filled cavity, the vestibule, into which the stapes abuts. The lower cavity is called the scala tympani; it connects with the scala vestibuli at the extreme tip of the cochlea (the helicotrema) and it connects to the middle ear cavity by another closed window, the fenestra rotunda.

When vibrations are set up in the fluid perilymph of the vestibule by the action of the stapes rocking in and out of the fenestra ovalis, they pass down the scala vestibuli via the helicotrema, into the scala tympani and end at the fenestra rotunda. The course of such vibrations

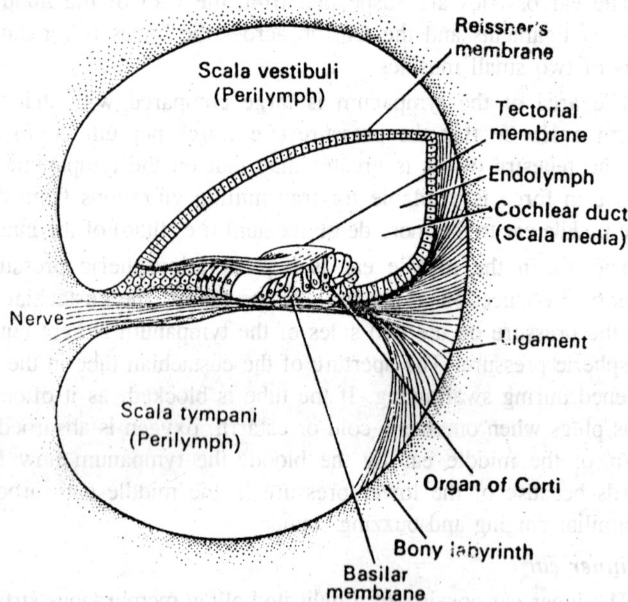

Fig. 15.32. Cochlear apparatus in T.S.

is indicated by the dotted lines. Fluid cannot be compressed so when the stapes rocks into the fenestra ovalis the pressure wave set up in the perilymph causes the fenestra rotunda to bulge into the middle ear. When the stapes moves out of the fenestra ovalis the fenestra rotunda moves inwards into the perilymph. The fenestra rotunda is a compensating mechanism which protects the delicate cochlea from the effects of pressure.

The upper membrane of the cochlea called Reissner's membrane is very delicate and can be disregarded in considering the sensory function of the ear. It is the lower membrane, called the basilar membrane which is the important one, and it is this membrane which we have compared to the piano sounding board. The basilar membrane is firmly attached at each side of the bony canal by fibrous tissue, that on the outer side being called the external spiral ligament. The membrane is widest at its apex; here the external spiral ligament is a very delicate structure compared with that at the base i.e. at the vestibular end of the basilar membrane. From these facts it is evident that the constituent 'strings' of the basilar membrane are longer and less taut at the apex than at the base. From our study of the relationship between the pitch of a note and the length and tension of the string

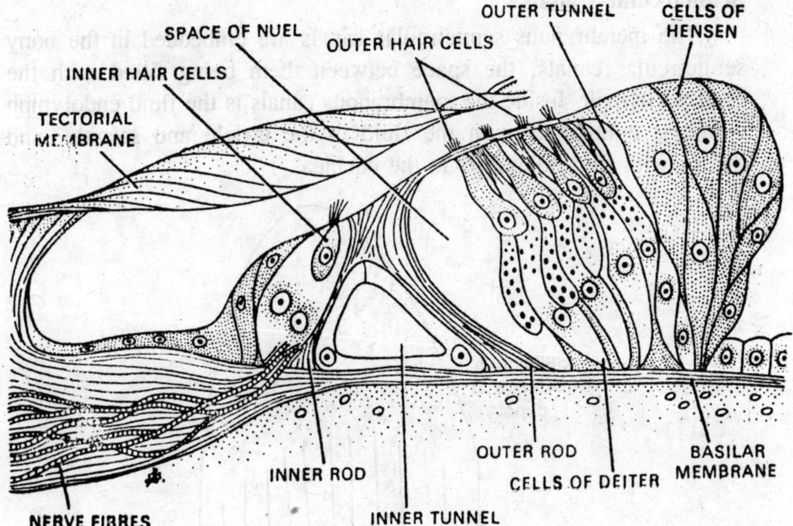

Fig. 15.33. Enlarged diagram of the organ of corti.

that produced it, we must conclude that notes of high pitch set the basal part of the basilar membrane in motion whereas notes of low pitch set the apical part of the membrane in motion whereas notes of low pitch set the apical part of the membrane vibrating. This has been confirmed experimentally.

The basilar membrane possesses rows of sense cells which because of the fine hairlike processes coming from their surfaces are called hair cells. These sensory hair cells are part of the organ of Corti. These processes are embedded in a ribbon of jelly called the tectorial membrane. The hair cells receive a very rich nerve supply from the auditory branch of the eighth cranial nerve. The sensory hair cells of a particular sector of the membrane are stimulated when that sector vibrates. Thus since particular parts of the membrane vibrate only to notes of a particular pitch, particular nerve cells are stimulated only when the ear receives notes of that pitch. This is the basis of the 'place theory' of hearing.

Balance and Orientation in Space

We have seen that the cochlea is the part of the membranous labyrinth concerned with hearing, now we must look at the remaining part of the labyrinth, consisting of the semicircular canals and the otolith organs (saccule and utricle).

Semicircular Canals

The membranous semicircular canals are embedded in the bony semicircular canals, the space between them being filled with the fluid perilymph. Inside the membranous canals is the fluid endolymph which is continuous with the fluid in the utricle and saccule, and thence with the endolymph in the cochlea.

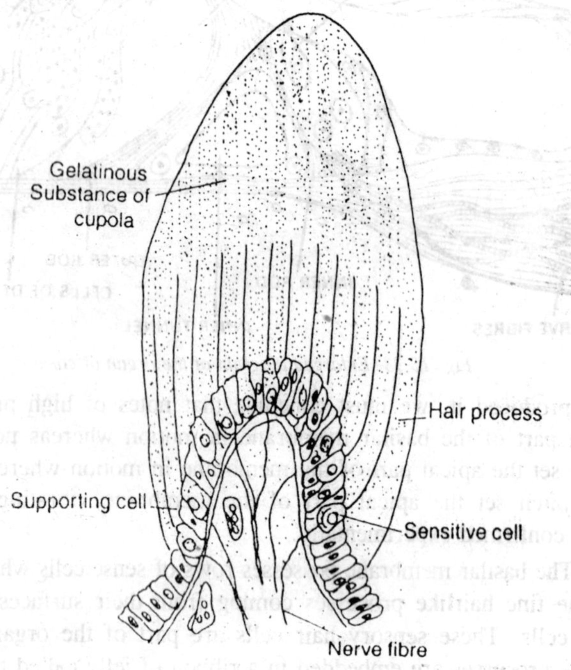

Fig. 15.34. The ampullary sense organ, the crista acoustica.

The canals lie in three planes which are at right angles to each other; there is a horizontal or lateral canal, a posterior vertical canal and an anterior vertical canal. At one end of each canal, where it passes into the utricle there is a swelling called the ampulla, containing a sense organ, the crista acoustica. The crista consists of a mound of cells containing sensory hair cells, the fine processes of which project from the mound to become embedded in a cone of jelly (the cupola). The sense cells are joined to nerve fibers of the eighth nerve.

Function of the semicircular canals

When the head moves from a resting position the endolymph in the semicircular canals is set in motion. According to the plane of movement of the head so the endolymph movement is restricted to a particular canal in that plane. Thus during and old-time waltz it is the horizontal semicircular canal in which the endolymph is mostly moving. There will be slight movements in the other two of course.

When the endolymph of a canal moves, the cupola in an ampulla is displaced and this stimulates the hair cells of the crista acoustica. From the pattern of stimuli coming from the three ampullae of each labyrinth the mammal is aware of the movement and the kind of movement of the head.

Otolith Organs

When the head is still the fluid in the semicircular canals is stationary and the sense organs in the ampullae are unstimulated. These provide no information about the position of the head at rest. This information comes from the sense organs in the utricle and perhaps the saccule. The sense organs of the utricle and saccule are called the maculae. The macula of the utricle consists of a thickening of the wall containing sensory hair cells surrounded by supporting cells. A collection of small crystals of calcium carbonate, called otoliths, adheres to the surface of the macula. The hair cells are supplied with nerve fibers and under the influence of gravity the hairs are distorted and the nerves stimulated due to the weight of the crystals. If the animal is upside down the crystals fall away from the hair processes and stimulation is reduced. From this sort of information the animal is aware of the position of its head in space at rest. Movement of the

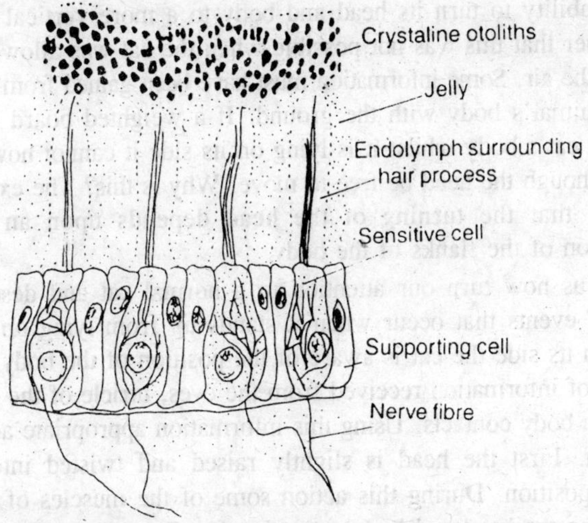

Fig. 15.35. Part of the macula sense organ of the utricle.

head also stimulates the maculae of course, even movement in a straight line will do so. (It would not stimulate the cristae of course).

Summary of the function of the labyrinth

(i) Cochlea - hearing.

(ii) Crista a coustica of the ampullae - sensitive to angular acceleration kinaesthetic sense.

(iii) Macula of the utridle - sensitive to gravity and linear acceleration static sense.

Other mechanisms for orientation. We have seen that the membranous labyrinth is a source of information about the position of the body in space but there are at least three others. The way that a mammal orientates itself in space can perhaps be best understood by giving a simplified account of some experimental work.

Role of the Eyes in Body Orientation

If a normal cat is held upside down by its legs and then dropped, it always lands on all four feet. However, if a blindfolded labyrinthectomized cat is dropped in a similar manner it falls on its back. When the blindfold is removed the labyrinthectomized cat lands on all four feet like a normal cat. It is obvious that the eyes have supplied information about the position in space.

Role of body contact in body orientation

A blindfolded labyrinthectomized cat lying on its side on the ground has the ability to turn its head and body to a more vertical position. Remember that this was not possible when the cat was allowed to fall through the air. Some information must have been gained from a contact of the animal's body with the ground. If a weighted board is placed upon the cat's body whilst it is lying on its side it cannot now turn its head, although the head be free to move. Why is this? The explanation must be that the turning of the head depends upon an unequal stimulation of the flanks of the body.

Let us now turn our attention to a normal cat and describe the series of events that occur when it stands up from lying on its side. Lying on its side the cat is aware of the position of the body in space because of information received from the eyes, utricle of the labyrinth and from body contacts. Using this information appropriate action can be taken. First the head is slightly raised and twisted into a near vertical position. During this action some of the muscles of the neck connecting the head and body are twisted and stretched. Stretching of the muscles stimulates proprioceptors within them, providing information about the position of the head in relation to the body. Appropriate action can now bring the head and body into alignment.

As soon as the head has started to move from the lying down position the sensory apparatus of the semicircular canals is brought into action providing the animal with information about the angular movement and acceleration of the head.

Summary of orientation mechanisms

We now know that the mammal gets information about the position and movement of the body in space from the following sources.

1. Membranous labyrinth; semicircular canals and utricle.
2. Eyes.
3. Body surfaces.
4. Proprioreceptors.

16

Reproductive System

The gonads of vertebrates produce gametes and, in most cases, steroid hormones. These hormones serve various functions, such as inducing or facilitating sexual behaviour and/or parental behaviour, preparing the reproductive tract for receiving the gemetes, caring for the zygote, and other functions.

Male Reproductive System

In most vertebrates, the testes are paired, but in some e.g., the cyclostomes, the left and right testes are fused, and in some teleost species only one tesitis develops, as in *Notopterus notopterus*. In most classes of vertebrates the sperm produced in the testes are conducted to the outside by a duct system; the cyclostomes and the salmonidae, however, such a duct system is lacking. In vertebrate species that do have a sperm duct system, this system may be separate from the urinary wolffian duct system as is the case in teleosts. In all other vertebrates, sperm are transported in tubules that originate as pronephric and mesonephric tubules.

The location of the testes varies in mammals. Giersberg and Rietschel (1968) distinguish the following five different types of arrangement.

1. The testes remain in the body cavity, and there is no gubernaculums as in the monotremes (duck-billede platypus, *Ornithorhynchus anatinus*, and the spiny anteaters, *Tachyglossus* sp., *Zaglossus* sp.) some primitive insectivore (the Macroscelididae or elephant shrews, and *Tenrec ecaudatus*), and the Hyraxes.
2. The testes remain in the body cavity, but a gubernaculums that later regresses is formed, as in the Sirenia (seacows), elephants, and the Edentata (solths and anteaters).

3. The testes move caudad, a gubermaculum and inguinal canal are formed but the testes do not enter the inguinal canal, as in the Cetacea (whales, dolphins, porpoises) and armadillos.
4. There is a gubernaculums, and the testes descend temporarily or permanently into a pouch of the cremaster muscle but there is no true scrotum. The testes descend temporarily in *Solenodon*, in moles, hedgehogs, shrews aardvarks (*Orycteropus afer*), and many rodents and in the lagomorphs (rabbis). The testes descend permanently in the phascolomidae (wombate), Tapiridae, Hippopotamidae, Rhnocerotidae, Manide (pangolins) Pinnipedia (seals, seal lions, walruses), some land carnivore (e.g., hyaenas, *Crocuta*) and some Chiroptera, e.g., the Vespertilionidae.
5. A gubernaculums is present and the testes decend through the inguinal canal either temporarily (usually during the breeding season) or permanently into the scrotum.

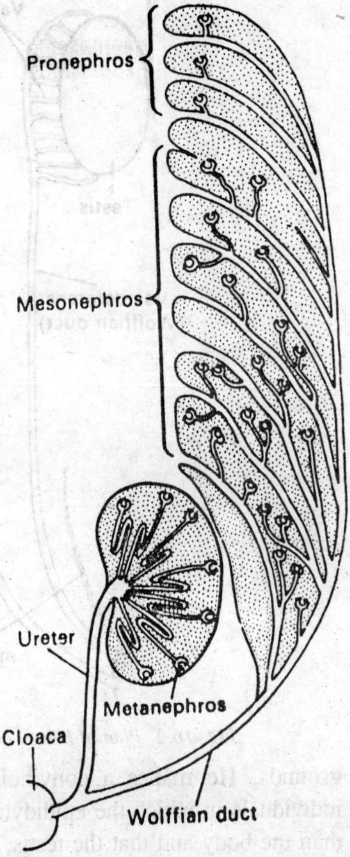

Fig. 16.1. Diagram to show serial development of pronephros, mesonephros and metanephros.

In most species in which the testes descend either into a cremaster muscle pouch or into a scrotum, the descent is permanent. In some species, the testes can return to the abdominal cavity or to the inguinal canal. Examples are: *Orycteropus* (aardvarks), some rodents (*Rattus*, *Sciurus*, *Tamias*, i.e., chipmunks) and chiroptera, and some primates (*Loris* and *Perodicticus*), according to Prasad (1974).

The phenomenon of testes in ascrotum, in which they are exposed to temperatures lower than the body, appears for the first time in mammals. Bedford (1978a,b) has tried to explain this on evolutionary

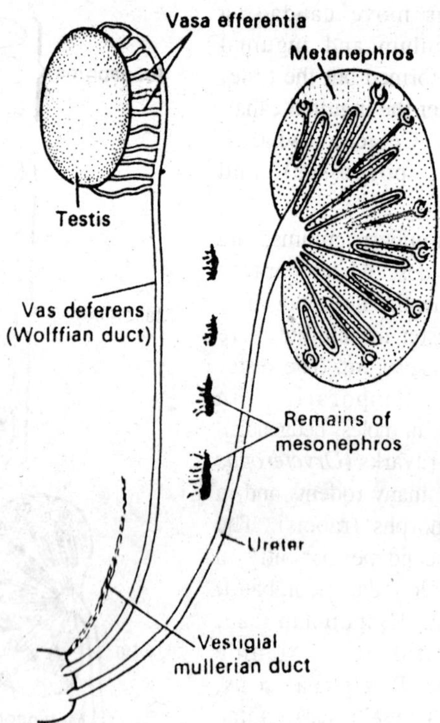

Fig. 16.2. Plan of kidney and genital ducts in male amniotes.

grounds. He makes a convincing argument that selection favoured individuals in which the epididymis was exposed to temperatures lower than the body and that the testis, which descended with the epididymis, may secondarily have lost the capacity to function at body temperature. There are however, mammals without a scrotum, in which the testes clearly function normally at body temperature. Except for marine mammals, scrotal mammals are polygynous species, in which the dominant male sires many more offspring than males lower in the dominance hierarchy. Ascrotal species, however, in general seem to be monogamous or live in family units. The high number of ejaculations by dominant polygynous males may therefore, have favored those males with epididymides in which sperm could be stored successfully. Such storage would probably be more successful at lower temperatures because metabolic rates would be lower; thus eventually, individuals with epididymides (and as a consequence testes) at temperatures lower than the body would have a selective advantage.

The apparent contradiction of polygynous ascrotal marine mammals may in fact be only an apparent contradiction. In the southern elephant seal (Mirounga leonina), the temperature of the testis is 6°C below body temperature apparently as a result of blood flow from the rear flippers. In some seals, e.g., *Callorhinus* and *Zalophus*, the testes are located in a scrotal pouch.

Many mammalian species have one or more well developed accessory reproductive glands. These glands are:

Ampullary Glands

These are derived from the wolffian ductsystem and have a simple cuboidal, or columnar epithelium; their secretions are high in ergothioneine.

Seminal Vesicles

More properly called vesicular glands, these are also derived from the wolffian duct system. They have a simple or pseudostratified epithelium and secrete fructose, citric acid, proteins, and ascorbic acid.

Paraurethral Glands

Also called glands of Littre, these are probably synonymous with the glands named paraprostate glands by McKeever (1970) and Holtz (1972). These glands originate from the epithelium of the urethra and are thus part of the urogenital sinus. They are usually small structures, the ducts of which drain into he urethra. According to Holtz (1972), there are two different types of paraprostate glands in the rabbit: (a) glands consisting of small clusters of fluid-filled vesicles, which, under microscopic examination, resemble the bulbourethral glands, and (b) small, compact, whitish glands, which are histologically identical with the prostate.

Proprostate Glands

These are found in the rabbit, and appear to be part of the prostate; however, they are separated from the prostate by a connective tissue system. These glands originate; as does the prostate, from endoderm of the urogenital sinus. Holtz (1972) emphasizes that some names given to these glands are not appropriate, e.g., coagulating gland (because the glands are not the source of the gel mass in the ejaculate), and prostate and glandula vesicularis (because they are not wolfian duct derivatives). The proprostate is an acinar gland, rich in smooth muscle and with a columnar epithelium. It secretes a white granular substance, which gives the gland a white color and has two large excretory ducts that lead to small urethral diverticula.

Prostate Gland

This originates as part of the urogenital sinus and has simple or pseudostratified columnar epithelium. The secretions contain enzymes, such as diastase, glucoronidase, fibrinolysin, acid phosphatase as well as citric acid. The gland has ducts that drain into the urethra.

Urethral Gland

This gland is found in some Chiroptera between the prostate and the openins of the ducts at the bulbourethral gland. It lies around the urethra, is covered by the muscles of the urethra, and its secretions drain via short ducts into the urethra.

Bulbourethral Glands (Cowpers Glands)

These glands are also part of the urogenital sinus. They have a simple columnar epithelium which secretes a viscous fluid containing sialoprotein and have ducts that drain into the urethra.

Bulbar Glands

Also called the *inferior bulbourethral glands*, these glands seem to be present in some of the sciuridae. They are glandular developments of the bullbourethral glands inside the corpus spongiosum. The secretions drain via a glandular duct (ductus penis) at the ventral side of the urethra.

Prostatic Utricle

Also called *uterus masculinus*, this is usually vestigial or even absent but can according to Prasad (1974), be well developed. It is derived from the Mullerian duct, and the ducts drain into the urethra.

Since this is a derivative of the Mullerian duct system might be expected to be sensitive to estrogen As estrogens have been suspected of inducing cancer of derivatives of the Mullerian duct system in women it may be prudent to consider seriously the wisdom of the use of estrogens in treating cancer of the prostate.

Pleputial Glands

These are modified sebaceous glands with a stratified, squamous epithelium and with ducts draining at the surface of the gland. They are enormously developed in the beaver.

Inguinal Glands

These are sebaceous glands and sweat glands found on the surface of the inguinal canal and in skin folds of the genital region. Holtz (1972) suggests the name *perineal glands* for these in the rabbit.

The effect of the presence of these accessory sex glands on fertility has been studied in the boar (*Sus scrofa*), which produces a large amount of fluid in the ejaculate (about 200 cc) 26 percent of which is contributed by the seminal vesicles, 56 percent by the prostate, and 19 percent by the bulbourethral glands. These large quantities of secretions make this species a very suitable animal for studying these gland. Removal of the seminal vesicles, or the bulbourethral glands, or both had no effect on fertility. The effect of removal of the prostate was not reported for the boar, but apparently removal of the prostate in humans does not affect fertility.

In rats, removal of the seminal vesicles and the prostate and coagulatin gland reduces fertility severally but removal of the following glands or combination of glands has no effect on fettility bulbouretheral glands, bulbourethral glands and seminal vesicles bulbourethral glands and coagulating gland seminal vesicles and prostate.

Prototheria

The penis of the Prototheria like that of the reptiles and birds is located on the floor or the clocaca and lies in a preputial fold of the ventral wall of the clocaca. The penis conducts the semaen but not the urine to the outside because the semen-urine duct gives off a branch to the clocaca before it enters the penis. The ductus deferens drains into the urogenital sinus, dorsal to the ureters not ventral as in the Metatheria and Eutheria. The penis contains fibrovascular tissue which is homologous with the corpus cavenosum penis of the Metatheria and Eutheria. The glands of the duck-billed platypus is bifurcated.

Metatheria

The penis is located in a cloacal fold which is surrounded, together with the anus, by a cloacla sphincter muscle, so that it is not visible generally except during erection. However, in *phascogale* (marsupial pocket misce) and *Dasyurus* (Australian native cats) the penis is free and not in a cloacal fold. The penis of the opossum Didelphis virginiana is bifurcated as is the penis of *Dasyurus* and *phascolomis* although in these last two genera, the bifurcation is not as marked. The penis of the Macropodidae and Tarsipes is definitely not forked.

In *Permelidae* (bandicoots), the urogenital sinus conducts only the semen to the outside; the urine is voided via the cloaca. In *Didelphis* and *Canenolestes* (bat opossum), the distal opening of the urinary duct lies at the base of the glands and the ductus deferens ends in the tip of the penis. In *Myrmecobius* (numbats), *Thylacinus* (pouchedwolf) and

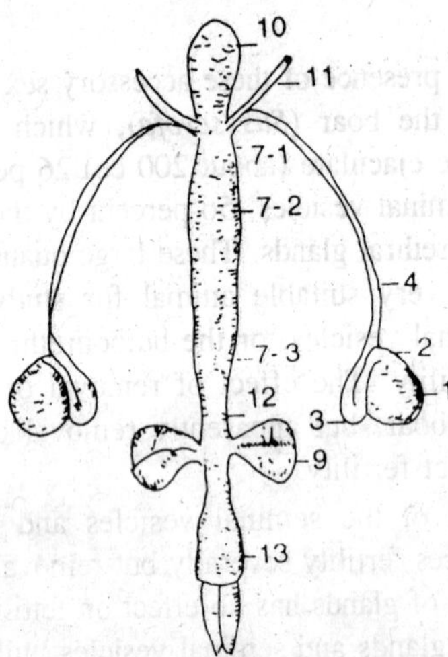

Fig. 16.3. Reproductive organs of the opossum. Note the bifurcated penis. 1–testis, 2–caput epididymis, 3–cauda epididymis, 4–ductus deferens, 5–ampullary gland, 6–vesicular glands, 7–prostate gland, 8–anterior lobe of prostate, 9–bulbourethral gland, 10–bladder, 11–ureter, 12–urethra, 13–penis.

Dasyurus, the urethra divides into two branches one branch going to one tip of the bifurcated glands, the other to the other tip.

Eutheria

The perineum divides the cloaca into the dorsocaudal rectum and the ventrocranial urogenital sinus, soi that the penis lies on the ventral side of the anus. The distance between anus and penis can be small, as in beavers (*Castor* sp.) and hares (*Lepus europeaus*) states that in beavers, the urogenital openings and the anus open into a cloaca a situation also found in the prototheria, Metatheria, Edentata and in the Insectivora, and *Neomys* sp. In other species, the penis lies with its preputial fold, near the umbilicus. The double skin fold, which surrounds the penis, partly or completely forms the prepuce which in the horse consists of two circular preputial folds which are telescoped so that the penis can be pulled inside the folds. The pars intrapreputialis penis, corresponds to the glands penis in humans.

Several types of penes have been distinguished.

1. *Indifferent type* (edentates and rodents). The penis is short and the corpus spongiosum penis is not covered by a tunica. The corpus cavernosum can be either fibrous tissue or tabecular erectile tissue.
2. *Fibro-elastic type* (reminants and whales). Usually the penis is long and fibrous and has a sigmoid flexure in the resting stage.

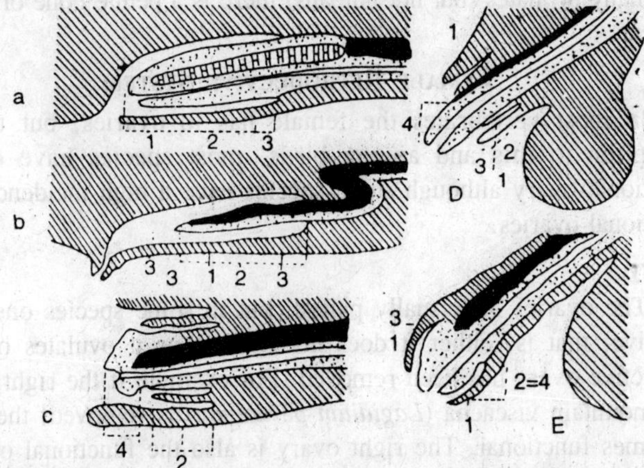

Fig. 16.4. Medial view of the penis of some mammals. A—seal (Phoca); B—bull (Bos taurus); C—horse (Equus caballus); D—long-tailed monkey (Cercopithecus sp.); E—human (homo sapiens). Black—corpus cavernosum penis; stippled—corpus spongiosum penis and corpus caernosum glands; heavy black lines—os penis in A. 1-preputium, 2-pars intrapreputialis penis, 3-preputial cavity, 4-glans penis.

There is little erectile tissue in the corpus cavernosum, the erection consisting of the straightening out of the sigmoid flexure. The corpus spongiosum is surrounded by a well-developed tunica.

3. *Vascular type* (perisodactyla, Carnivora, Primates, and some Insectivora). The penis if not is erection is long and flexible. The corpus cavernosum penis and corpus spongiosum urethra are surrounded by a fibrous capsule the tunica albuginea. Erection results from the corpora cavernosa and corpora spongiosa filling with blood.

4. *Intermediary type* (elephants, and the sirenia). This type combines features of types 2 and 3. According to Slijper as cited by Prasad (1974), there is a general correlation between the type of penis and the length of coitus Animals with a fibro-elastic type penis have a short coitus the ones with the indifferent type have a relatively short coitus the ones with the intermediary type have a relatively short coitus the ones with the intermediary type have a longer coitus than the other two types but shorter than animals with the vascular type in which coitus may be fairly long. However, coitus in swine which have an intermediate type penis, may last as long as 30 minutes!

In most species of Chiroptera, Rodentia, Carniovora, Insectivora and many primates (but not humans) there is a penile bone or baculum present.

Female Reproductive System

In most vertebrates the female has to ovaries, but there are exception to this and among birds, most species have only one functional ovary although some species have a high incidence of two functional ovaries.

Ovary

The ovaries are usually paired, but in some species one may be inactive; that is, either it does not ovulate or it ovulates only after the active ovary has been removed. For example if the right ovary of the mountain viscacha (*Lagidium peruanum*) is removed, the left one becomes functional. The right ovary is also the functional one in the following bats: *Miniopterus schreibersi*, *Rhinolophus* sp., *Tadarida*

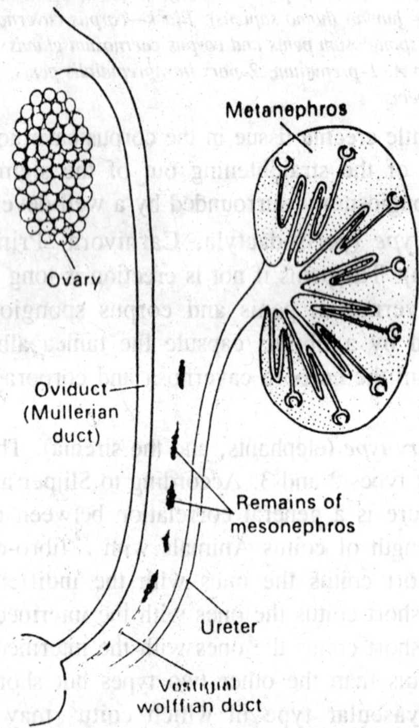

Fig. 16.5. Plan of kidney and genital ducts in female amniotes.

cyanophala and *Molossus* ater. The left ovary is dominant in the bat *Miniopterus natalensis* and in the waterbuck *kobus defassa*.

Table 16.1. Occurrence of ovarian bursa in mammals

Animal	*Anatomy of bursa*				
	1	2	3	4	5
Prototheria					
Tachyglossus	+				
Ornithorhynchus				+	
Methatheria					
Phalangeridae				+	
Phascolomidae		+			
Macroopoidae				+	
Eutheria					
Insectivora					
Tenrecinae	+				
Erinaceinae	+				
Soricinae	+				
Talpidae	+				
Chiroptera	+				
Primates					
Lemuridae			+		
Hominidea			+	+	
Edentata					
Bracdypodidae	+				
Dasypodidae		+			
Lagomorpha					
Lepus europaeus					
Oryctolagus cuniculus		+			
Rodentia					
Sciuridae		+			
Castoridae		+			
Cricetidae	+				
Muridae	+				
Gliridae					
Dipodidae	+				
Hystricidae		+			

Caviidae			+		
Dasyproctidae					+
Cetacea					+
Carnivora					
Canidae	+				
Ursidae	+				
Procyonidae	+				
Mustelidae	+				
Viverridae	+	+			
Hyaenidae		+			
Felidae		+	+		+
Pinnipedia		+			
Hyracoidea					
Procaviidae		+			
Perissodactyla					
Equidae				+	+
Artiodactyla					
Suidae		+			
Camelidae	+				
Cervidae			+	+	
Bovidae		+			

Each ovary is suspended from the dorsal body wall by the mesovarium or secondarily by the broad ligament. In many species the ovary lies in a membranous sac called the *ovarian bursa*, which is a fold of the mesosalpinx, the mesentery of the oviduct. The function of the bursa has not been determined.

The ovary consists of the stroma, the germinal epithelium, the germ cells and their associated structures, interstitial cells, and various vestigial structures. The germinal epithelium, which covers the ovary, may be either directly on the ovarian stroma (e.g., in sherws) or it may be on a fibrous ovarian tunica albuginea. It may consist of squamous or low columnar cells. The germ cells and their associated structures include oocytes, follicles in different stages of development, connective tissue, corpora lutea, old corpora lutea called *corpora albicantia*, and interstitial tissue.

Follicles

These can be characterized as small medium and large follicles, according to Pedersen and peters (1968). The main distinctions are

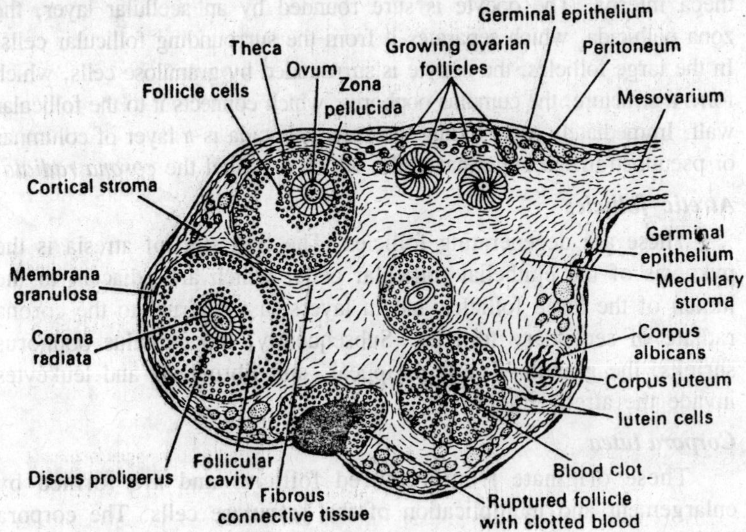

Fig. 16.6. T.S. of the ovary.

made on the basis of oocyte size number of layers of follicular cells, and the presence of follicular fluid. The follicular wall has a granulosa layer, a theca *externa*, and a theca *interna*. The oocyte is surrounded

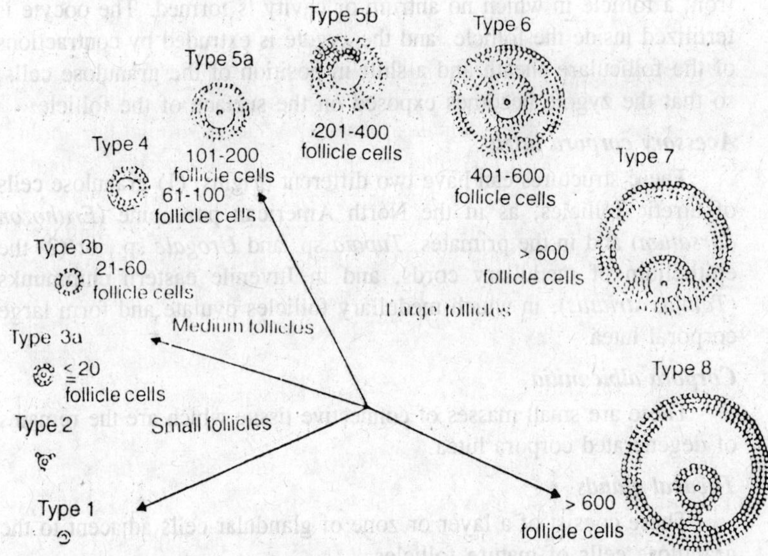

Fig. 16.7. Classification of ovarian follicles.

by an acellular wall has a granulose layer a theca externa, and a theca interna. The oocyte is sure rounded by an acellular layer, the zona pellucida, which separates it from the surrounding follicular cells. In the large follicles, the oocyte is surrounded by granulose cells, which form a structure, the cumulus oophorus, which connects it to the follicular wall. Immediately surrounding the zona pellucida is a layer of columnar or pseudostratified columnar cells, which is called the *corona radiata*.

Atretic follicles

These are degenerating follicles. The first sign of atresia is the pyknosis of the epithelial follicular cells, which are adjacent to the lumen of the large follicles and in layers just external to the corona radiate of secondary follicles. Subsequently, the cumulus oophorus shrinks, the granulosa cells disappear, and fibroblasts and leukeytes invade the atretic follicle.

Corpora lutea

These originate from ruptured follicles and are formed by enlargement and multiplication of the granulose cells. The corpora lutea consist of a solid mass of polyhedra cells. Mossman and Duke (1973a) proposed the term "luteal gland" in preference to corpous luteum, but the use of the term *corpus luteum* is of such long standing that substitution of a new term seems to have little chance of adoption. The corpora lutea of the Madagascar hedgehog (*Setifer setosus*) develop from a follicle in which no antrum or cavity is formed. The oocyte is fertilized inside the follicle, and the zygote is extruded by contractions of the folliculare sheath and a shift in position of the granulose cells, so that the zygote becomes exposed on the surface of the follicle.

Acessory corpora lutea

These structures can have two different origins: (1) granulose cells of atretic follicles, as in the North American porcupine (*Erithozon dorsatum*) and in the primates, *Tupaia* sp. and *Urogale* sp., or (2) the epithelium of medullary cords, and in Juvenile eastern chipmunks (*Tamias striatus*), in which medullary follicles ovulate and form large corporal lutea.

Corpora albicanitia

These are small masses of connective tissue which are the remains of degenerated corpora lutea.

Theacal glands

These consist of a layer or zone of glandular cells adjacent to the granulose cells of mature follicles.

Interstitial glands

These may originate from the ovarian stroma, medullary cord cells, and the theca interna of atretic follicles. Mossman and Duke (1973b) stress the inaccuracy of calling these structures "*luteinized follicles*."

Some structures in the ovary are vestigial. The *rate ovarii* plays a role in oögenesis. It is a regular network of spaces lined with a simple low columnar epithelium. It is well developed in the tarsier, *Tarsius* sp. (a primate) and in pangolins, *Manis* sp. The *epoophoron* consists of vestiges of efferent ductules and is homologous with the epididymis of the male. *Gartner's duct* or the duct of the epoophoron is a vestige of the ductus deferens. The *paraoophoron* consists of the more rostral and the more caudal mesonephric tubules - or ductuli aberrantes. It is homologous with the (also vestigial) paradidymis of the male.

The ovary receives blood from the ovarian artery, a branch of the dorsal aortam, and blood leaves the ovary via the ovarian vein, which drains into the caudal vena cava. Some branches of the ovarian artery and vein anastomose with ovarian branches of the uterine blood vessels.

Oviduct

The oviduct is suspended by the mesosalpinx on one side; the other side is attached to the mesotubarium superius, which connects the oviduct to the fimbria. The fimbria consist of a fringe of irregular processes. The relationship between oviduct and mesenteries can vary considerabley among species. Beck an Boots (1974) give a comprehensive summary of the various relationships in a large number of species.

Starting at the cranial end, the oviduct consists of: (1) infundiblum, with a fimbriated, funnel-like opening, (2) the ampulla, (3) the isthmus, (4) the utero-tubal junction. The wall of the oviduct consists of several layers: The *tunica serosa*, the outermost layer is a thin layer of connective tissue covered by a single layer of squamous epithelium. The *tunica muscularis* consists of smooth muscle, which may be arranged in four distinct patterns, i.e., (1) a predominant circular layer of smooth muscle without distinguishable longitudinal layers, (2) thin longitudinal layers on both sides of a thicker circular layer, (3) a thick circular layer with an outer longitudinal muscle layer, and (4) a circular layer with an inner longitudinal muscle layer. The *tunica mucosa*, the innermost layer consists of a lamina propria of connective tissue and a lamina epithelialis, which has ciliated cells and three types of nonciliated cells: secretory cells, wedge-type cells called "peg cells," and basal cells along the base of the epithelium.

In species such as pigs, mice, rats, humans and rhesus monkeys of ciliated cells depends on the stage of the reproductive cycle; however, the occurrence of cyclic changes in cilia is controversial.

The antomy of the utero-tubal junction varies among species. Beck and Boots (1974) described ten different morphological types of utero-tubal junctions. Type 1-3 lack mucosal projections of oviductal or uterine origin. Type 4-10 have specialized mucosal projections of oviductal or uterine origin. The reader is referred to this excellent review by Beck and Boots (1974) for more details. The oocytes or the zygotes are transported rapidly through the utero-tubal junction, which may play a role in sperm transport.

Uterus

The uterus is suspended by the mesometrium. There are different types of uteri. In the prototheria, the uteri are paired and open into the urogenital sinus; no vagina is present. In the metatheria, or marsupials there are 2 vaginae, 2 cervices, 2 uteri. Sharman (1976) has illustrated some of the variations of this general plan. The kangaroo (*Macropus* sp.) has an open midline vagina after the first partiurition, so that it resembles the Eutherian midline vagina, but it has been formed in a different manner.

Among the Eutheria one finds four kinds of uteri: (1) The duplex uterus found in the rabbit has 1 vagina, 2 cervices, and 2 uterine horns. (2) The bicornuate uterus found in the pig, has 1 vagina, 1 cervix, and 2 uterine horns. (3) The bipartite uterus found in cattle, sheep and horses has 1 vagina, 1 cervix and either a small or a prominent uterine body, and 2 separate uterine horns. (4) The simplex uterus found in most primates, including humans, has 1 vagina, 1 cervix, and 1 uterine body. In humans, various anomalies have been found; for example, all the three different types of uteri listed for the Eutheria, a double uterus and vagina as a result of failure of the Mullerian ducts to fuse, a double uterus and vagina, and a subseptate uterus.

Small variations are found in the four different types of uteri. For example, the sloth *Bradypus* has a simplex uterus, one cervix, but a paired vagina as the result of a sagittal septum; the armadillo (*dasypus* sp.) has a simplex uterus with one cervix opening into the urogenital sinus.

Sharman (1976) has pointed out that the peculiar vagina of marsupials is the result of the position of the ureters of marsupials, which grow forward *between* the wolffian and Mullerian ducts. This

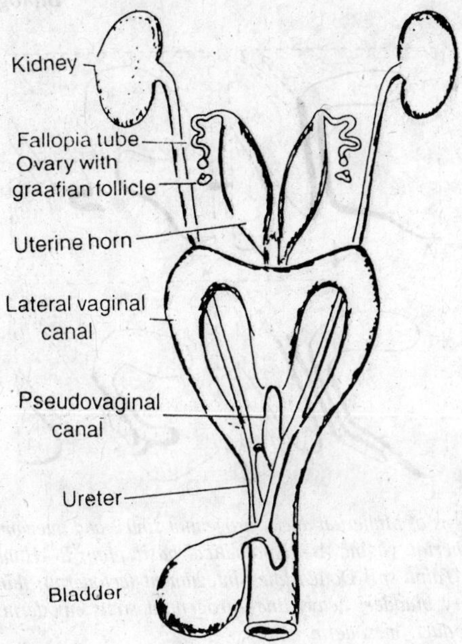

Fig. 16.8. Peculiar anatomy of the female reproductive tract in marsupials. Spermatozoa ascend by way of the lateral vaginal canals, and the fetus is born through the central pseudovaginal canal, which is occluded at other times.

prevents the fusion of the vagina in the midline (because of the intrusion of the ureters). The lateral vagina serve for transport of the spermatozoa, but the birth of the young occurs via the midline birth canal, which connects the uterine openings *directly* with the urogenital sinus.

Cervix

The cervix is a sphincter at the posterior end of the uterus. Its wall is thick, and the inner wall has ridges. In consists of a serosa, an outer longitudinal and an inner circular muscle, a submucoa, mucosa consisting of columnar epithelium at the cranial end and squamous epithelium at the caudal end and many goblet cells that secrete mucus. The consistency of this secretion varies during the menstrual cycle and is sometimes used to diagnose the time of ovulations in women, as is the crystallization pattern of the mucus.

Vagina

The relation of the Mullerian ducts, urogenital sinus, and the integument in the formation of the vagina is illustrated in Figure. The vaginal wall consists of a serosa, an outer longitudinal and an inner

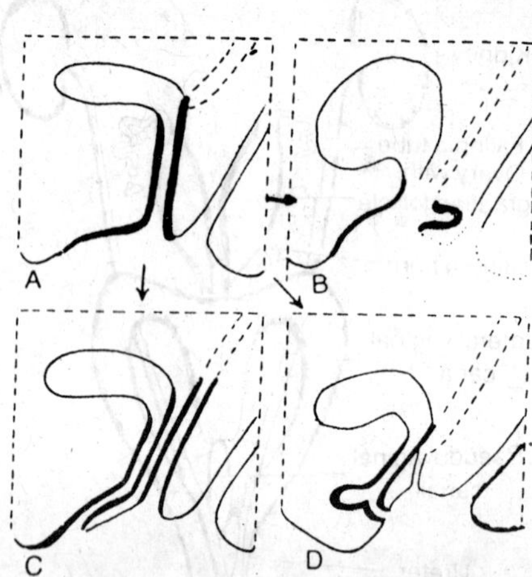

Fig. 16.9. Contributions of Mullerian ducts, urogenital sinus, and integument in formation of the eutherian vagina A—Hypothetical basic plan; B—Human; C—Mouse; D—Mole (Talpa sp.) Dotted line—integument (ectoderm); thin line—intestine and urinary bladder; heavy line—urogenital sinus entoderm; dashed line—Mullerian duct, mesoderm.

circular muscle layer, a submucosa, and a mucosa consisting of stratified squmous epithelium.

In some laboratory species, especially the rat, mouse, and hamster, the stage of the estrous cycle is correlated with differences in the histological appearance of the vaginal epithelium, so that vaginal smears

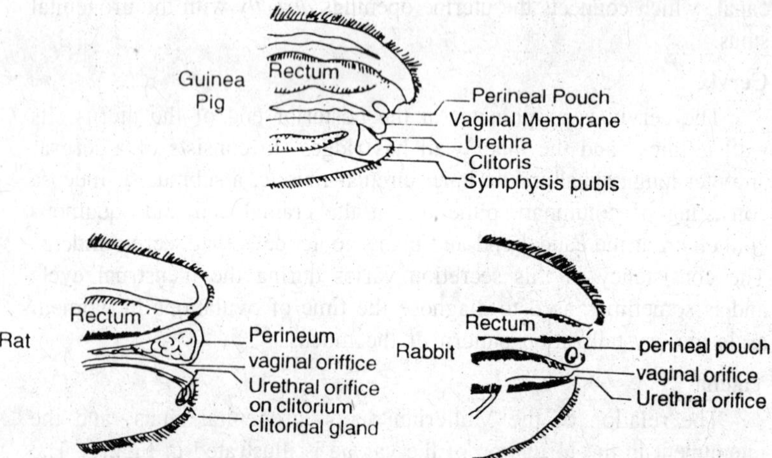

Fig. 16.10. Location of the urethral and vaginal orifices in the abdomen of the guinea pig rat and rabbit.

can be used to diagnose the stage of the estrous cycle. During proestrus epithelial cells predominate; during estrus there are some remnants of epithelial cells and many large cornified cells; during metestrus there are many cornified and epithelial cells and leukocytes; during diestrus few epithelial cells remain and few to many leukocytes are found.

Prostate

Although the prostate is mostly a male secondary sexual gland, a female prostate is well developed in some species. In laboratory rats, the "Witschi rat," developed by selective breeding, has a well-developed prostate.

Table 16.2. Part of prostate found in female mammals.

Species	*Part of prostate*
Insectivora	
Hemicentetes semispinosus	Dorsal-lateral
Erinaceus europaeus	Dorsal-lateral
Talpa europaea	Ventral
Chiroptera	
Taphozous	Dorsal
Nycteris luteola	Ventral
Coleura atre	Dorsal
Primates	
Macaca mulatto	Dorsal (?)
Cercopithecus aethiops	Dorsal (?)
Homo sapines	Dorsal
Lagomorpha	
Oryctolagus cuniculus	Dorsal (?)
Sylvilagus floridanus	Dorsal
Rodentia	
Arvicola sapidus	Ventral (?)
Apodemus sylvaticus	Ventral
Arvicanthis cinerus	Ventral
Rattus norvegicus	Ventral
R. rattus	Ventral
Mastomys erythroleucus	Ventral
Mus musculus	Ventral

External Genitalia

The external genitalia consist of the labia majora, labia minora, and the clitoris, but only the human has true labia. Generally, the

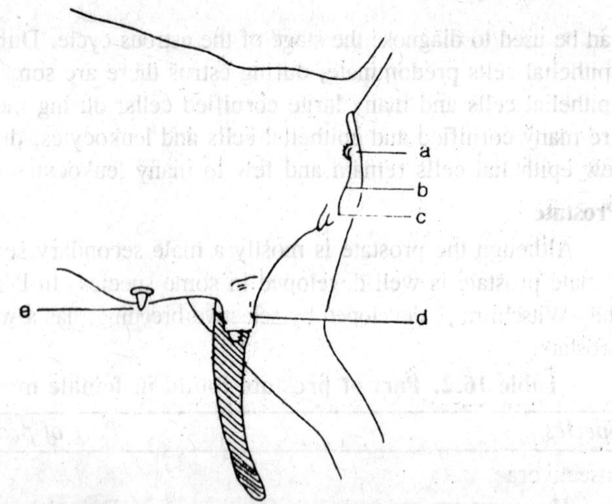

Fig. 16.11. Exterior female genitalia of hyaena crocuta. a—anus, b—perineum, c—false scrotum, d—clitoris not in erection and erection (hatched), e—teat.

clitoris is poorly developed unless the fetus has been exposed to larger amounts of androgens than normal e.g., in the adrenogenital syndrome, in the freemartin, or after experimental androgen treatment. However, in the hyaena (*Hyaena crocuta*), the clitoris is very well developed. This extreme development and the anatomy make it almost impossible to distinguish a male from a female, especially since the female also has a false scrotum.

Sexual Skins

Some primates have a so-called sexual skin, which is contiguous with the external genitalia. It consists of a highly vascularized edematous or pigmented dermis. Its color and vascularization vary during the reproductive cycle.

INDEX